AF832077

Ecological Studies

Analysis and Synthesis

Volume 225

Series editors

Martyn M. Caldwell
Logan, UT, USA

Sandra Díaz
Cordoba, Argentina

Gerhard Heldmaier
Marburg, Germany

Robert B. Jackson
Durham, NC, USA

Otto L. Lange
Würzburg, Germany

Delphis F. Levia
Newark, DE, USA

Harold A. Mooney
Stanford, CA, USA

Ernst-Detlef Schulze
Jena, Germany

Ulrich Sommer
Kiel, Germany

Ecological Studies is Springer's premier book series treating all aspects of ecology. These volumes, either authored or edited collections, appear several times each year. They are intended to analyze and synthesize our understanding of natural and managed ecosystems and their constituent organisms and resources at different scales from the biosphere to communities, populations, individual organisms and molecular interactions. Many volumes constitute case studies illustrating and synthesizing ecological principles for an intended audience of scientists, students, environmental managers and policy experts. Recent volumes address biodiversity, global change, landscape ecology, air pollution, ecosystem analysis, microbial ecology, ecophysiology and molecular ecology.

More information about this series at http://www.springer.com/series/86

Farshid S. Ahrestani · Mahesh Sankaran
Editors

The Ecology of Large Herbivores in South and Southeast Asia

Springer

Editors
Farshid S. Ahrestani
Department of Ecology, Evolution and Environmental Biology
Columbia University
New York, NY
USA

and

Frontier Wildlife Conservation
Mumbai
India

Mahesh Sankaran
Faculty of Biological Sciences
University of Leeds
Leeds
UK

and

National Centre for Biological Sciences
Bangalore
India

ISSN 0070-8356 ISSN 2196-971X (electronic)
Ecological Studies
ISBN 978-94-017-7568-7 ISBN 978-94-017-7570-0 (eBook)
DOI 10.1007/978-94-017-7570-0

Library of Congress Control Number: 2016932749

© Springer Science+Business Media Dordrecht 2016
This work is subject to copyright. All rights are reserved by the Publisher, whether the whole or part of the material is concerned, specifically the rights of translation, reprinting, reuse of illustrations, recitation, broadcasting, reproduction on microfilms or in any other physical way, and transmission or information storage and retrieval, electronic adaptation, computer software, or by similar or dissimilar methodology now known or hereafter developed.
The use of general descriptive names, registered names, trademarks, service marks, etc. in this publication does not imply, even in the absence of a specific statement, that such names are exempt from the relevant protective laws and regulations and therefore free for general use.
The publisher, the authors and the editors are safe to assume that the advice and information in this book are believed to be true and accurate at the date of publication. Neither the publisher nor the authors or the editors give a warranty, express or implied, with respect to the material contained herein or for any errors or omissions that may have been made.

Printed on acid-free paper

This Springer imprint is published by Springer Nature
The registered company is Springer Science+Business Media B.V. Dordrecht

Contents

Contributors

Farshid S. Ahrestani Department of Ecology, Evolution and Environmental Biology, Columbia University, New York, NY, USA

Sumanta Bagchi Centre for Ecological Sciences, Indian Institute of Science, Bangalore, India

Yash Veer Bhatnagar The Snow Leopard Trust, Seattle, WA, USA; Nature Conservation Foundation, Mysore, India

Faysal Bibi Department of Mammalogy, American Museum of Natural History, New York, NY, USA; Museum für Naturkunde, Leibniz Institute for Evolution and Biodiversity Science, Berlin, Germany

Jedediah F. Brodie Biodiversity Research Centre, University of British Columbia, Vancouver, BC, Canada

Richard T. Corlett Xishuangbanna Tropical Botanical Garden, Chinese Academy of Sciences, Menglun, Mengla, Yunnan, China

Alys Granados Biodiversity Research Centre, University of British Columbia, Vancouver, BC, Canada; Department of Zoology, University of British Columbia, Vancouver, BC, Canada

Thomas N.E. Gray WWF Greater Mekong, Phnom Penh, Cambodia

Ignas M.A. Heitkönig Resource Ecology Group, Wageningen University, Wageningen, The Netherlands

Kavita Isvaran Centre for Ecological Sciences, Indian Institute of Science, Bengaluru, India

Yadvendradev V. Jhala Wildlife Institute of India, Dehradun, Uttarakhand, India

Hisashi Matsubayashi Department of Forest Science, Tokyo University of Agriculture, Setagaya, Japan

Kim McConkey School of Natural Sciences and Engineering, National Institute of Advanced Studies, Indian Institute of Science Campus, Bangalore, India

Grégoire Métais CR2P, Paléobiodiversité et Paléoenvironnements, UMR 7207 (CNRS, MNHN, UPMC), Sorbonne Université, Muséum National d'Histoire Naturelle, Paris Cedex 05, France

Charudutt Mishra The Snow Leopard Trust, Seattle, WA, USA; Nature Conservation Foundation, Mysore, India

Channa Phan WWF Cambodia Country Program, Phnom Penh, Cambodia

Soumya Prasad School of Life Sciences, Jawaharlal Nehru University, JNU New Campus, New Delhi, India; Nature Science Initiative, Dehradun, India

Herbert H.T. Prins Resource Ecology Group, Wageningen University, Wageningen, The Netherlands

Sovanna Prum WWF Cambodia Country Program, Phnom Penh, Cambodia; Forestry Administration, Ministry of Agriculture Forestry and Fisheries, Khann Daun Penh, Phnom Penh, Cambodia

Mahesh Sankaran Ecology & Evolution Group, National Centre for Biological Sciences, Tata Institute of Fundamental Research, Bangalore, India; School of Biology, University of Leeds, Leeds, UK

Sachin Sridhara James Cook University, Cairns, QLD, Australia

Kulbhushansingh R. Suryawanshi The Snow Leopard Trust, Seattle, WA, USA; Nature Conservation Foundation, Mysore, India

Chapter 1
Introduction: The Large Herbivores of South and Southeast Asia—A Prominent but Neglected Guild

Farshid S. Ahrestani and Mahesh Sankaran

Abstract The large herbivores of South and Southeast Asia comprise an ancient and diverse guild with a long history of association with humans. To this day, our knowledge of the ecology of these herbivores, and the ecological roles they play in ecosystems, remains largely inadequate. In this edited collection of chapters, we attempt to synthesize and integrate ongoing research on large herbivores in the region, and highlight directions for future research that is critical for the management of large herbivores in South and Southeast Asia. Given the dramatic population declines and range contractions of large herbivores in the region, there is an urgent need for scientifically based management efforts to ensure both their continued persistence as well as the integrity of the ecological services they provide in ecosystems.

Keywords Artiodactyla · Biogeographic realms · Perisodactyla · Species richness and diversity · Ungulate ecology

As primary consumers, large (>5 kg) terrestrial mammalian herbivores are an integral component of nearly every ecosystem on Earth (Duncan et al. 2006). Besides their obvious and irreplaceable role in supporting viable populations of large carnivores, large herbivores also play other critical roles in ecosystems as regulators of energy and nutrient cycles, modulators of plant community composition and grassland-woodland transitions, and as agents of seed dispersal (Janzen

---

F.S. Ahrestani (✉)
Department of Ecology, Evolution and Environmental Biology, Columbia University, New York, NY 10027, USA
e-mail: farshid.ahrestani@gmail.com

M. Sankaran
Ecology & Evolution Group, National Centre for Biological Sciences, Tata Institute of Fundamental Research, GKVK Campus, Bellary Road, Bangalore 560065, India
e-mail: mahesh@ncbs.res.in

M. Sankaran
School of Biology, University of Leeds, Leeds LS2 9JT, UK

© Springer Science+Business Media Dordrecht 2016

F.S. Ahrestani and M. Sankaran (eds.), *The Ecology of Large Herbivores in South and Southeast Asia*, Ecological Studies 225,
DOI 10.1007/978-94-017-7570-0_1

1984; Olff and Ritchie 1998; Olff et al. 2002; Augustine and Mcnaughton 2004; Couvreur et al. 2004; Burns et al. 2009). Large herbivores are also recognized as 'keystone' species in many ecosystems, influencing not only the structure of vegetation but also the richness, abundance and distribution of several other taxa ranging from arthropods to birds (Duncan et al. 2006; Pringle et al. 2007; Palmer et al. 2008; Greenwald et al. 2008; Banks et al. 2010; Goheen et al. 2010; Foster et al. 2014). Their loss can thus have effects that cascade through ecosystems (Ripple and Beschta 2006; Beschta and Ripple 2012). Given that the majority of large herbivore species, particularly those in developing tropical countries, are undergoing dramatic population declines and range contractions, there is an urgent need for scientifically based management efforts to ensure both their continued persistence as well as the integrity of the ecological services they provide in ecosystems (Ripple et al. 2015).

For the general public, and many scientists alike, the term 'large herbivore' is synonymous with Africa, bringing to mind images of wide expanses of savannas teeming with ungulates. Indeed, Africa with 94[1] large herbivore species—the majority of which inhabit the Afrotropic biogeographic realm (Udvardy 1975)—is the most species rich of all continents. However, with 83[1] species (Table 1.1), the countries of South and Southeast Asia (SSEA; Fig. 1.1), which collectively comprise the bulk of Indo-Malayan biogeographic realm (Udvardy 1975), support more than twice as many large herbivore species per unit area when compared to Africa (8.22 vs. 3.11 species per million km^2; also see Chap. 2). Further, many protected areas in South Asia also support biomass densities of large mammalian herbivores that are comparable to those found in many flagship parks in Africa (Karanth and Sunquist 1992). And yet, despite the global significance of SSEA in terms of its large mammalian herbivore diversity and abundance, our knowledge of the herbivores that inhabit its ecosystems, and the ecological roles they play in them, remains woefully inadequate.

The large herbivore guild of SSEA is both diverse and ancient. Evolutionary radiations of large herbivores occurred in Asia millions of years ago (see Chap. 2), and large herbivores in SSEA are today found across a wide range of habitats and environmental gradients, from deserts and arid lands to wet evergreen tropical forests, and from the plains to over 5000 m asl. Some species such as the sambar *Rusa unicolor*—arguably the most widely distributed large herbivore in SSEA—are found in a variety of habitats across wide precipitation, temperature, and altitudinal gradients, while there are species that are found on islands in Southeast Asia with highly restricted ranges. For example, no more than 300 tamaraw *Bubalus mindorensis* are left in the wild on Mindoro Island, Philippines.

Large herbivores in SSEA also have a long history of association with humans, perhaps more so than anywhere else on Earth. Species such as elephant, rhinoceros, gaur, water buffalo, markhor and ibex figure prominently as motifs on seals of the

[1]The species list for large herbivores was drawn primarily from Wilson and Reeder (2005) while incorporating widely accepted updates by the IUCN.

Table 1.1 List of the large herbivore species found in South and Southeast Asia, their IUCN status, and their presence (X) in the different countries in the region

Species (Common name)	IUCN	SRI	AFG	PAK	IND	NPL	BNG	BHU	MYA	THD	LAO	CAM	VTN	MAL	IDO	PHI
Order Cetartiodactyla (Artiodactyla)																
Family Bovidae																
Subfamily Bovinae																
Bos gaurus (gaur) Domestic form: *Bos frontalis* (gayal)	V				X	X	X	X	X	X	X	X	X	X		
Bos javanicus (banteng) Domestic form: *Bos javanicus*	E								X	X	X	X	X	X	X	
Bos mutus (wild yak) Domestic form: *Bos grunniens*	V				X	X										
Bos sauveli (kouprey)	CE											X				
Boselaphus tragocamelus (nilgai)	LC			X	X	X										
Bubalus arnee (water buffalo) Domestic form: *Bubalus bubalis*	E				X	X		X	X	X		X				
Bubalus depressicornis and *quarlesi* (lowland and mountain anoa)	E														X	
Bubalus mindorensis (tamaraw)	CE															X
Pseudoryx nghetinhensis (saola)	CE										X		X			
Tetracerus quadricornis (four-horned antelope, chousingha)	V				X	X										
Subfamily Antilopinae																
Antilope cervicapra (blackbuck)	NT				X											
Budorcas taxicolor (takin)	V				X			X	X							
Capra aegagrus (wild goat) Domestic form: *Capra hircus*	V		X	X												

(continued)

Table 1.1 (continued)

Species (Common name)	IUCN	SRI	AFG	PAK	IND	NPL	BNG	BHU	MYA	THD	LAO	CAM	VTN	MAL	IDO	PHI
Capra falconeri (markhor)	E		X	X	X											
Capra sibirica (Himalayan ibex)	LC		X	X	X											
Capricornis milneedwardsii (Chinese Serow)	NT								X	X	X	X	X			
Capricornis rubidus (red serow)	NT				X		X		X	X						
Capricornis sumatraensis (southern serow)	V									X				X	X	
Capricornis thar (Himalayan serow)	NT				X	X	X	X	X							
Gazella bennettii (Indian gazelle/chinkara)	LC		X	X	X											
Gazella subgutturosa (goittered gazelle)	V		X	X												
Hemitragus jemlahicus (Himalayan tahr)	NT				X	X		X								
Naemorhedus baileyi (red goral)	V				X				X							
Naemorhedus goral (gray goral)	NT			X	X	X		X								
Naemorhedus griseus (Chinese goral)	V				X				X	X	X		X			
Nilgiritragus hylocrius (Nilgiri tahr)	E			X												
Ovis ammon (argali sheep)	NT		X	X	X											
Ovis orientalis (urial) Domestic form: *Ovis aries*	V		X	X	X											
Pantholops hodgsonii (Tibetan antelope, chiru)	E				X											
Pseudois nayaur (bharal, Himalayan blue sheep)	LC			X	X	X		X	X							

(continued)

Table 1.1 (continued)

Species (Common name)	IUCN	SRI	AFG	PAK	IND	NPL	BNG	BHU	MYA	THD	LAO	CAM	VTN	MAL	IDO	PHI
Family Cervidae **Subfamily** Cervinae																
Axis axis (chital)	LC	X			X	X	X	X								
Axis calamianensis (Calamian deer)	E															X
Axis kuhlii (Bawean deer)	CE														X	
Axis porcinus (hog deer)	E			X	X	X	X	X	X							
Cervus elaphus (red deer, wapiti, American elk)	LC		X	X	X											
Muntiacus atherodes (Bornean yellow muntjac)	LC														X	
Muntiacus crinifrons (black muntjac)	V								X							
Muntiacus feae (Fea's muntjac)	DD								X	X						
Muntiacus muntjak (red muntjac)	LC	X			X	X	X	X	X	X	X	X	X	X	X	
Muntiacus puhoatensis (Pu Hoat muntjac)	DD										X		X			
Muntiacus putaoensis (leaf deer)	DD				X				X							
Muntiacus rooseveltorum (Roosevelt's muntjac)	DD										X		X			
Muntiacus truongsonensis (Annamite muntjac)	DD										X		X			
Muntiacus vuquangensis (giant muntjac)	E										X	X	X			
Rucervus duvaucelii (barasingha, swamp deer)	V				X	X										
Rucervus eldii (Eld's deer, thamin)	E				X				X		X	X				
Rusa alfredi (Philippine spotted deer)	E															X
Rusa marianna (Philippine sambar)	V															X

(continued)

Table 1.1 (continued)

Species (Common name)	IUCN	SRI	AFG	PAK	IND	NPL	BNG	BHU	MYA	THD	LAO	CAM	VTN	MAL	IDO	PHI
Rusa timorensis (rusa)	V														X	
Rusa unicolor (sambar)	V	X			X	X	X	X	X	X	X	X	X	X	X	
Family Tragulidae																
Moschiola indica (Indian chevrotain)	LC				X	X										
Moschiola meminna (Sri Lankan chevrotain)	LC	X														
Tragulus javanicus (Javan chevrotain)	DD														X	
Tragulus kanchil (lesser Oriental chevrotain)	LC								X	X	X	X	X	X	X	
Tragulus napu (greater Oriental chevrotain)	LC								X	X				X	X	
Tragulus nigricans (Balabac chevrotain)	E															X
Tragulus versicolor (silver-backed chevrotain)	DD												X			
Tragulus williamsoni (Yunnan chevrotain)	DD									X						
Family Moschidae																
Moschus berezovskii (Forest musk deer)	E												X			
Moschus chrysogaster (Himalayan musk deer)	E				X	X		X								
Moschus cupreus (Kashmir musk deer)	E		X	X	X											
Moschus fuscus (black musk deer)	E				X	X		X	X							
Moschus leucogaster (white-bellied musk deer)	E				X	X		X								
Family Suidae																

(continued)

Table 1.1 (continued)

Species (Common name)	IUCN	SRI	AFG	PAK	IND	NPL	BNG	BHU	MYA	THD	LAO	CAM	VTN	MAL	IDO	PHI
Babyrousa babyrussa (golden babirusa)	V														X	
Babyrousa celebensis (Sulawesi babirusa)	V														X	
Babyrousa togeanensis (Togian babirusa)	E														X	
Porcula salvanius (pygmy hog)	CE				X			X								
Sus ahoenobarbus (Palawan bearded pig)	V															X
Sus barbatus (bearded pig)	V													X	X	
Sus bucculentus (Heude's pig, Vietnam warty pig)	DD												X			
Sus cebifrons (Visayan warty pig)	CE															X
Sus celebensis (Sulawesi warty pig)	NT														X	
Sus oliveri (Mindoro warty pig)	E															X
Sus philippensis (Philippine warty pig)	V															X
Sus scrofa (Eurasian wild pig) *Domestic form*: *Sus domesticus*	LC	X	X	X	X	X	X	X	X	X	X	X	X	X	X	
Sus verrucosus (Javan warty pig)	E														X	
Order Perissodactyla																
Family Equidae																
Equus hemionus (Asiatic wild ass, kulan)	E			X	X											
Equus kiang (Tibetan wild ass, kiang)	LC			X	X	X										
Family Rhinocerotidae																
Dicerorhinus sumatrensis	CE														X	

(continued)

Table 1.1 (continued)

Species (Common name)	IUCN	SRI	AFG	PAK	IND	NPL	BNG	BHU	MYA	THD	LAO	CAM	VTN	MAL	IDO	PHI
(Sumatran rhinoceros)																
Rhinoceros sondaicus (Javan rhinoceros)	CE														X	
Rhinoceros unicornis (Indian rhinoceros)	V				X	X										
Family Tapiridae																
Tapirus indicus (Asian tapir)	E													X	X	
Order Proboscidea																
Family Elephantidae																
Elephas maximus (Asian elephant)	E	X			X	X	X	X	X	X	X	X	X	X	X	

The countries in the region are: *SRI* Sri Lanka; *AFG* Afghanistan; *PAK* Pakistan; *IND* India; *NPL* Nepal; *BNG* Bangladesh; *BHU* Bhutan; *MYA* Myanmar; *THD* Thailand; *LAO* Laos; *CAM* Cambodia; *VTN* Vietnam; *MAL* Malaysia; *IDO* Indonesia; *PHI* The Philippines

IUCN status: *LC* Least Concern; *V* Vulnerable; *NT* Near Threatened; *E* Endangered; *CE* Critically Endangered; *DD* Data Deficient

Note The species list was drawn primarily from Wilson and Reeder (2005) while incorporating widely accepted updates by the IUCN. We confined the Phylogeny of the Orders and their subunits in accordance with Hassanin et al. (2012)

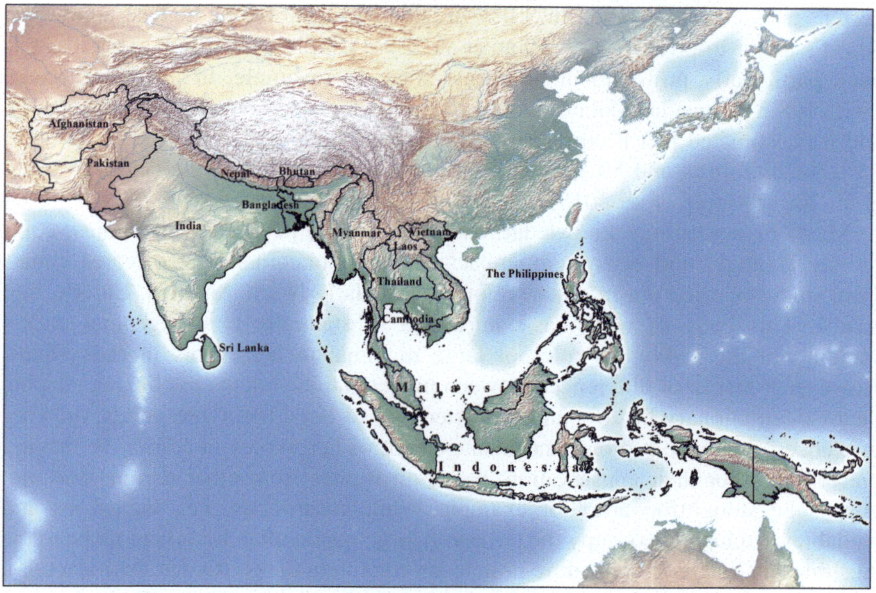

Fig. 1.1 Map showing the countries of South and Southeast Asia. (*Map credit* Sivakumar Sankar and Srinivas Vaidyanathan)

Indus valley civilization, dating back more than 5000 years. The Indus Valley region in Northwest India and Pakistan and westward to the Fertile Crescent (Southwest Asia) is also considered to be one of the principal centers of domestication of cattle (which resulted in the cattle breed *Bos indicus),* pigs, sheep and wild goats, while East Asia (China and countries south of it) is believed to have been a center for the domestication of buffalo, pigs and yak (Bruford et al. 2003; Chessa et al. 2009). Other large herbivores that have been domesticated or tamed in SSEA include the banteng *Bos javanicus*, Asian elephant *Elephas maximus*, and gaur *Bos gaurus*. Large herbivores have been, and continue to be an integral component of the fabric, culture, folklore, and religious practices of the region, and have also played important roles in ancient warfare in India and Central Asia. Despite this antiquity of association with humans, SSEA has the distinction of having one of the world's most recently discovered large herbivore species—the saola *Pseudoryx nghetinhensis*, a long-horned antelope native to dense forests in Laos and Vietnam was discovered for the first time by conservationists in 1992 (Dung et al. 1993).

Habitat loss, land-use change and hunting are pervasive threats to large herbivore populations worldwide, and nearly 60 % of all extant large herbivore species are today threatened with extinction (Ripple et al. 2015). The problem is particularly pronounced in SSEA, which is home to the highest number of threatened large herbivore species globally (Ripple et al. 2015). Over the past century, SSEA has lost nearly 50 % its forest cover (Houghton 1994; Richards and Flint 1994), and the current rates of deforestation in SSEA are considered to be the highest among the world's major tropical regions (Zhao et al. 2006; Ripple et al. 2015). This loss has been particularly profound in Southeast Asia; for example, in just one decade, 1972–1982, peninsular Malaysia lost 18 % of its forested area (Brown et al. 1991), and it is estimated that if current trends continue, Southeast Asia could lose up to 75 % of its forests and 50 % of its biodiversity by the end of the century (Sodhi et al. 2004; Ripple et al. 2015).

Much of our current knowledge on the behavior, population dynamics, community ecology, and ecosystem impacts of large mammalian herbivores comes from studies carried out in other regions, particularly North America, Europe, and Africa. This body of work has contributed substantially to our general understanding of the ecological roles of large mammalian herbivores in ecosystems. However, because of differences between biogeographic realms in climate, edaphic factors, evolutionary history, and characteristics of large herbivore guilds, drawing inferences about the ecology and management of herbivores in SSEA, or making prescriptions for their conservation, based on studies from other regions may not always be straightforward. For example, while macropods, i.e., kangaroos and wallabies, exclusively occupy the terrestrial large mammalian herbivore niche in the Australasian biogeographical region, even-toed ungulates (Order Artiodactyla), odd-toed ungulates (Order Perisodactyla), and elephants (Order Probodscidae) dominate the large herbivore assemblages in the rest of the world. The range ofbody masses, as well as the modal body mass of the herbivore guild varies across biogeographic zones and continents. Compared to ungulates, macropods that dominate the Australasian realm have a more restricted body size range, as well as a lower modal body mass (Fritz and Loison 2006). SSEA and Africa are fairly similar in terms of their herbivore body mass range (SSEA:5–3500 kg (chevrotain—elephant), Africa: 5–5000 kg (chevrotain/dik-dik—elephant)), but the range of body weights of ungulates in North America, South America, and Europe are more restricted (see Fritz and Loisson 2006). Although SSEA and Africa support herbivore assemblages with similar body mass ranges, a key feature that distinguishes their large herbivore guilds is that 30 % of the species found in SSEA are deer species, while only one of the 94 large herbivore species found in Africa is a deer.

As both human population growth and the loss of natural habitat continue unabated, designing appropriate management strategies for the conservation of large herbivores in SSEA based on a sound knowledge of their ecology is becoming increasingly critical. Our objective in putting this edited volume together was twofold: (i) to showcase some of the current ongoing work on large herbivore ecology in SSEA and (ii) to highlight gaps in our knowledge base that catalyze future research on the ecology of large herbivores in SSEA. There is, however, a paucity of knowledge on large herbivores from SSEA. For example, a quick search on the ISI Web of Science database on published research on large herbivores in Asia unearthed 86 relevant peer reviewed publications between 2005 and February 2015, in comparison to 409 publications from the same period in Africa, where the large herbivores of SSEA find their closest analogues (Fig. 1.2; see figure title for details). Although by no means comprehensive, our search is nevertheless

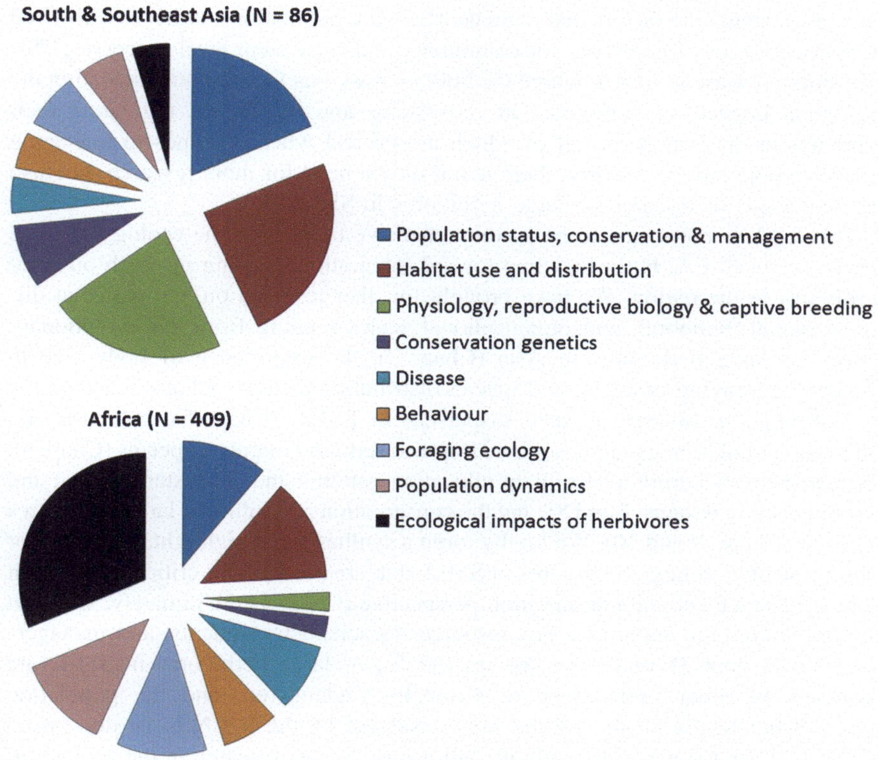

Fig. 1.2 A comparison of research efforts on native large herbivore species in South and Southeast Asia (SSEA) and Africa between January 2010 and February 2015. Data for the graphs were generated based on searching the ISI Web of Science database using the search string "topic = (ungulate OR herbivore OR grazing OR browsing) AND (*continent* OR *countries*) NOT (arthropod* OR fish* OR insect* OR butterfly*), where '*continent*' and '*countries*' represent Asia or Africa and their constituent countries, respectively. We excluded all studies that focused exclusively on domestic herbivores, those that only simulated herbivory under either field or greenhouse conditions, and those that focused primarily on the chemical analysis of browse species. For Asia, we only included studies that were carried out within the 15 countries (see Fig. 1.1 and Table 1.1) that comprise South and Southeast Asia. Studies were subjectively classified into different focal research areas based on the primary objectives of the study. Our final dataset included 86 studies from SSEA and 409 from Africa. While we are aware that the search is by no means comprehensive, we believe it is nevertheless reflective of the relative efforts expended on research on native large herbivore species in the two regions

illustrative of the dearth of research on herbivores in SSEA. Importantly, there are also critical differences in the nature of the research being carried out, and the kinds of questions being addressed, in both regions. The bulk of work in Asia focuses on quantifying population sizes (primarily for endangered species), habitat use, and distribution patterns of species, and studying the physiology and reproductive biology of species in captivity (Fig. 1.2). Only a small fraction of studies in SSEA have investigated the factors that regulate herbivore populations, or have considered the consequences of herbivory for community and ecosystem level processes. This is in stark contrast to Africa, where the bulk of work has focused on elucidating the ecological impacts of herbivores in ecosystems and the factors regulating their dynamics in the long term, all of which are critical when it comes to managing herbivore populations. Clearly, there is not only a need for more research, but also different kinds of research on large herbivores in SSEA.

This book is not meant to be a comprehensive treatise on the ecology of large herbivores in SSEA, but a first step toward integrating ongoing research on large herbivores in the region. We have brought together research on herbivores at different spatial, temporal, and organizational scales ranging from the evolutionary history of large herbivores in Asia (Chap. 2), the importance of body size in influencing foraging strategies and species distribution patterns (Chaps. 3 and 4), the role of large herbivores as seed dispersers in SSEA (Chap. 5), behavior and long-term population dynamics of a widespread semiarid ungulate species (Chap. 6), consequences of herbivory for plant allocation patterns and ecosystem carbon and nutrient cycling (Chaps. 7 and 8), and the conservation and future of large herbivores in SSEA (Chaps. 9 and 10). We finally forge a synthesis and highlight directions for future research on large herbivores in SSEA that are particularly critical and urgent from a management and conservation perspective (Chap. 11). Ultimately, we want that this volume to serve as a key resource for scientists, students, and managers alike. Given that 31 of the 83 species (37 %) of large herbivores in SSEA are classified as either Endangered or Critically Endangered, and the population trends of practically all the species are considered by the IUCN to be decreasing (Table 1.1), we hope that this volume will inspire further research of the ecology of this prominent but understudied guild in SSEA.

References

Augustine DJ, Mcnaughton SJ (2004) Regulation of shrub dynamics by native browsing ungulates on East African rangeland. J Appl Ecol 41:45–58

Banks JE, Jackson C, Hannon LM et al (2010) The cascading effects of elephant presence/absence on arthropods and an Afrotropical thrush in Arabuko-Sokoke Forest, Kenya. Afr J Ecol 48:1030–1038

Beschta RL, Ripple WJ (2012) The role of large predators in maintaining riparian plant communities and river morphology. Geomorphology 157:88–98

Brown S, Gillespie AJ, Lugo AE (1991) Biomass of tropical forests of south and southeast Asia. Can J For Res 21:111–117

Bruford MW, Bradley DG, Luikart G (2003) DNA markers reveal the complexity of livestock domestication. Nat Rev Genet 4:900–910

Burns CE, Collins SL, Smith MD (2009) Plant community response to loss of large herbivores: comparing consequences in a South African and a North American grassland. Biodivers Conserv 18:2327–2342

Chessa B, Pereira F, Arnaud F et al (2009) Revealing the history of sheep domestication using retrovirus integrations. Science 324:532–536

Couvreur M, Christiaen B, Verheyen K et al (2004) Large herbivores as mobile links between isolated nature reserves through adhesive seed dispersal. Appl Veg Sci 7:229–236

Duncan P, Danell K, Bergstrom R et al (2006) Introduction to large herbivore ecology. In: (ed) Large Herbivore Ecology, Ecosystem Dynamics and Conservation, Cambridge University Press, New York, p 1-18

Dung VV, Giao PM, Chinh NN et al (1993) A new species of living bovid from Vietnam. Nature 363:443–445

Foster CN, Barton PS, Lindenmayer DB (2014) Effects of large native herbivores on other animals. J Appl Ecol 51:929–938

Fritz H, Loison A (2006) Large herbivores across biomes. In: Danell K, Duncan P, Bergstrom R, Pastor J (eds) Large herbivore ecology ecosystem dynamics and conservation. Cambridge University Press, Cambridge, pp 19–49

Goheen JR, Palmer TM, Keesing F et al (2010) Large herbivores facilitate savanna tree establishment via diverse and indirect pathways. J Anim Ecol 79:372–382

Greenwald KR, Petit LJ, Waite TA (2008) Indirect effects of a keystone herbivore elevate local animal diversity. J Wildl Manag 72:1318–1321

Hassanin A, Delsuc F, Ropiquet A et al (2012) Pattern and timing of diversification of Cetartiodactyla (Mammalia, Laurasiatheria), as revealed by a comprehensive analysis of mitochondrial genomes. CR Biol 335:32–50

Houghton R (1994) The worldwide extent of land-use change. Bioscience 44:305–313

Janzen DH (1984) Dispersal of small seeds by big herbivores: foliage is the fruit. American Naturalist 123:338–353

Karanth KU, Sunquist ME (1992) Population structure, density and biomass of large herbivores in the tropical forests of Nagarahole, India. J Trop Ecol 8:21–35

Olff H, Ritchie ME (1998) Effects of herbivores on grassland plant diversity. Trends Ecol Evol 13:261–265

Olff H, Ritchie ME, Prins HHT (2002) Global environmental controls of diversity in large herbivores. Nature 415:901–904

Palmer TM, Stanton ML, Young TP et al (2008) Breakdown of an Ant-Plant Mutualism Follows the Loss of Large Herbivores from an African Savanna. Science 319:192–195

Pringle RM, Young TP, Rubenstein DI et al (2007) Herbivore-initiated interaction cascades and their modulation by productivity in an African savanna. Proc Natl Acad Sci 104:193–197

Richards JF, Flint EP (1994) A century of land-use change in South and Southeast Asia. In: (ed) Effects of Land-Use Change on Atmospheric CO2 Concentrations, Springer, pp 15–66

Ripple WJ, Beschta RL (2006) Linking a cougar decline, trophic cascade, and catastrophic regime shift in Zion National Park. Biol Conserv 133:397–408

Ripple WJ, Newsome TM, Wolf C et al (2015) Collapse of the world's largest herbivores. Sci Adv. doi: 10.1126/sciadv.1400103

Sodhi NS, Koh LP, Brook BW et al (2004) Southeast Asian biodiversity: an impending disaster. Trends Ecol Evol 19:654–660

Udvardy MDF (1975) A classification of the biogeographical provinces of the world. Morges (Switzerland): International Union of Conservation of Nature and Natural Resources. IUCN Occasional Paper no. 18

Wilson DE, Reeder DM (2005) Mammal species of the world—a taxonomic and geographic reference. Johns Hopkins University Press, Baltimore, MD, USA

Zhao S, Peng C, Jiang H et al (2006) Land use change in Asia and the ecological consequences. Ecol Res 21:890–896

Chapter 2
Evolutionary History of the Large Herbivores of South and Southeast Asia (Indomalayan Realm)

Faysal Bibi and Grégoire Métais

Abstract Modern day South and Southeast Asia falls almost entirely within a single biogeographic region, the Indomalayan Realm. Here, we review the Cenozoic geological and environmental history of the Indomalayan Realm, and the evolutionary history of the large herbivorous mammals that have inhabited it. For the most part bounded by major physical features, the cohesiveness of the Indomalayan Realm in its mammalian faunas can be recognized as far back as the middle Miocene or even earlier. Many of the extant large herbivores of the Indomalayan Realm have a diverse fossil record in this part of the world, though a few, such as cervids, are relative newcomers. Many extant clades that are not currently present in the Indomalayan Realm had records in the region up to the Pleistocene, including giraffids, hippopotamids, and reduncin antelopes. The island archipelago of Southeast Asia in particular witnessed radiative speciation of numerous clades including proboscideans and ruminants, at least through the climatic cycles of the Pleistocene, if not earlier. If there is a single common thread to the evolutionary history of Indomalaya's large herbivores, it may be the loss of taxonomic diversity, with much greater taxonomic representation recorded at numerous times in the past in almost all large herbivore clades. Today, diversity loss continues at the hands of anthropogenic, rather than natural, environmental causes, which threaten to violently curtail millions of years of evolutionary heritage.

F. Bibi (✉)
Department of Mammalogy, American Museum of Natural History, 200 Central Park West, New York, NY 10024, USA
e-mail: Faysal.Bibi@mfn-berlin.de

F. Bibi
Museum für Naturkunde, Leibniz Institute for Evolution and Biodiversity Science, Invalidenstrasse 43, Berlin 10115, Germany

G. Métais
CR2P, Paléobiodiversité et Paléoenvironnements, UMR 7207 (CNRS, MNHN, UPMC), Sorbonne Université, Muséum National d'Histoire Naturelle, 8 rue Buffon, CP 38, 75231, Paris Cedex 05, France
e-mail: gregoire.metais@mnhn.fr

© Springer Science+Business Media Dordrecht 2016
F.S. Ahrestani and M. Sankaran (eds.), *The Ecology of Large Herbivores in South and Southeast Asia*, Ecological Studies 225,
DOI 10.1007/978-94-017-7570-0_2

Keywords Cenozoic · Climate · Conservation · Evolution · Paleobiogeography · Paleoecology

2.1 Introduction

> The region is of comparatively small extent, but it has a very diversified surface, and is proportionately very rich.

So begins Wallace's (1876) chapter on the Oriental Region, a geographic area characterized by a cohesive assemblage of birds, mammals, reptiles, and insects that differentiates it from the rest of Asia to the north, Australasia to the southeast, and Africa and Arabia to the west. As the smallest of the world's six main continental biogeographic regions, the Indomalayan Realm (as it is now better known, Udvardy 1975) is not to be underestimated as a center of global biodiversity, both present and past. In terms of large herbivores, the Indomalayan Realm is three times as rich in species as the Neotropics, which is twice as large in area, and almost as rich in species as the Afrotropics, which is three times as large in area. Its smaller size is also no indication of its animal diversity—three of the world's most recently discovered large mammals, the kouprey, the saola, and the large antlered muntjac have come from the Indomalayan area. The saola was only described in 1992 (Dung et al. 1993), indicating that even in the late twentieth century, large animals entirely unknown to the outside world were still to be found in little explored stretches of Southeast Asian forests. More recent but no less spectacular, the discovery of the living Laotian rodent *Laonastes aenigmamus*, nowadays restricted to the rugged Annamite Mountains (Lao People's Democratic Republic) that turns out to be a surviving member of a clade (Diatomyidae) that was formerly believed to have been extinct for more than 11 million years (Dawson et al. 2006).

Though names have changed, Wallace's faunal regions have remained much the same as when he defined them over 100 years ago (though see Holt et al. 2013, for a recent update, including newly defined Saharo-Arabian and Sino-Japanese realms, and an Oriental-Oceanian boundary east of Sulawesi). With the notable exception of Afghanistan, modern South and Southeast Asia (SSEA) falls almost entirely within the Indomalayan Realm (Fig. 2.1). In discussing the late Cenozoic evolution of this region's large herbivores, the distinctiveness of the Indomalayan faunas is a principal theme, namely because the physical and climatic barriers to dispersal that account for this region's cohesive character appear to have been in place since at least the middle Miocene, if not even earlier (e.g., Wang et al. 2013).

The fossil record of the Indomalayan region is as impressive as its living faunas. A large number of fossil sites are known from throughout SSEA that record the Eocene to Holocene evolution of the modern physical, vegetational, and faunal characteristics of the region. Many of the extant large herbivores in SSEA have fossil records indicating a long period of presence and evolution in the region (e.g., bovids, tragulids). Others, though well established in SSEA today, are relative newcomers to the region (e.g., cervids). Yet other clades have far more complex histories than indicated

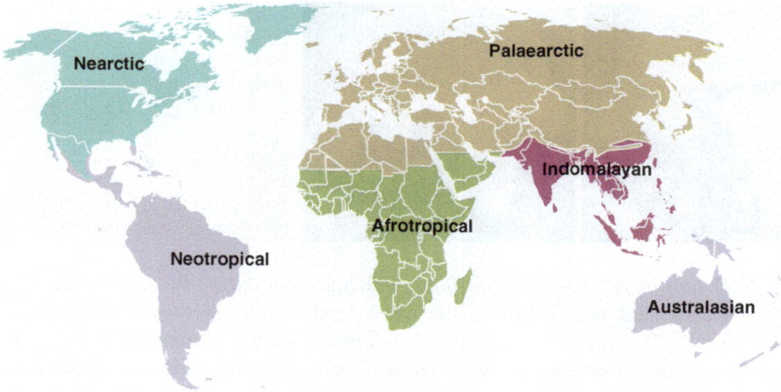

Fig. 2.1 Terrestrial Realms of the world. SSEA falls almost entirely within the Indomalayan Realm. Exact realm boundaries differ among different sources (modified from WikiMedia Commons, user *Lokal_Profil*, CC-BY-SA-2.5)

by their living representatives (e.g., proboscideans), and there are fossil taxa that are entirely absent from the region today (e.g., giraffids, hippopotamids).

The evolutionary history of the large herbivores of SSEA is a large topic, and no single paper can do justice to the immense wealth of geological, paleontological, archaeological, zoological, and genomic information on the subject. In what follows, we present some basic geological background to the development of the modern SSEA landscape (Sect. 2.2); followed by some of the broader scenarios of large herbivore evolution (Sect. 2.3), particularly regarding the emergence of present day climatic and geological configurations and the continuous interplay of environmental change and faunal evolution in the region (Sect. 2.4); an introduction to some of the main fossil deposits of the region yielding large herbivores (Sect. 2.5); and summaries of the main evolutionary features of the large herbivore clades of the Indomalayan Realm, both living and extinct (Sect. 2.6). We end with a brief consideration of SSEA's past, present, and future biodiversity (Sects. 2.7–2.8).

2.2 Geological Background

2.2.1 Physical Boundaries

The Indomalayan Realm is bounded by prominent physical barriers, topographical features resulting from the collision of the Indian Subcontinent with Eurasia during the early Cenozoic and the subsequent uplift of the Tibetan Plateau. The Indomalayan and Palaearctic Realms are separated by the Himalayan mountain chain, its tectonic tributaries to the west, and a complex transitional zone around the Yangtze River (Xie et al. 2002), that was established synchronously with the uplift of the eastern Tibetan Plateau (Zheng et al. 2013). The main geographical,

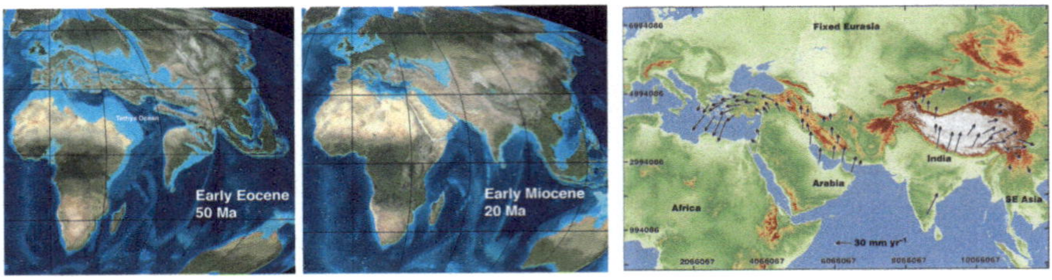

Fig. 2.2 Collision of Afro-Arabia and India with Laurasia during the Cenozoic, with closure of the Tethys Ocean and the formation of stable land bridges between southern and northern continents by 20 Ma, and the continuing development of the Himalayan-Alpine mountain belt thereafter (paleomaps from Blakey 2008, tectonic map from Edwards and Grasemann 2009)

topographical, and structural feature of SSEA is the Himalayan mountain chain, being the largest and highest mountainous zone on the planet, the Cenozoic history of which has greatly influenced the landscapes, climate, and evolutionary history of mammalian communities of the region. The Himalayas form part of the 10,000 km-long Alpine-Himalayan belt, which includes the European Alps, the Zagros Mountains of Iran, and the Indo-Burmese mountains, stretching from Spain to Indonesia, and that formed as a result of the closure of the Neotethys Ocean over the past 180 million years (Fig. 2.2).

2.2.2 Closure of the Neotethys Sea and the Formation of the Mountain Chains of Southern Asia

The closure of the Neotethys is the result of the breakup of the supercontinent of Gondwana from the early Jurassic onwards, and the associated northward movement of the Indian and Afro-Arabian plates, which began to collide with Eurasia by the middle Paleogene, or around 35 Ma (Allen and Armstrong 2008). The Alpine-Himalayan collision system, resulting from the closure of the Neotethys (its western part represented by the closed residual Mediterranean Sea) consists of high mountain chains and great plateaus that were formed during the Cenozoic. The Turkish–Iranian and Tibetan are the two great plateaus that were uplifted as a result of the closure of the Neotethys and subsequent continental collision (Şengör and Kidd 1979). The final closure of the Neotethys Ocean during the early Neogene triggered intense crustal deformation, with faulting and folding to accommodate major compressional forces that resulted in the high elevations and sharp topographic relief of much of southern and central Asia.

While the Alps, Zagros, Himalaya, and Indo-Burmese mountain chains are products of the same broad geodynamic process (i.e., closure of the Neotethys), the mode and timing of this process vary along the wide belt. Here, we focus on the two main areas where the disappearance of the Neotethys Ocean and the subsequent intercontinental collision have greatly shaped the topography and climate of southern and southeastern Asia (Popov et al. 2004).

2.2.3 Collision of the Afro-Arabian Continent with Eurasia

The collision of the Afro-Arabia plate with Eurasia has shaped the Western border of the Indomalayan Province and conditioned the land connection with Eurasia. The Indo-Pakistan plate is bounded by the Chaman-Ornach-Nal left lateral transform fault system and the Makran accretionary prism related to the northward subduction of the Arabian Sea lithosphere (Bender and Raza 1995). This collision brought about the gradual closure of the Neotethys Ocean through northward subduction of oceanic lithosphere, producing the Urumieh-Dokhtar volcanic arc in Iran (Dewey et al. 1973) and the Chagaï volcanic arc along the Pakistan and Afghanistan borders (Nicholson et al. 2010). Around the same time, at about 25 Ma, a new oceanic basin, the Red Sea, was formed as Arabia began to separate from Africa, and seismically.

Reconstructing the history of the closure of the Neotethys requires various sources of data that sometimes lead to contradictory conclusions, and geological structures such as the Zagros orogen or Anatolian crustal faults can be variously interpreted. Consequently, controversy surrounds the exact sequence of events and the timing of continental collision. Plate kinematic reconstruction (Van Hinsbergen et al. 2009) shows that during the early Cenozoic (ca. 56 Ma) the Tethyan oceanic basin was ca. 1300 km wide, while the Tethys Ocean itself was wider still as it covered much of the Arabian carbonate platform (Fig. 2.2). The Tethyan oceanic basin gradually narrowed through the Cenozoic, eventually closing and disconnecting the Indian and Atlantic Oceans from each other. The exact timing of this disconnection, however, is controversial. Oceanographic, plate tectonic, and climatic modeling studies as well as stable neodynium isotope data indicate the persistence of a marine connection between the Indian and Atlantic oceans through a residual Tethyan seaway well into the middle Miocene (Stille et al. 1996; McQuarrie et al. 2003). Geological evidence, however, indicates an initial collision of the submerged Arabian and Eurasian plate margins during the Late Eocene (ca. 35 Ma, Allen and Armstrong 2008), and marine faunal evidence indicates the progressive creation of separate Mediterranean and Indian Ocean marine realms shortly thereafter (Harzhauser et al. 2002, 2007). Both bivalve and gastropod assemblages show biogeographical divergence and two separate Tethyan sub-provinces during the Oligocene, with a greatly increased endemism during the early Miocene (Harzhauser et al. 2002, 2007). The collision of submerged continental margins did not immediately result in the formation of continental land bridges between Arabia and Asia. Late Oligocene mammals from Ethiopia indicate the seaway was still a barrier to terrestrial mammals at 27 Ma (Kappelman et al. 2003), but a major influx of Eurasian mammals into Africa around the Oligocene–Miocene boundary (23 Ma) indicates that a secure land connection between Afro-Arabia and Eurasia had developed by this time (Bernor et al. 1987).

The closure of the Neotethys resulted in the disruption and reorganization of global oceanic current systems, and this certainly contributed to the elevated global temperatures that characterized the Middle Miocene Climatic Optimum (MMCO) event at 17–15 Ma (Flower and Kennett 1994). This interval of global warming, detected in both marine and continental sediments, was marked by a 3–4 °C increase in global temperature (You et al. 2009).

2.2.4 Collision of the Indian Subcontinent with Eurasia

The Himalayas provides a classic example of the collision of two continental plates (e.g., Molnar and Tapponnier 1977; Yin 2010). The northward movement of the Indian plate can be reconstructed through magnetic anomalies on the Indian Ocean floor (Patriat and Achache 1984). Continental collision of India into Asia from the Eocene to the present day has resulted in major crustal shortening and thickening that has produced the Himalayan and Tibetan Plateau.

The tectonic evolution of the Indian plate, which traveled around 9000 km in 160 million years, is largely the story of its separation from Gondwana and its subsequent collision with the Kohistan-Ladakh Arc and Asia. By the start of the Cenozoic (65 Ma), the Indian Subcontinent had become an island continent entirely separated from Africa and Antarctica. Blazing northwards at a geologically rapid rate of ca. 20 cm/year, the Indian plate began its collision with Asia during the Early Eocene (ca. 50 Ma), slowing its rate down to 5 cm/year thereafter (Copley et al. 2010; Chatterjee et al. 2013). Most data suggest that the India-Eurasia continental collision began near the Paleocene-Eocene transition (ca. 55 Ma, Beck et al. 1995), and sedimentary deposits eroded from the rising Himalayas date back to the Middle–Late Eocene epoch (49–34 Ma), providing evidence for early mountain building (Najman et al. 2001; Zhuang et al. 2015). With the continuing convergence of India with Asia through the Paleogene, the northern edge of the Indian plate was fractured, thrusted, and uplifted, resulting in the formation of the Himalayas. The Higher, Lesser, and Sub-Himalayas are thus slices of the old Indian shield that have been stacked over each other along a series of southward verging thrusts. Fully, one third of the original Indian plate was subducted beneath Asia during the initial phases of collision and now lies buried beneath the Tibetan Plateau (Guillot et al. 2003). For roughly a million square kilometers beneath the Tibetan Plateau, the crust is about twice as thick as the average continental crust (Mattauer et al. 1999; Van der Voo et al. 1999).

2.2.5 Climatic Implications of Himalayan Uplift

Over long time scales, geography and topography and atmospheric circulation interact to produce major climatic patterns (Ruddiman 1997, e.g. Molnar et al. 2010), and the Indomalayan region is one in which the interactions between physical processes and biotic feedbacks are plain to see. The South and East Asian monsoon systems are major climatic features of the Indomalayan region that have shaped vegetation and biodiversity over many millions of years. Monsoons are

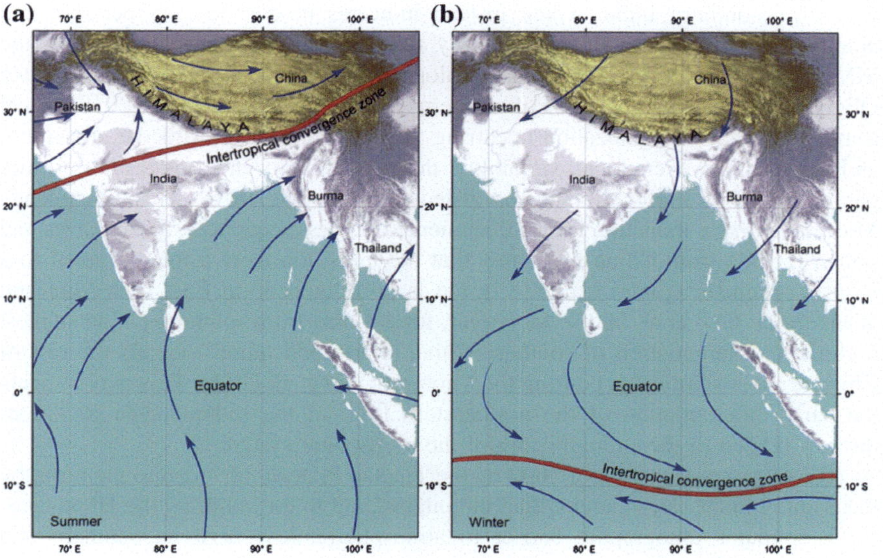

Fig. 2.3 The climate of southern and SE Asia is dominated by monsoon winds, which are related to the uplift of the Himalaya-Tibetan Plateau. **a** Summer winds from the southwest carry moisture from the Indian Ocean and bring heavy rains from June to September. **b** Winter winds blow from the northeast and carry little moisture. Temperatures remain high as the Himalayan mountains prevent cold air from passing southwards into the Indian Subcontinent (from Chatterjee et al. 2013)

strong onshore and offshore winds that are produced by differences in heat capacity between land and water (Fig. 2.3). In winter, the Himalayas and the Tibetan Plateau become cold, which cools the air above them. This cool, dense air descends and flows toward the Indian Ocean and displaces the warmer air there. In summer, the mountains and plateaus heat up, causing the air above them to rise and draw strong winds, the summer monsoons, in from the Indian Ocean. Sweeping northward from the Bay of Bengal, the Indian summer monsoon strikes the eastern Himalayas with great force, producing a tropical depression and torrential rain. This seasonal climatic pattern has profoundly shaped landscape and vegetation, and therefore the habitats of animals in the region.

When this monsoonal regime got started, and how it varied through the Cenozoic remain open questions. The age of inception of the South and East Asian monsoons remains controversial but a variety of environmental indicators suggest that the East Asian monsoon, at least, was established during the late Eocene (Licht et al. 2014) with an intensification by the start of the Miocene (ca. 24 Ma, Sun and Wang 2005), probably correlated with a period of major uplift. Interestingly, the marine climatic record suggests a breakdown of Indian monsoonal seasonality over the Himalayas and southern China between 17 and 15 Ma ago during the Mid-Miocene Climatic Optimum (MMCO) (You et al. 2009). However, this temporary disruption of Indian monsoonal intensity is not reflected in palynological data from southern India, which instead indicates modern-like temperature and precipitation cyclicity with only slightly reduced rainfall seasonality during the MMCO (Reuter et al. 2013). Palynological assemblages from different late Oligocene to Miocene localities of northern Thailand (Sepulchre et al. 2010) reveal a major climatic reorganization occurring near the Oligocene–Miocene transition, and significant vegetational variability during the Miocene, suggesting further parameters controlling climate and paleo-environments there. This Oligocene–Miocene climatic transition is also documented in loess sequences, fossil leaves and pollen records from China that show that the transition from a mainly zonal to a monsoon climate system occurred in the latest Oligocene in East Asia (Sun and Wang 2005; Guo et al. 2008). Moreover, fossil floras from southern China suggest a climatic differentiation of southern China (warm and humid) versus the rest of China (greater seasonality) during the Miocene (Yao et al. 2011). This may indicate the early differentiation of the northeastern limit of the Indomalayan Realm as directly related to the strengthening of the monsoonal system.

At an average elevation of 5000 m, the Tibetan Plateau, bounded to the north by the Kunlun Shan, Altyn, and Qilian mountains, and to the south by the Himalayas, forms a major topographic barrier to atmospheric flow between southern and Central Asia. The mountains and highlands forming the Tibetan Plateau dramatically affect continental albedo and atmospheric energy balance, stimulate intense seasonality on their southern slopes, and produce a massive rain shadow on their leeward (northern) slopes. The Tibetan Plateau has been an important contributor to long-term Cenozoic cooling, at both local and global scales, as well as to Northern Hemispheric glaciation (Hay et al. 2002). Rapid uplift of the surface topography of the Higher Himalaya by 8 Ma coincides with a substantial increase in precipitation and erosion of rising mountains as revealed from deep-sea sediments in the Bay of Bengal (Ganges-Brahmaputra drainage) and the Arabian Sea (Indus drainage) (Prell and Kutzbach 1992). According to Clift (2008), the intensification of monsoons 8 Ma ago (producing increasing erosion) may have been triggered when the Tibetan Plateau reached a threshold height of 2000 m. This period also records enhanced aridity in the Asian interior, and the onset of Indian and East Asian monsoons (An et al. 2001; Patnaik et al. 2012). Further uplift of the Himalayas and the Tibetan Plateau in the Pliocene also resulted in enhanced aridity in the Asian interior (Gobi and Mongolian deserts) (Ramstein et al. 1997; An et al. 2001), as evidenced by the deposition of extensive eolian (loess) deposits in central and eastern China at about 2.5 million years ago (An et al. 2001).

Though the inception and strengthening of the Indian-Asian monsoon system is classically linked to the uplift of the Tibetan Plateau during the Neogene (Prell and Kutzbach 1992; Clift and Plumb 2008), simulations show that a monsoon-like climatic regime existed during the Eocene regardless of the exhumation of the Tibetan Plateau (Huber and Goldner 2012). Models of early monsoonal dynamics have highlighted the role of several parameters that are often neglected in previous reconstructions of the monsoon's history, such as sea-surface temperature and greenhouse gas concentration (Huber and Goldner 2012; Licht et al. 2014).

Likewise, the opening and closure of oceanic gateways as a result of continental plate motions and related changes of ocean-atmosphere circulation probably had a major effect on global climate. For example Ramstein et al. (1997) and Fluteau et al. (1999) stressed the combined role of the closure of the Tethys and the retreat of the Paratethyan epicontinental sea (both occurring around the time of the Eocene–Oligocene transition) in the formation of large-scale patterns of monsoonal climate, at least as much as the influence of the uplift of the Tibetan Plateau (see also Zhang et al. 2007). Further complicating the picture is the fact that the early (Paleogene) history of the Asian monsoon is relatively poorly documented and additional data are needed to test large-scale paleoclimatic interpretations based on model simulations. It remains clear, however, that atmospheric circulation in the absence of the Himalayas and Tibetan Plateau would have been very different from what it is today.

2.3 Environment and Evolution

2.3.1 Modern Climates and Herbivore Habitats

The relationships of temperature, precipitation, and seasonality go a long way toward explaining the vegetational characteristics of any region on earth. Except at high altitudes, modern SSEA is characterized by the presence of a warm climate year-round. Precipitation in its northern areas from (Pakistan to Bangladesh and Myanmar) is monsoonal and markedly seasonal, with rains falling mainly in summer (June–September) and much of the rest of the year remaining relatively dry. The southern (equatorial) portions of SSEA (mainly Sri Lanka and SE Asia) have less strongly seasonal rainfall patterns, with some places receiving more rainfall during the northern winter (November–February). Much of the vegetational and faunal characteristics of southern Asia from Pakistan to Bangladesh are in turn a product of the monsoonal climate regime, resulting especially in a dominance of drier, more xeric habitats (deciduous forests, woodlands, grasslands) and a much lower proportion of humid forests than in equatorial and southeastern Asia.

Some of the large herbivore species of SSEA can be found living in a wide range of habitats, perhaps most tolerant among these being sambar deer (*Rusa unicolor*) and elephant (*Elephas maximus*). Many species, however, are restricted to humid forested habitats such as the tapir (*Tapirus indicus*), bovids like serows (*Capricornis* spp.) and the saola (*Pseudoryx nghetinhensis*), and cervids such as muntjacs (*Muntiacus* spp.), hog deer (*Axis porcinus*), and tufted deer (*Elaphodus*). The grasslands and dry forests of the Indian Peninsula are home to a range of species more tolerant of drier, more open habitats. Among these are axis deer (*Axis axis*), nilgai (*Boselaphus tragocamelus*), blackbuck (*Antilope cervicapra*), and chinkara (*Gazella bennettii*), the last being particularly well adapted to semi-arid and arid habitats.

2.3.2 The Late Miocene Reduction of Forested Habitats

The fossil record indicates that for much of the Cenozoic, humid forest habitats, today restricted to equatorial regions, covered much of the subtropical and temperate regions of the world. Global temperature reconstructions based on the oxygen isotopic composition of benthic foraminifera retrieved from oceanic drill cores indicate that global temperatures have been gradually decreasing since about 15 Ma (Zachos et al. 2001), or just after an especially warm period known as the Mid-Miocene Climatic Optimum. In SSEA, as in much of the world, the early and middle Miocene fossil record indicates that environments were much more humid and forested than today. The switch to drier, more open habitats is particularly well recorded in the Siwaliks of northern Pakistan, where the fossil ape *Sivapithecus* and other forest taxa had gone extinct by 8.1 Ma (e.g. Nelson 2007) as humid forest habitats fairly rapidly gave way to drier, deciduous forests and grasslands.

Through their finely resolved sedimentary and fossil sequences, the Siwalik deposits of the Potwar Plateau provide one of the world's best records of increasing aridity and increasing seasonality of terrestrial environments since 10 Ma. In particular, the Siwalik geochemical and faunal evidence indicates a transition from wet monsoonal forests before 8.5 Ma, to dry monsoonal forests at 7.0 Ma, to savannah habitats at 6.0 Ma (Badgley et al. 2008), with increasing proportions of grassland species into the Pliocene and Pleistocene (Patnaik 2015). The Siwaliks provided the first paleosol carbon isotopic data indicating a major vegetational replacement of C_3 vegetation habitats by C_4 vegetation in the late Miocene between about 8 and 6 Ma (Quade et al. 1989, 1992; Quade and Cerling 1995), a phenomenon that was later demonstrated to have occurred at many low-latitude locations around the world (Cerling et al. 1997). Barry et al. (2002) found elevated levels of speciation and extinction during this period of major vegetational change (Fig. 2.4). Badgley et al.

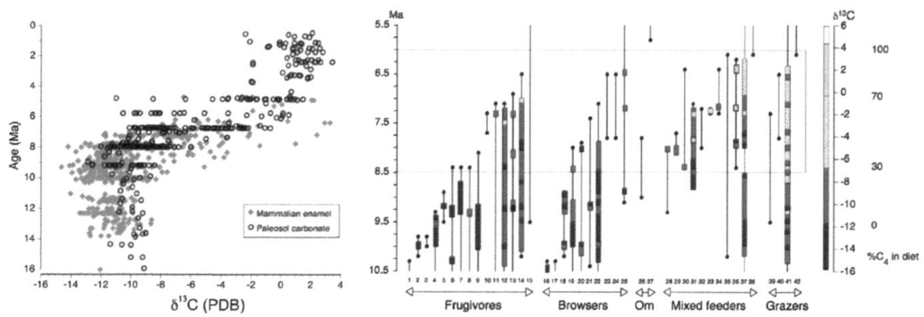

Fig. 2.4 At *left*, the carbon isotopic shift indicating rapid expansion of C_4 (tropical grass) habitats in northern Pakistan between 8 and 6 Ma as recorded in herbivore tooth enamel and in paleosol carbonates. At *right*, coincident with C_4 vegetation expansion, Siwalik faunas record a loss of many frugivorous and browsing lineages, and an increase in mixed feeders and grazers. Both panels from Badgley et al. (2008)

(2008) determined that the expansion of grassland habitats negatively impacted forest herbivores the most, with 15 out of 17 browsing and frugivorous lineages going extinct during this time, and while grazing and mixed feeding herbivores also suffered extinctions, these were balanced by an almost equal number of immigrants that preserved this dietary niche. Palynological data from the sub-Himalayan region in Nepal similarly provides evidence for the replacement of forests by grasslands between 8 and 6.5 Ma (Hoorn et al. 2000). A similar loss of humid forest habitats, termed as the 'mid-Vallesian crisis,' is recorded slightly earlier in southern Europe (e.g., Agustí et al. 1997, 2003), implicating global-scale climatic changes in the loss of forest habitats through the late Miocene.

2.3.3 The Pleistocene and Holocene: Extinctions and Insularity

Another time of great extinction among SSEA large herbivores is the late Pleistocene. The late Pleistocene to early Holocene transition (about 14–10,000 years BP) witnessed the extinction of many species of large herbivores across the globe. In the Americas and Australasia the extinctions were particularly severe, and in these regions and northern Eurasia all species of 1000 kg or greater were lost (Stuart 1999). In contrast, SSEA and sub-Saharan Africa appear to have been less affected, with elephants and rhinoceroses surviving (though just barely) in both regions. Though debate continually rages, the main causes of the late Pleistocene megaherbivore extinctions were probably a combination of human overkill and climate change, but the evidence surrounding extinction events, as well as their timing, varies from continent to continent. In general, there is a good correspondence between the first appearance of humans and the onset of mega herbivore extinctions in North America (11.5–11 ka), South America (15–10 ka), and Australasia (72–44 ka) (Martin 1967; Alroy 2001; Barnosky et al. 2004; Sandom et al. 2014). In contrast, humans or their tool making ancestors (e.g., *Homo erectus*) evolved in Africa, and have also lived in Europe and Asia, including SSEA, since the early Pleistocene, or around 1.8 Ma (Swisher et al. 1994). A popular view holds that the extinctions in sub-Saharan Africa and SSEA were less severe on account of a much older coevolution of humans and fauna (Martin 1966), but human overkill does not explain all the observed extinction patterns (MacPhee 1999; Barnosky et al. 2004; Louys et al. 2007; Campos et al. 2010). Factors such as disease might also have played a role (MacPhee and Marx 1997).

Detailed assessment of extinction severity in the late Pleistocene of SSEA is complicated by a relatively low number of late Pleistocene sites, and poor chronological control at most of these. However, Louys et al.'s (2007) faunal lists show that, though many large herbivore species did disappear, extinctions in the late Pleistocene of Southeast Asia may not necessarily have been more severe than

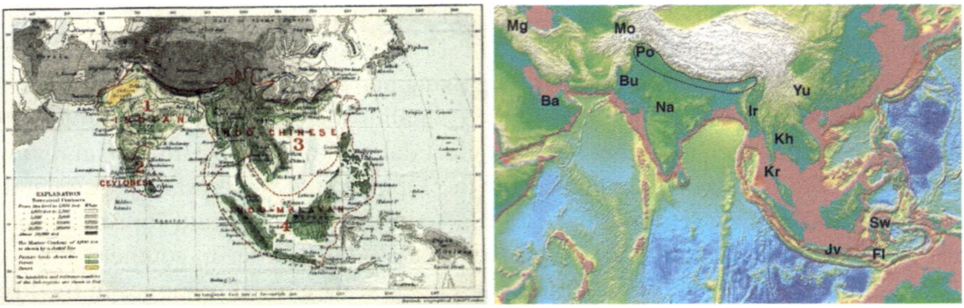

Fig. 2.5 *Left*, the Indomalayan (Oriental) Realm and its provinces, as described by Wallace (1876). *Right*, digital elevation map of SSEA showing the shallowness of the seaway around the Indonesian islands, and the locations of some of the fossil sites mentioned in the text. *Ba* Baynunah, United Arab Emirates; *Bu* Bugti Hills, Pakistan; *Fl* Flores, Indonesia; *Ir* Irrawady Basin, Myanmar; *Jv* Java, Indonesia; *Kh* Khorat, Thailand; *Kr* Krabi Basin, Thailand; *Mg* Marageh, Iran; *Mo* Molayan and Taghar, Afghanistan; *Na* Narbada Valley, India ; *Sw* Sulawesi, Indonesia; *Po* Potwar Plateau, Pakistan, and encircled in *dotted line* is the extent of the Siwalik deposits of the Indian Subcontinent; *Yu* Yunnan Province, China

during preceding intervals of time. This brings up the question of whether SSEA, like sub-Saharan Africa, experienced a pronounced late Pleistocene extinction pulse at all or whether it was all just ‘evolution as usual.’ If anything, as Louys et al. point out (also Woodruff 2010), fluctuating sea levels throughout the Pleistocene (especially over the last 800 kyr) probably had a much stronger influence in shaping the extant faunas (herbivores included) of southeastern Asia than the last glacial to interglacial transition at the Pleistocene-Holocene boundary (or even a massive volcanic eruption for that matter, Louys 2007, 2012a, b).

Today, the Malaysian Peninsula, Sumatra, Java, and Bali are not separated by any significant stretch of ocean. The great distances between Borneo and the Philippines are deceptive because the ocean separating these islands from each other and from the Asian mainland is shallow, and is in fact an inundated continental shelf rather than a true oceanic basin (Fig. 2.5). With a sea level drop of just 40–50 m the Sunda shelf would be joined such that one could walk from the Malaysian Peninsula to Java and Borneo, and a drop of 120 m would extend an almost continuous dry land connection to parts of the Philippines (Voris 2000; Tougard 2001; Woodruff 2010), thereby providing regular and easy access for the dispersal of large mammals into the Southeast Asian island archipelago at numerous times during the Pleistocene.

Global sea level rises and falls cyclically (mainly tracking astronomical cycles), but the intensity (amplitude) of these cycles has changed through time. Today, sea levels are at a maximum level that has not changed significantly over the past 4 myr (Fig. 2.2 in Woodruff 2010). However, minimum sea levels have over time become progressively lower, especially with increases in astronomic cyclic variability (causing global cooling and increases in polar ice volume). Cyclical drops of 50 m or more probably began to take place around 2.7 Ma, reaching 100 m or more on several occasions since *ca* 0.8 Ma (Woodruff 2010). Regardless of tectonic uplift, much of the Southeast Asian archipelago was repeatedly connected to the Asian mainland throughout the Pleistocene, providing an easy route for colonization of these islands. Van den Bergh (1999) and van den Bergh et al. (2001a, b) review the routes and timing of mammalian dispersal between mainland SE Asia and the island archipelago, specifically in the direction of Java. These authors state that, in accordance with the sea level record, early dispersals took place during the Early Pleistocene (after 2.5 Ma), and increased with the formation of land bridges at 0.8 Ma, which was also a time of a major faunal turnover between Early Pleistocene and Middle Pleistocene island faunas.

Indeed, many of the large herbivore species that are restricted to islands appear to have originated from a mainland ancestor during the Pleistocene. Among these are the 'dwarf' buffaloes, namely the tamaraw of Mindoro Island (*Bubalus mindorensis*), the mountain and lowland anoas of Sulawesi and Butung islands (*B. quarlesi* and *B. depressicornis*), and the extinct *Bubalus cebuensis* of Cebu (Croft et al. 2006) among other fossil species. Genomic age estimates (Tanaka et al. 1996; Bibi 2013; Hassanin et al. 2013) suggest that the divergence date of the water buffalo (*Bubalus bubalis*) and the lowland anoa was probably between 2 and 1 Ma, implying an early Pleistocene age for the first colonization of these islands by the mainland buffalo. This is in agreement with the fossil record, the oldest indications of buffalo from the modern SE Asian archipelago being *Bubalus palaeokerabau* at 1.2 Ma from Ci Saat in Java (Louys et al. 2007). The Javan Rusa deer (*Rusa timorensis*) of Java and Bali is estimated to have diverged from the sambar (*R. unicolor*) during the middle Pleistocene, or less than 1 Ma (Pitra et al. 2004; Gilbert et al. 2006; Hassanin et al. 2012), suggesting a younger and separate dispersal event than that of the dwarf buffaloes.

Van den Bergh (1999) and van den Bergh et al. (2001a, b), following Thenius (1970), point out one seeming exception to the mainly Pleistocene timing of large mammal dispersal into the Southeast Asian archipelago. The *Babyrousa* lineage, they suggest, may have crossed to Sulawesi and nearby islands from the mainland Sunda Shelf as early as the mid-Oligocene, when a land bridge connection with the mainland may have last existed before the Pleistocene. This would make the babirusa the oldest continuous inhabitant among the large mammals of the Southeast Asian islands. Unfortunately, nothing at all is known of the fossil record of the babirusa prior to the Pleistocene (Hooijer 1948a, b; Hardjasasmita 1987; Groves 2001a, b). Molecular estimates suggesting that *Babyrousa* diverged from Suinae between 18 and 9 Ma (Gongora et al. 2011) might support Miocene colonization of SE Asian islands. The evolutionary and geographic origins of the babirusa lineage, whether on the Asian mainland, on Sulawesi, or on other islands, will have to await further fossil discoveries.

2.4 Paleobiogeography

2.4.1 Boundaries and Cohesiveness of the Indomalayan Realm

Much of the evolutionary history of mammals in SSEA reflects a basic biogeographic similarity to the geographic limits of modern Realms dating back to at least the late Miocene. Sclater (1858) and Wallace (1876) classified the world's terrestrial animal faunas into six major zoogeographic regions, with SSEA united as a single Oriental Region (Fig. 2.5), today more commonly known as the Indomalayan Realm (Udvardy 1975). Through several name changes, the basic boundaries of Wallace's biogeographic divisions have remained relatively unchanged in over 100 years (Fig. 2.1). The northwestern and northern limits of the Indomalayan Realm are bounded by the Sulaiman, Hindu Kush, and Himalayan mountain ranges. Comparison of late Miocene faunas from the Kabul basin in Afghanistan indicates almost no similarity to contemporaneous faunas from the Siwaliks only ~300 km to the southeast in northern Pakistan (Brunet et al. 1984; Beden and Brunet 1986; Sen et al. 1997). In contrast, Miocene faunas from Pakistan to Thailand and southern China all share many taxa in common, documenting the coherency of the Indomalayan Realm and its isolation from Central Asia to the north for at least the majority of the Neogene. The reasons for biogeographic differentiation in the Miocene may be similar to those today, namely elevated mountain chains and associated latitudinal climatic and environmental differences. Environmental conditions in the Molayan fauna (ca. 8–7 Ma) of the Kabul Basin indicate the presence there of a relatively open and dry sclerophyllous shrubland, but one that was composed entirely of C_3 vegetation (Zazzo et al. 2002; Merceron et al. 2004), in stark contrast to the C_4 grasses that had begun to take hold at this time in northern Pakistan and SSEA. Today, C_4 grasses are widespread in areas that are hot and receive a summer rainy season, such as tropical seasonal and monsoonal environments. In contrast, C_3 grasses dominate in areas receiving winter rains (e.g., Mediterranean biomes), and at higher latitudes and elevations.

To the northeast, the boundaries of the Indomalayan Realm do not overlap cleanly with major physical or political boundaries. While China for the most part harbors vegetation and animal assemblages different than those in SSEA, biogeographic zonation distinguishes southern China (i.e., south of about 29° N) as part of the Indomalayan Realm rather than the Palaearctic Realm to the north. A similar configuration is known at least as far back as the late Miocene, when faunas from sites in Yunnan Province are unlike contemporaneous sites from central and northern China, and are instead similar to those from the Siwaliks and Myanmar (Flynn and Qi 1982; Wei 1987; Qiu and Qiu 1995; Ni and Qiu 2002; Dong and Qi 2013). Climatic reconstructions for the Xiaolongtan fossil flora (Xia et al. 2009) indicate that temperatures in southern China were about the same in that part of the late Miocene as today, though precipitation may have been quite a bit higher.

The southern boundaries of the Indomalayan Realm are delimited by the Indian and Pacific Oceans. Perhaps the most significant changes along this front have taken place on the Sunda Shelf (Malaysia, Indonesia, Philippines), as its low-lying crust has been variably exposed or submerged, and its islands and archipelagos connected or isolated, by global sea level drops or rises. SSEA is connected to the Australian continent through a porous network of islands that make up the Southeast Asian archipelago but the faunas of these continental regions are starkly different. The transitional nature of the mammals and birds across this island archipelago has historically not been so clean-cut, and 'too many lines' have been proposed in attempts to categorize it (Simpson 1977). Many authorities have put the boundary between the Indomalayan and Australian Realms along what is termed Wallace's Line, a division that passes most famously between the islands of Bali and Lombok which are less than 40 km apart, while others have preferred a demarcation to the east of Sulawesi (Udvardy 1975; Holt et al. 2013).

2.4.2 Late Paleogene and Neogene Relationships with Africa

The Indomalayan Realm's closest modern comparison lies with Africa and specifically with the Afrotropical Realm, which covers sub-Saharan Africa and the southern portion of the Arabian Peninsula. Today, these two regions share obvious similarities in their large herbivore faunas, such as the presence of rhinocerotids, tragulids, elephants, as well as a high diversity of bovids. These similarities extend back in time into the Miocene, having been greater in the late Miocene than they are today (Thomas 1984; Bibi 2011). However, prior to the Miocene, Africa and Eurasia were separated by the Tethys Ocean throughout the entire Cenozoic. Rare occurrences of Oligocene and earliest Miocene assemblages in SSEA and Africa provide a picture of the past years of isolation, and the first instances of faunal exchange between these two landmasses.

For a long time, most of our knowledge of Afro-Arabian mammalian communities in the earliest Tertiary has come from the Fayum depression of northwestern Egypt. Vertebrate paleontological research in the Fayum Depression began in 1879, with Georg Schweinfurth's recovery of whale and fish fossils on the island of Geziret el-Qarn in Birket Qarun (Schweinfurth 1886). Since then, the marine Eocene formations underlying the famous continental series have yielded additional cetacean remains (Gingerich 2008). The age of the Fayum fossil fauna has been a matter of some controversy (Kappelman et al. 1992; Seiffert 2006), but it seems that the terrestrial deposits of the Fayum Depression include the Eocene–Oligocene transition. The mammal fauna from the Fayum Depression has been intensively studied since the early twentieth century, and it shows a dominance of African endemics (e.g., afrotheres), as might be expected at this early age. A few Asian elements are present, however, and include primates (Seiffert 2012), marsupials (Hooker et al. 2008), and anthracotheriid artiodactyls (Ducrocq 1997).

Another important late Paleogene African fauna is that from the late Oligocene of Chilga, Ethiopia (Kappelman et al. 2003). This fauna documents a time interval poorly represented in Afro-Arabia between archaic, endemic Afro-Arabian faunas of the late Eocene and early Oligocene as known from the Fayum Depression, and early Miocene replacement faunas, principally documented in eastern Africa, containing Eurasian ungulate migrants such as ruminants, suids, and rhinos. The Chilga fauna includes taxa belonging predominantly to the paenungulate orders (Hyracoidea, Embrithopoda, and Proboscidea) and is globally comparable to the Fayum fauna (Kappelman et al. 2003). Most notably, and together with the late Oligocene elephantimorphs of Eritrea (Shoshani et al. 2006), the Chilga fauna includes the oldest known deinotheres and elephantoids, whose prior fossil records did not extend before the early Miocene (Sanders et al. 2004). The Chilga fauna indicates the continuing biogeographic separation of Afro-Arabia from Eurasia into the latest Paleogene as a result of the Neotethys Sea between these landmasses, and the lack of formation of stable land bridges between the regions at this time. However, while exhibiting a high degree of endemism, Paleogene African faunas do record episodic trans-Tethyan faunal exchanges between Eurasia and Afro-Arabia during the Paleogene and even earlier, during the Mesozoic (see Gheerbrant and Rage 2006).

A great faunal exchange took place between Eurasia and Afro-Arabia in the early Miocene, much as would take place between North and South Americas during the late Pliocene (e.g., MacFadden 2006). In this scenario, the Oligocene to Miocene fossil sequences of Pakistan (see below) plays a critical role in understanding the timing and nature of faunal exchanges. In southern Asia, early Palaeogene vertebrate faunas were for a long time only documented by a few localities in northern Pakistan that yielded fragmentary fossils (Zu Dehm and Oettingen-Spielberg 1958). Until the reassessment of the age of the Chitarwata Formation (Welcomme et al. 2001), the Oligocene was considered as lacking in the entire Indian continent. As such, the Bugti fauna provides a unique window for documenting the transition between the early middle Eocene faunas from the Indo-Pakistani region (e.g., Thewissen et al. 2001a, b), and the Neogene Siwaliks faunas of the Potwar Plateau.

The Proboscidean Datum Event (PDE), supposedly recording the first appearance of proboscideans in Eurasia and the establishment of terrestrial land connections between Afro-Arabia and Eurasia, was long considered as a reliable biochronologic marker dated at 17.5 Ma. Introduced by Madden and Van Couvering (1976), the PDE became less a datum than an interval when the fossil record of Eurasian proboscideans became better known. Tassy (1990) cast serious doubt on the consistency of such a datum, and suggests a diachrony between the first appearance of elephants in Europe (where the PDE is around 18 Ma) and the rest of Asia. Beyond the PDE, the introduction of African large mammals into Eurasian faunas was certainly multi-phased, and depended on both biotic (e.g., dispersal abilities) and abiotic (e.g., sea-level lowering) factors. A thorough analysis of the Oligocene–Miocene sequence exposed in the Sulaiman Range of Pakistan reveals first occurrences of African migrants much earlier than expected by previous ideas, namely during the late Oligocene (Antoine et al. 2003a, b; Barry et al. 2005).

Faunal interchanges between Africa and Eurasia were therefore diachronous, occurring over an interval of time rather than as a single event, and were probably determined by many different biological and physical parameters. The exchange of faunas between Africa and Eurasia in the late Paleogene and early Negoene was a critical event in the evolutionary history of the ungulate faunas of both continental areas, though it appears to have impacted African faunas more severely than the Eurasian ones. Indeed, many endemic African mammal clades were reduced or even totally gone by the middle Miocene, replaced by incoming Eurasian taxa such as rhinos, bovids, and suids (e.g., Ungar et al. 2012 and therein).

Though similarities can always be found at higher (family) taxonomic levels, it is interesting that the Afrotropical and Indomalayan Realm rarely shared the same species among them. Among fossil bovids, which are very abundant and diverse, only a small handful of occurrences of the same or similar species can be counted in the late Miocene, though this itself is still significantly more than could be counted for the Pliocene or Pleistocene (Bibi 2011). Perhaps the clearest example of a fauna that combined both African and Indomalayan species is that of the late Miocene Baynunah Formation in the United Arab Emirates (Whybrow and Hill 1999). This fauna is mainly African in composition, but interestingly includes Indomalayan species that are never recorded in Africa such as the bovid *Pachyportax latidens* and the suid *Propotamochoerus hysudricus* (Bishop and Hill 1999; Gentry 1999; Hill et al. 2012; Bibi et al. 2013). The Baynunah fauna therefore offers a glimpse into late Miocene biogeographic configurations that reflect a greater overlap of biogeographic boundaries between the Indomalayan and Afrotropical realms at this time.

The Pliocene and Pleistocene probably saw more dispersal of large herbivores into SSEA from Africa than in the opposite direction. This includes the Asian elephant, *Elephas maximus*, the lineage of which (*Elephas* spp.) first evolved in Africa in the early Pliocene and dispersed into SSEA by the late Pliocene. This and other examples of late Neogene continental dispersal are discussed below.

2.5 The Major Fossiliferous Deposits of SSEA

2.5.1 The Bugti Hills, Pakistan

The succession of fossiliferous Tertiary units exposed in Pakistan makes this area one of the most critical for the study of mammalian evolution on the western border of the Indomalyan Realm. The Sulaiman and Kirthar Ranges are north–south mountains resulting from Plio-Pleistocene deformation of Mesozoic and Cenozoic sedimentary rocks. These Mesozoic units had accumulated on the Indian platform of the eastern Tethys Ocean prior to Indian collision. The early Cenozoic sequences record the onset of continental collision, and the thick late Paleogene and Neogene sequences record the emergence of the area and the establishment of the Himalayan

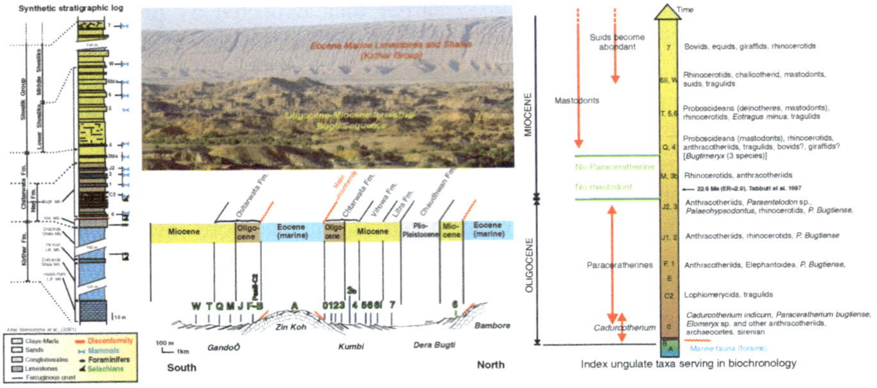

Fig. 2.6 Biostratigraphic framework of the Bugti Hills, which documents about 30 million years of terrestrial ecosystem evolution in the Indian Subcontinent. The photo is taken facing north and captures the Gandoi area with the Zin Koh Anticline in the background. The biostratigraphic chart on the *left* shows the distribution of fossiliferous horizons in the sequence (letters for the Gandoi Syncline, numbers for the Dera Bugti Syncline), and proposed biochronologic correlations with the Siwaliks and the Zinda Pir area. At *right*, large herbivore fossil taxa indicative of biochronology (adapted from Welcomme et al. 2001; Métais et al. 2009b; Antoine et al. 2013)

foreland and the Indus drainage. The fossiliferous sequence recording the transition from marine to terrestrial deposition is exposed on the eastern side of the Sulaiman Range, and is best exposed in the Bugti Hills, a region ruled by the Baloch Bugti tribe (Fig. 2.6). The Bugti Hills have been known for their fossil discoveries since 1846 when Vickary, a soldier in the British Army, first reported 'large bones,' which were subsequently identified and described by Blanford (1883). The conglomerates and sandstones yielding fossil vertebrates were then ascribed to the 'Upper Nari Formation,' and soon became famous for their 'large fossil mammals' described in several reports (Pilgrim 1907, 1908, 1910a, b, 1912; Forster-Cooper 1913, 1915, 1924). These fluvial deposits and their fossil content, including the giant rhino *Paraceratherium (=Indricotherium, Baluchitherium)* were then considered to be pre-Siwalik in placement and early Miocene in age (Pilgrim 1912). The stratigraphy and the time slices represented in the Bugti Series, however, turned out to be more complex than previously evaluated, with the sedimentary sequence there documenting an interval (with numerous gaps) ranging from the middle Oligocene to the Pleistocene. The classical early Miocene 'Bugti fauna' is actually made up of different fossiliferous horizons yielding distinct mammal assemblages that were erroneously considered as a single unit (Welcomme et al. 2001). No magnetostratigraphic study has yet been undertaken in the Bugti Hills and relative ages continue to be provided only by mammalian biochronology. The only radiometric date available is a fission track date of 22.6 ± 2.9 Ma from a tuffaceaous horizon in the vicinity of Lundo (Tabbutt et al. 1997), a locality situated stratigraphically at the top of the Bugti Member of the Oligocene to Miocene Chitarwata Formation (Welcomme et al. 2001; Métais et al. 2009; Antoine et al.

2013). The biostratigraphic (Downing et al. 1993; Lindsay et al. 2005) and magnetostratigraphic (Friedman et al. 1992) work carried out in the nearby Zinda Pir Dome (150 km NE of the Bugti region) has also helped provide important age correlations. The Bugti sequence is divided into formations that can be recognized across the Sulaiman foothills (Fig. 2.6). These include the Chitarwata Formation (Oligocene-early Miocene), Vihowa (early middle Miocene, and correlative with the Kamlial Formation, the lowermost unit of the Siwaliks in the Potwar Plateau), Litra (late Miocene, roughly Chinji-Nagri equivalent), and Chaudhwan (Plio-Pleistocene). Ungulate mammals are particularly well represented in these successive faunas and most of them serve as key index taxa for establishing biochronological correlations. For example among ruminants, tragulids and lophiomerycids are common in the Chitarwata Formation, while bovids and giraffids become more common in the Vihowa Formation (Ginsburg et al. 2001; Métais et al. 2003; Barry et al. 2005; Métais et al. 2009b).

2.5.2 The Siwalik Hills of Pakistan and India

Located in northeastern Pakistan and northwestern India, along the southern flanks of the Himalayas, the Siwalik Hills present one of the largest, most complete, and richest sequences of Neogene fossil mammals anywhere in the world. These foothills of the Himalayan Range essentially consist of Miocene to Pleistocene molassic deposits derived from the erosion of the rising Himalaya, and preserve a fairly continuous record of deposition by drainages ancestral to the modern Indus, Ganges, and Brahmaputra (Friend et al. 2001). Siwalik sediments consist of conglomerates, sandstones, and shales with a thickness of around 5000 m, with vertebrate fossils occurring as dense aggregates of disarticulated elements, hundreds or even thousands in number, filling small channels within braided river systems.

Lyell (1830, pp. 7–8) relates that the fourteenth century Persian historian Ferishta noted the discovery of 'bones of elephants and men, some of them petrified' in the Siwaliks area in the process of waterway excavations ordered by the emperor Feroz Shah the Third (also in Murchison 1868, p. 4). Siwalik fossils may have also provided inspiration for some of the stories in the Mahabharata, the Indian literary epic assembled in the first millennium B.C.E. (van der Geer et al. 2008).

Siwalik fossils are among the oldest studied fossils, with scientific descriptions of these appearing before the mid-nineteenth century by the naturalist Hugh Falconer and his colleague, the engineer Proby T. Cautley. The Siwaliks have since then been the target of numerous local and international scientific expeditions and continue to produce what is certainly the best representation of Miocene terrestrial biotic evolution for the entire SSEA region. The Siwaliks sequence has been divided into several geological formations (Fig. 2.7). These include the Kamlial (18.3–14.2 Ma), Chinji (14.2–11.2 Ma), Nagri (11.2–9.0 Ma), Dhok Pathan (10.1–ca. 3.5 Ma), Tatrot (3.5–3.3 Ma), and Pinjor (2.5–0.6 Ma) formations (Barry et al. 2002; Nanda 2002). Formational boundaries, however, appear to be time

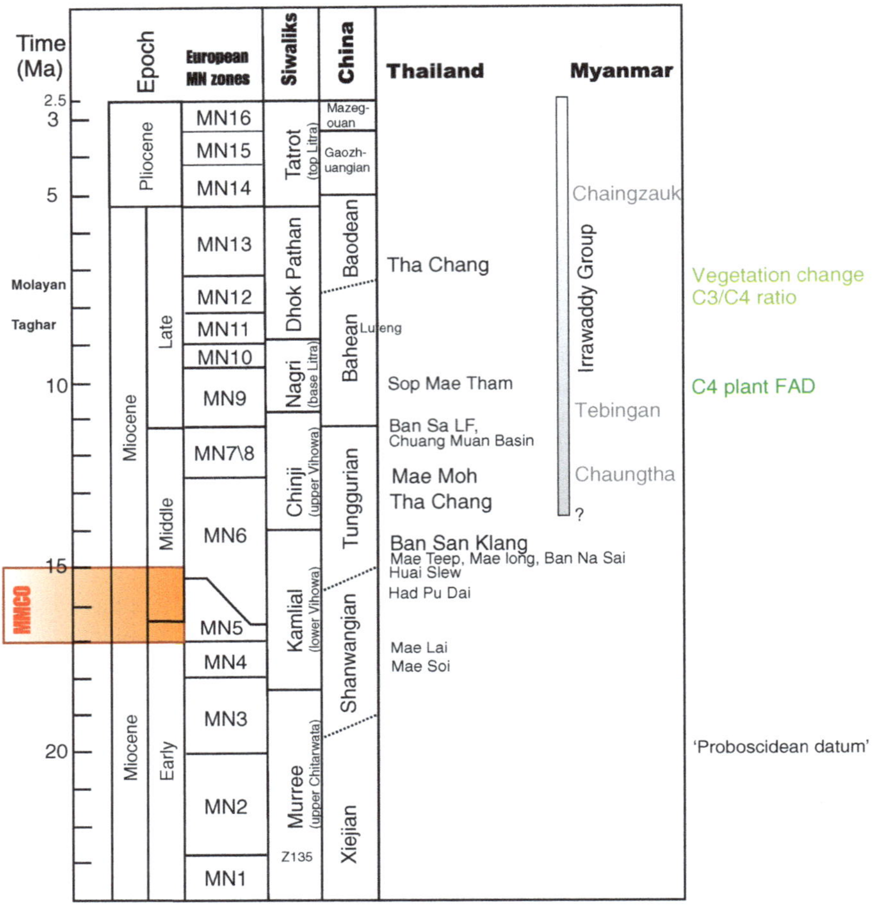

Fig. 2.7 Chronology and correlation of key SSEA Neogene localities with large herbivores. Tentative correlation with the European MN zones is also provided. MMCO designates the Middle Micoene Climatic Optimum, which marks an attenuation of the Asian Monsoon regime for about 2 Ma (based on Barry et al. 2002; Sen 1998; Kunimatsu et al. 2004; Chavasseau et al. 2006; Ogino et al. 2011; Antoine et al. 2013; Qiu et al. 2013)

transgressive in places, and many other formations are named from overlapping time intervals. Oftentimes fossils are referred to similarly named 'faunas' (e.g., Chinji fauna, Tatrot fauna), regardless of the actual formation of origin.

Tripathi (1986) reviewed the history of geological study up to that time, but the most synthetic recent summaries of the geology, chronology, and faunal composition of the Miocene Siwaliks are provided by Barry et al. (2002, 2013), Flynn et al. (2013), and Patnaik (2013). Barry et al. (2002) presented precise ages for a large sequence of sites spanning from 10.7 to 5.8 Ma using magnetostratigraphy, which then allowed for large-scale analyses of faunal diversity and ecological dynamics (Barry et al. 2002; Badgley et al. 2008). A very large number of publications describe Siwalik fossils, and many of these have been cited here in sections below summarizing the history of different clades of fossil herbivores.

Among the many fossil taxa recovered there, the Siwaliks have received attention for the occurrence of a hominoid, *Sivapithecus*. Much of the recent work from the Siwaliks has gone into understanding the habitat preferences of *Sivapithecus*, and the reasons for its extinction there by 8.1 Ma, which appear to be a result of climatic changes and a major loss of forest habitats at that time (Nelson 2007; Badgley et al. 2008).

2.5.3 Paleogene and Neogene Fossil Deposits of Myanmar

The present drainage of the Irrawaddy River and its tributaries forms a large basin extending from north to south through central Myanmar (Fig. 2.8). The formation and sedimentary infilling of the Irrawaddy Basin is closely related to the collision of India with the southern margin of Asia, and the erosion of the resulting uplifted terrain (Bender 1983). From the late middle Eocene, the Irrawaddy Basin has been intermittently flooded by the sea, but the large majority of its deposits are fluvial in origin (Bender 1983). Neogene continental deposits are well exposed in the Irrawaddy basin, and late Paleogene deposits in the Chindwin basin (a tributary of the Irrawaddy). Climatically, the north–south extending Indo-Burmese Ranges (1500–4000 m) impede the passage of humid summer monsoonal winds from the southwest into central Myanmar, producing arid conditions in this area. At present, central Myanmar is a semi-arid zone with an annual precipitation of 600–1000 mm and a mean annual temperature of around 30 °C (Bender 1983). In contrast, the western and southern parts of Myanmar receive 3000–4000 mm of rainfall annually and are covered with tropical rainforests. At their northern extent, the Indo-Burman Ranges are continuous with the Himalayan-Tibetan Plateau.

The late Middle Eocene Pondaung Formation is about 2000 m thick, and the lower part of the upper member yielding fossil vertebrates is comprised of fluvio-deltaic deposits (Soe et al. 2002). The Pondaung fossil fauna is one of the richest Paleogene mammalian faunas in SSEA. Mammalian fossils from this formation were first described by Pilgrim and Cotter (1916). Among the Pondaung mammals, artiodactyls and perissodactyls dominate, with anthracotheres, small ruminant-like artiodactyls, and primitive rhinos and tapiromorphs, along with carnivores, primates, and rodents (Colbert 1938; Ducrocq et al. 2000; Tsubamoto et al. 2002, 2005; Métais 2006; Métais et al. 2007). Interestingly, the Pondaung Formation has yielded a diacodexeid-like artiodactyl (Ducrocq et al. 2015), thus shifting up the youngest occurence of these basal artiodactyls fairly common from the early Eocene deposits of Holarctic areas and India, and suggesting that Southeast Asia may have played as early as Eocene the role of refuge pretty much as it does now. The paleoenvironment of the Pondaung fauna is reconstructed as mangroves and dry dipterocarp forest (Licht et al. 2015) under a with monsoonal regime, supplied by large rivers that were located not far from

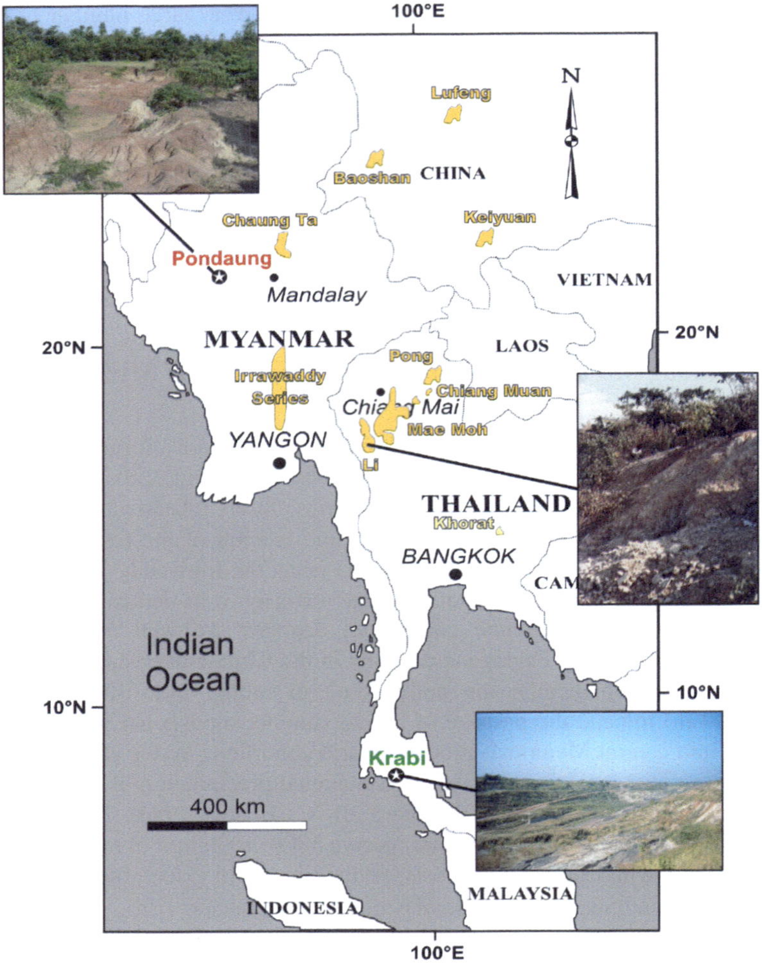

Fig. 2.8 Tertiary basins in Myanmar, Thailand, and southern China. Photos of Saba Pondaung, Pondaung Formation (G. Métais), an outcrop in the Li Basin (L. Ginsburg), and the Krabi Coal Seam (S. Ducrocq)

the coast of the eastern Tethyan Sea. At this time there are no herbivorous species with truly hypsodont cheek teeth (as in living horses), also reflecting a general absence of dry and open habitats.

The Neogene and Quaternary fluvial deposits of the Irrawaddy Formation are well exposed along the banks of the modern Irrawaddy River, can reach thickness of 2000 m, and span the Middle Miocene to the Pleistocene (Fig. 2.7) (Bender 1983). Mammalian fossils from these sediments have been documented since the late nineteenth century (e.g., Lydekker 1876a, b) and were loosely correlated with those of the Siwaliks of the Indian subcontinent through biostratigraphy (Colbert 1938, 1943).

The lithology of the Irrawaddy series is quite homogeneous, and fossil vertebrates are to be found scattered in thick repeating sandy units. Biochronological correlations with Neogene faunas of Thailand and the Siwaliks of Pakistan sometimes provide the only age indication. For example, several taxa described from the locality of Chaungtha also occur in the Chinji Formation (14–11 Ma, Barry et al. 2002) of the Siwaliks and the Mae Moh Formation of northern Thailand (Chavasseau et al. 2006). This suggests that, like today, SE Asia and Pakistan during the middle Miocene belonged to the same biogeographical region and perhaps shared a similar forested environment under a monsoon-like climate regime (Ducrocq 1994). The abundance of silicified fossil wood of Dipterocarpaceae (Bande and Prakash 1986) in the 'Lower Irrawaddy' series suggests the existence of tropical rainforests during the early Neogene.

Late Miocene faunas of the Irrawady Formation are dated to between 10.4 and 8.8 Ma on the basis of biochronology and magnetostratigraphy (Jaeger et al. 2012). Stable isotope data on the fossil equid *Hipparion* indicate an evergreen forest in a C_3-vegetation environment, suggesting humid conditions there, prior to the replacement of C_3 forested habitats by C_4 grasslands between 8 and 6 Ma (discussed above).

Several 'Irrawaddy faunas' have been reported from the Chaingzauk area, and correlations with the Plio-Pleistocene Tatrot-Pinjor faunas or with the latest Miocene (ca. 6–5 Ma) upper Dhok Pathan faunas of the Siwaliks have been proposed (Fig. 2.7) (Colbert 1938). The Chaingzauk faunas are abundant in ungulates, with rhinocerotids, bovids, hippopotamids, tragulids, anthracotheres, monkeys, suids, and proboscideans. Carbon and oxygen isotopic investigation indicates that mammals inhabited a transitional environment between woodland and grassland (Zin-Maung-Maung-Thein et al. 2011) and that C_4 grassland habitat expansion had taken place by the latest Miocene in Myanmar, as it did in the Siwaliks of Pakistan. The present day arid environment of central Myanmar (the driest region in SE Asia) must have developed gradually during the late Miocene to early Pliocene, and is probably related to the continuing uplift of the Indo-Burmese Ranges and the Himalaya-Tibetan Plateau.

2.5.4 *Paleogene and Neogene of Thailand*

The Cenozoic mammal localities of Thailand range chronologically from late Eocene to Pleistocene, and geographically from the peninsular region to the north of the country (Figs. 2.7 and 2.8). A number of small basins form part of a NW–SE rift zone that crosses Myanmar and Laos and ends in the gulf of Thailand. The Tertiary rift basins of central Thailand are delimited to the west by the Sagaing fault zone and to the east by the Red River fault zone (Morley 2002). These major strike-slip faults resulted from tectonic activity associated with the Himalayan orogeny that forced the area of Indochina to move to the southeast relative to the remainder of Asia (Tapponnier et al. 1986). During the Cenozoic, the subsidence of these small pull-apart basins and subsequent infilling with fluvio-lacustrine sediments resulted in the deposition of thick sedimentary sequences that are accessible largely due to mining activities.

The coalmine of Krabi (Fig. 2.8) has yielded one of the most diverse assemblage of fossil mammals from Southeast Asia. Mammalian biostratigraphy indicates a late Eocene age (Ducrocq et al. 1995), while magnetostratigraphic work suggests the Krabi section is earliest Oligocene, between 34 and 31 Ma in age (Benammi et al. 2001). The Krabi fauna, comprising over 30 species (Ducrocq et al. 2006) is generally considered as a reference fauna in South Asia. The main coal seam, which has yielded most of the fossil mammals, is sandwiched between overlying lacustrine sediments and underlying fluvial sediments. The abundant unionid shells, crocodiles remains, and freshwater turtles suggest fluvio-lacustrine and wetland environments for the Krabi fauna. Despite the high diversity of the large mammal fauna, micromammal remains are very scarce, and most vertebrate remains belong to medium and large herbivorous mammals typical of swampy environments. Ungulate mammals, especially artiodactyls, are fairly well represented at Krabi, with anthracotheriids (five species in three genera) forming the majority of the fossil mammals collected there. Perissodactyls are usually an important component of Asian Paleogene mammal faunas (Russell and Zhai 1987), but the Krabi fauna is dominated by artiodactyls (about 15 different species), with only two rhinocerotoid perissodactyls (Ducrocq et al. 2006). The low diversity of perissodactyls in Krabi could either reflect an incompleteness of the Paleogene fossil record in southern Asia, or a greater diversity of artiodactyls in lower latitudes relative to a greater diversity of perissodactyls in higher latitudes in Central Asia at this time. The greater diversity of dental patterns for artiodactyls in Krabi (ruminants, anthracotheres, suoids) seems to be analogous to that of perissodactyls from most other Paleogene Asian localities. Beyond anthracotheres, artiodactyls are represented by tragulid and lophiomerycid ruminants (Métais et al. 2001), a suid (Ducrocq et al. 1998), and a tayasuid (Ducrocq 1994). The body mass cenogram of the Krabi fauna (Ducrocq et al. 1995) is quite similar to that of extant faunas living in humid tropical forests (Legendre 1989), suggesting a closed canopy forest habitat.

The middle Miocene localities of Thailand are situated in the northwestern part of the country. These basins are filled mainly by fluvio-lacustrine Neogene deposits, and intercalated lignite seams are commercially exploited. Different basins have been investigated and it seems that most mammal localities from northern Thailand occur in a time span ranging from about late early to early late Miocene (Von Koenigswald 1959; Ginsburg 1984; Ducrocq et al. 1994), overlapping with the Kamlial and Chinji faunal zones of the Siwaliks in Pakistan. About 45 species of mammals (from chiropterans to proboscideans) are reported from the different basins of northern Thailand (Ducrocq et al. 1995).

The principal fossiliferous localities are situated in the basins of Li, Mae Moh, Chiang Muan, and Pong (Fig. 2.8). Magnetostratigraphic studies have been carried out on different sequences of the Mae Moh and Chiang Muan Basins and provide further constraint for dating mammal localities (Benammi et al. 2002; Coster et al. 2010). The Mae Moh coal mine has produced a diversified late middle Miocene fauna (Suraprasit et al. 2015), including the early bovid *Eotragus*, and lagomerycids *Lagomeryx* and *Stephanocemas* (Suraprasit et al. 2014). The Li Mae Long fauna in the Li Basin is the most diversified fauna, the early middle Miocene age of which is constrained by micromammals (Mein and Ginsburg 1997). Large mammals are represented by the proboscidean *Stegolophodon*, the rhinocerotid *Gaindatherium*, the lagomerycid *Stephanocemas rucha*, the tragulid *Siamotragulus haripounchai*, the anthracothere *Brachyodus* cf. *onoideus*, and the suid *Conohyus sindiensis* (Mein and Ginsburg 1997; Ducrocq et al. 2003). This fauna has been correlated with the European Mammal Zone MN4 (late early Miocene), making this the oldest Neogene fauna from northern Thailand.

The Chiang Muan Basin has yielded a younger fauna which is supposedly late middle Miocene in age (Chaimanee et al. 2003; Pickford et al. 2004; Suganuma et al. 2006) and probably correlative with the upper Chinji-lower Nagri faunas of the Siwaliks. The fauna from the Ban Sa locality in the Chiang Muan basin includes a tetralophodont mastodon, rhinocerotids, tragulids, bovids, suids, and hominoid primates. The Neogene Vietnamese site of Hangmon (Ginsburg et al. 1992; Covert et al. 2001) and the presumed contemporaneous Chinese faunas, such as at Lengshuigou, Tunggur, Tsaidam or Lufeng, do not provide meaningful comparisons with the Thai locality because of their distinctly different mammalian assemblages, suggesting a degree of provincialism in the Chinese ruminant faunas of this time.

Several fossil vertebrate localities of late Miocene to Pliocene age have also been reported in Nakhon Ratchasima Province (Khorat), northeastern Thailand (Chaimanee et al. 2004; Hanta et al. 2008; Claude et al. 2011). The ungulate mammal fauna includes proboscideans (*Deinotherium indicum*, *Gomphotherium* sp., *Stegolophodon* sp.), an anthracotheriid, a suid (*Hippopotamodon sivalensis*), rhinocerotids (*Chilotherium palaeosinense*, *Brachypotherium perimense*, *Alicornops complanatum*), bovids, giraffids (*Sivatherium* sp.) and some hipparionine equids. This assemblage seems to correlate to the upper Nagri and lower Dhok Pathan Formations of the Siwaliks and indicates a late Miocene age, or between about 9 and 7 Ma (Barry et al. 2002).

2.5.5 Afghanistan

Though part of SSEA as defined politically, Afghanistan's faunas resemble those of Central Asia, eastern Europe, and the Middle East more so than those of the Indomalayan Realm. Afghanistan is accordingly classified within the Palaearctic (or Saharo-Arabian) Realm (Holt et al. 2013), and the distinction between its faunas and those of the Indomalayan Region (even nearby Pakistan) is visible at least as far back as the late Miocene (Brunet et al. 1984).

The known fossil record of Cenozoic mammals in Afghanistan is limited to a handful of Miocene-Pliocene localities in intramontane basins. A few small mammal remains were first reported by Lang and Lavocat (1968) from the Bamian Basin (150 km west–northwest of Kabul), and revised by Flynn (1983). Raufi and Sickenberg (1973) reported on mammal remains from the Sarobi and Jalalabad Basins, about 60 km East of Kabul. Between 1976 and 1979, French paleontological expeditions prospected different basins southeast of Kabul where the late Neogene

Lataband series (Mennessier 1961) was exposed, and several fossil mammal localities of late Miocene to Pliocene age were discovered (Heintz et al. 1978a, b). Among these, the site of Molayan bears the richest assemblage so far recovered from Afghanistan, both in terms of taxonomic diversity and number of specimens (Heintz et al. 1978a, b; Brunet et al. 1984; Sen 2001). Faunal studies suggest a Middle Turolian (ca.8–6.5 Ma) age for Molayan (Sen 1998) while Taghar, another locality stratigraphically lower than Molayan, has yielded large mammals that all indicate an early Turolian age (ca. 9–8 Ma) (Sen et al. 1997). Geographically, Molayan and Taghar are situated close to the border between two paleobiogeographical provinces, the Greco-Iranian and the Siwalik provinces (de Bonis et al. 1992). However, the mammalian faunas from these two Afghan localities are firmly Greco-Iranian in composition, with hardly a species in common with the Potwar Plateau only some 300 km to the southeast (Brunet et al. 1984). Both Molayan and Taghar have yielded diverse ungulate faunas including Bovidae, Giraffidae, Equidae, Chalicotheridae, Rhinocerotidae, Gomphoteriidae, Deinotheriidae, and Suidae. Cervid remains are also known from Taghar, which is interesting since cervids have not been recorded from the late Miocene and early Pliocene faunas south of the Himalayas (Heintz and Brunet 1982, and see below). Studies of microwear and stable isotopes of fossil bovids from Molayan indicate an open and partly forested alluvial plain forming hilly landscapes, where evergreen sclerophyllous shrubs and C_3 grasses constituted the majority of the vegetation (Zazzo et al. 2002; Merceron et al. 2004).

2.5.6 Southern China and Northern Vietnam

While not within the boundaries of political SSEA, southern China and northern Vietnam forms the north-easternmost part of the Indomalayan Realm. Fossil sites at least as old as the late Eocene record the evolution of Indomalayan mammals here since at least that time, indicating similarities to contemporaneous SSEA faunal assemblages to the south, and differences from Chinese and Central Asian assemblages to the north. During the late Eocene up yo the Miocene, numerous pull-apart basins formed on the southern Chinese microplate and northern Vietnam along the strike-slip Red River faults system (Pubellier et al. 2008). It resulted thick sequences of terrestrial (often coal bearing) sediments that have produced faunas mostly documented by larges mammals. The Naduo and Gongkang Formations in the Bose Basin of southern China probably straddle the Eocene-Oligocene transition, and their faunas is domminated by anthracotheres, tapiroids, earliest suoids (Liu 2001), and at a lesser extent ruminants (Wang et al. 2007 and reference therein). The Naduong Basin of nothern Vietnam has also yielded a rich fauna and flora reported by Böhme et al. (2011, 2013). The large mammals includes anthracotheriids (Böhme et al. 2013; Ducrocq et al. 2015), the rhinocerotid Epiaceratherium naduongense, and traguloids. The fossil mammals from Na Duong show unexpected similarities with European (Böhme et al. 2013) and to some extent to African faunas (Ducrocq et al. 2015). The Na Duong Formation probably deposited within a swamp ecosystem with both aquatic and non-aquatic environments. Fossil plants suggest a forest of Dipterocarpaceae, and the sedimentary sequence exposed in the Na Duong basin pleads for tropical humid condition without a dry season (Böhme et al. 2013). During the Neogene, the main fossil sites of southern China are those of Yunnan Province, mainly in the Xiaolongtan, Lufeng, Yuanmou, and Zhaotong Basins (Dong and Qi 2013; Wang et al. 2013), while the Early Miocene Hang Mon fauna represent the only and still poorly documented record of Miocene mammals in Vietnam (Ginsburg et al. 1992). Numerous localities of Pleistocene are also known in southern China (Li 1981; Ji 1982; Huang and Zhang 2006).

2.5.7 Southeast Asian Islands

The fossil record of the Southeast Asian archipelago is primarily Pleistocene in age. Fossil sites are known from the islands of Java, Sumatra, Flores, Borneo, Sulawesi, and several of the Philippine islands, among others (e.g., Croft et al. 2006; Louys and Meijaard 2010). The fossil record of these islands tells a story of insular evolution of mainland mammal lineages, with waves of dispersal into the islands associated with the major and cyclical sea level drops characteristic of the Pleistocene (discussed above).

2.6 Evolutionary History of the Large Herbivores of SSEA

2.6.1 Artiodactyla

Artiodactyls (or cetartiodactyls) are the most diverse clade of extant large mammals, occurring in continental areas from the equator to high latitudes, as well as both marine and freshwater habitats around the world. Artiodactyls such as oxen, goat, sheep, pigs, and camels played a pivotal role in the Neolithic development of human settlements and agricultural and cultural practices, and are to this day a critical source of productivity and food for humanity.

This clade has been the focus of much scientific attention over the past few decades, particularly as genomic sequencing techniques came to be applied to phylogenetic reconstruction. The classical concept of Artiodactyla as proposed by Richard Owen in 1848 for terrestrial, even toed ('cloven hoofed') ungulates (i.e., pigs, hippos, camels, and ruminants) was challenged by molecular phylogenies based on RNA, cytochrome *b*, or other molecular markers (Fig. 2.9). These phylogenies offered numerous hypotheses of relationships within Artiodactyla that were all distinct from those classically admitted by comparative anatomical approaches since the mid nineteenth century. Most surprisingly, molecular phylogenies suggested that cetaceans (whales and dolphins) were closely related to extant hippos, and thus nested deep within Artiodactyla (e.g. Montgelard et al. 1997).

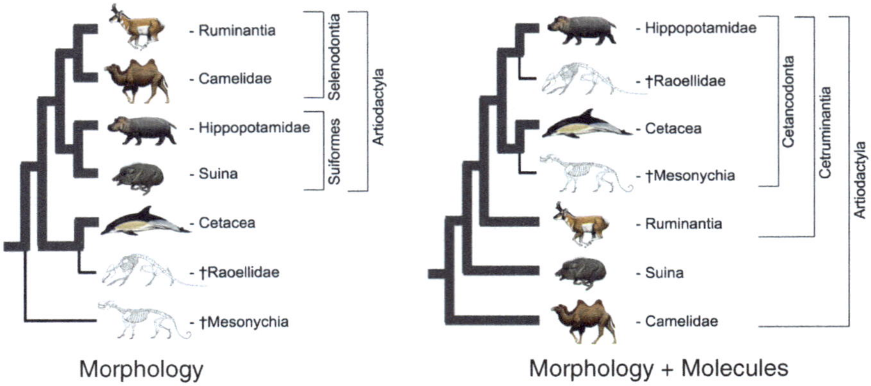

Fig. 2.9 The phylogenetic tree of artiodactyls produced by morphological characters versus a tree obtained from analysis of morphological and molecular characters combined. Both approaches recognize the relationship of whales to artiodactyls, but are conflicted on relationships among the artiodactyl clades (modified from Spaulding et al. 2009)

These provocative topologies in support of a 'whippo' clade were vindicated by the paleontological discovery of a primitive fossil whale ankle bone with a double-trochlea astragalus, a definitive anatomical feature of artiodactyls (Gingerich et al. 2001; Thewissen et al. 2001a, b). The exact evolutionary origins of Cetacea within Artiodactyla, however, remain uncertain, particularly relative to numerous poorly understood early fossil artiodactyls. Raoellids, a small group of early to middle Eocene amphibious artiodactyls endemic to India and Pakistan, display morphological and histological features suggestive of a degree of adaptation to aquatic life (Thewissen et al. 2007), making them possibly ancestral to Cetacea (Geisler and Uhen 2005). Several inconsistencies remain between molecular and morphological phylogenies, especially regarding the relationships of tylopods and ruminants within Artiodactyla. Molecular analyses also support a closer relationship of 'whippos' to ruminants than to pigs or camels (Fig. 2.9), a clade named Cetruminantia (e.g., Hassanin et al. 2012). Further paleontological work is needed to better illuminate the early history of these groups, using multidisciplinary approaches to decode the evolutionary history of morphological changes and their potential functional and phylogenetic significance.

The relationships of artiodactyls among placental mammals are also a matter of debate, often accentuated by this clade's scanty early Paleogene fossil record, and the difficulty in securely determining the position of extinct basal groups to extant crown clades. Molecular (Madsen et al. 2001; Meredith et al. 2011) as well as combined morphological-molecular phylogenies (O'Leary et al. 2013) suggest that artiodactyls are part of the clade Laurasiatheria that also includes carnivores, bats, pangolins, and perissodactyls. Laurasiatheres and archontes (rodents and primates) are thought to originate in the Holarctic area sometime between the early Cretaceous and the Paleocene. Beyond topologies of phylogenetic trees, divergence time (i.e., inter- or intra-ordinal diversification) is often a matter of controversy between paleontologists and molecular biologists. Molecular biologists use many different approaches and inevitably rely on the fossil record to calibrate molecular evolutionary rates and derive molecular age estimates, often with contradictory results.

The fossil record provides slightly less room for chronological ambiguity. Artiodactyls first appear in the fossil record in the earliest Eocene (ca. 55 Ma) of Eurasia, India, and North America (Rose 2006). They are mostly documented by the genus *Diacodexis*, although several enigmatic and poorly documented forms from the late Paleocene to the early Eocene of China may indicate higher taxonomic diversity in Asia near the Paleocene–Eocene boundary (Tong and Wang 2006). During the Middle Eocene, several groups of basal artiodactyls (usually named dichobunoids) diversified concurrently in the three main Holarctic areas. However, with the exception of tylopods (including camelids), which first appeared and diversified in North America (Honey et al. 1998), the earliest representatives of the lineages leading to the crown artiodactyl clades are to be found in Southeast Asia (Webb and Taylor 1980; Ducrocq 1994, 1997; Métais et al. 2001, 2007).

The modern dominance of artiodactyls among Asian herbivore faunas was not always as such in the past. The prevalence of perissodactyls over artiodactyls in the Eocene faunas of Asia is clearly perceptible, both in terms of taxonomic diversity and demographic abundance. The average size of artiodactyls (most of which were ecologically similar to extant rabbits) was much smaller than that of perissodactyls, suggesting that these early ungulates were ecologically divergent and not sharing the same food resources. During the Oligocene and Neogene, however, artiodactyls became larger and more diverse while the diversity of perissodactyls decreased. This reverse, continental scale trend between the two main groups of Asian ungulates appears to have been the result of vegetational changes that favored the digestive efficiency of ruminants (foregut fermenters) relative to perissodactyls (hindgut fermenters) (Janis 1989).

Suidae

SSEA has the highest diversity of suid species in the world today, represented by several species of *Sus*, *Porcula salvanus*, and the peculiar babirusa (*Babyrousa* spp.) recently divided among three species. It is interesting that most babirusa species are restricted to particular islands of Southeast Asia, suggesting that their differentiation must have been related to processes of island formation (tectonic motions and sea level changes).

Suids are omnivores that rely on plant-based diets, but opportunistically incorporate a large range of small animals. Unlike most herbivores that possess teeth with long enamel ridges (lophodont, selenodont), suids tend to have teeth with rounded cusps (bunodont). Humans and bears are two other taxa with similarly bunodont teeth, a functional similarity that gives away the omnivorous diets of these three otherwise distantly related taxa. Suiodea (alternately Suina or Suiformes) is the clade uniting pigs (Suidae), peccaries (Tayassuidae), and extinct fossil taxa that are closely related to these living groups. The earliest fossil suoids are known from the late middle or late Eocene of SSEA, suggesting that the origins of both suids and tayassuids are to be found in the Eocene of SSEA (Rose 2006; Harris and Liu 2007; Orliac et al. 2010a, b). Basal suoids have been described from the Eocene Krabi Basin of Thailand (Ducrocq 1994; Ducrocq et al. 1998; Orliac et al. 2011). One of these, *Egatochoerus*, was initially described as a tayassuid, but later reassigned to the base of Suiodea (Orliac et al. 2010a, b).

The earliest suids may be early Miocene Eurasian hyotheriines (Orliac et al. 2010a, b) though the taxonomy of Suidae below the family level has a complicated and unsettled history (Harris and Liu 2007). Other extinct fossil suid clades from the Neogene include the Tetraconodontinae and Listriodontinae. Living suids, however, may all be assigned to a single subfamily, Suinae, with the exception of the babirusa (Gongora et al. 2011). The earliest fossils of Suinae appear in the middle Miocene of both SSEA (Chinji) and Europe (at about 13 Ma or younger) (Harris and Liu 2007). A notable fossil suine from the late Miocene of SSEA is *Propotamochoerus hysudricus*, which is recorded in the Siwaliks of northern Pakistan from 10.2 to 6.8 Ma (Badgley et al. 2008). This long-lived species is known only from SSEA with the notable exception of an occurrence in the late Miocene Baynunah fauna of the United Arab Emirates (Bishop and Hill 1999; Bibi et al. 2013). The earliest fossils attributed to *Sus* are from the earliest Pliocene of SSEA and Europe (Harris and Liu 2007).

Molecular phylogenies estimate the divergence of Suidae and Tayassuidae at 47–23 Ma, Suinae and *Babyroussa* at 18–9 Ma, and the origin of crown Suinae at 15–7 Ma (Gongora et al. 2011). These ranges appear broadly compatible with the fossil record (Orliac et al. 2010a, b), though Thenius (1970) proposed an Oligocene or older split of the babirusa from other pigs. The babirusa, with its bizzare horn-like canines, has an enigmatic evolutionary history. Besides a few Pleistocene occurrences, the babirusa has no fossil record, and it has been proposed that its colonization of Southeast Asian islands may have been as old as the Oligocene, in contrast to the other mammalian lineages inhabiting the Southeast Asian archipelago, most of which are believed to have dispersed there in the Pleistocene (van den Bergh et al. 2001a, b).

Schizochoerus gandakasensis, recorded from 10.1 to 8.7 Ma in the Siwaliks (Badgley et al. 2008) is a member of an enigmatic group of pig and peccary-like species generally referred to as palaeochoerids (Boisserie et al. 2010; Orliac et al. 2010a, b). There is no consensus as to the position of the different 'palaeochoerid' taxa, but a recent analysis suggests *Schizocheorus* may be an early suid, with other 'palaeochoerids' lying closer to peccaries (Orliac et al. 2010a, b). *Pecarichoerus* from the early and middle Miocene Siwaliks (Colbert 1933) was first thought to be a peccary but was later placed among the 'palaeochoerids' (Orliac et al. 2010a, b).

Hippopotamidae

Though today restricted to Africa, hippopotamids were previously more widely distributed across Eurasia, with a presence in SSEA from the late Miocene to the late Pleistocene. Until recently, many Asian and African fossil hippopotamids, as well as the living west African pygmy hippo, were classified in the genus *Hexaprotodon*. A recent review (Boisserie 2005) determined that *Hexaprotodon,* first named on the basis of fossil material from the Siwaliks by Falconer and

Cautley (1836a, b, c), is in fact a restricted clade of extinct SSEA hippopotamids. The earliest SSEA hippo is *Hexaprotodon sivalensis* (Fig. 2.10), which is recorded at 5.9–3.5 Ma in the Siwaliks (Badgley et al. 2008). The lineage survives into the

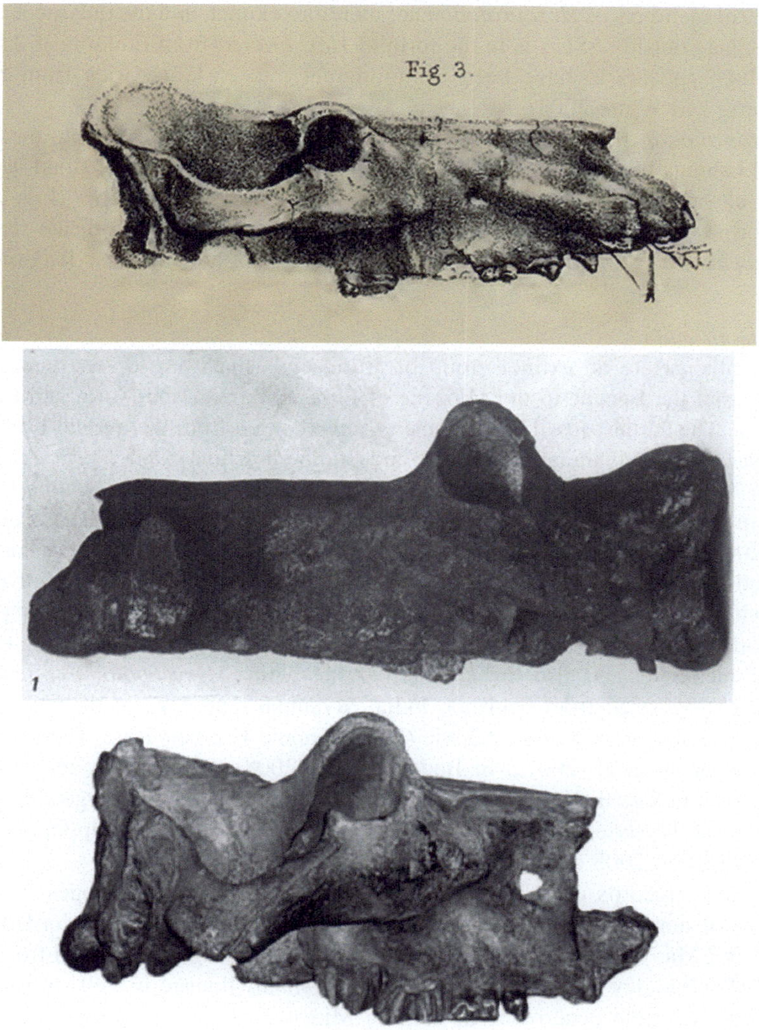

Fig. 2.10 Fossil hippopotamids from southern Asia. From top: *Hexaprotodon sivalensis* from the latest Miocene to Pliocene of India and Pakistan (from Falconer in Murchison 1868, p. 2), *Hex. palaeindicus* from Pleistocene India (from Boisserie 2005 Fig. 2.8), and *Hexaprotodon* sp. from Pleistocene Indonesia (from Hooijer 1950, p. 2). Note the evolution of elevated orbits in the younger specimens, a feature indicative of a semi-aquatic lifestyle as in modern hippos, evolved in parallel in Asian *Hexaprotodon* and African *Hippopotamus* clades (Boisserie 2005). Crania are not to scale

late Pleistocene of Central India as *Hex. palaeindicus*, and possibly as a different species in Indonesia (Colbert 1935a, b; Hooijer 1950; Boisserie 2005; Chauhan 2008; Patnaik et al. 2009). While the living African pygmy hippo has often been classified as *Hexaprotodon liberiensis*, numerous morphological differences indicate it is better assigned to a separate genus, *Choeropsis liberiensis* (Boisserie 2005). All members of *Hexaprotodon* are therefore extinct, and the only occurrence of this clade outside SSEA is in the form of *Hex. bruneti* from Ethiopia at 2.5 Ma, and which appears to have been an immigrant into eastern Africa from SSEA (Boisserie and White 2004).

Hexaprotodon sivalensis possessed relatively low orbits that indicate more terrestrial habitats than in living hippos. By Pleistocene times, the orbits had become highly elevated above the skull profile, as seen in *Hex. palaeindicus* (Fig. 2.10). High orbits are found in living *Hippopotamus amphibius*, and indicate that the Pleistocene SSEA hippo was also well adapted to a semi-aquatic lifestyle (Boisserie 2005).

Anthracotheriidae

Anthracotheres are an extinct group of artiodactyls that were diverse and widespread from the Eocene to the Miocene of Eurasia, Africa, and North and Central America. The earliest fossil anthracotheres are recorded from the middle Eocene of southeast Asia (Myanmar, Vietnam, and southern China) and North America, though the family's origins are certainly older and probably in southern Asia (Pilgrim 1928; Tsubamoto et al. 2002; Ducrocq and Lihoreau 2006a, b; Lihoreau et al. 2007; Tsubamoto 2010; Böhme et al. 2013). A fossil anthracothere has been reported from the Eocene of Timor Island (Von Koenigswald 1967), but it is a somewhat anomalous occurrence and its original provenience has been put to question (Ducrocq 1996; Lihoreau et al. 2007a, b). The last worldwide occurrence of anthracotheres is also in southern Asia, with *Merycopotamus dissimilis* (Fig. 2.11) recorded from Pakistan, India, Myanmar, and Java, with its youngest records in Pakistan at around 2.5 Ma (Steensma and Hussain 1992; Dennell et al. 2006; Lihoreau et al. 2007a, b; Badgley et al. 2008) and possibly as young as 1.7 Ma (Dennell et al. 2006). Certain anthracotheres achieved very large sizes and a semi-aquatic lifestyle, and this group may have given rise to the Hippopotamidae (Boisserie et al. 2005 and references therein).

The main records of SSEA anthracotheres come from two lineages from the Siwaliks of northern Pakistan. There, the *Microbunodon* lineage is recorded from 13.8 to 9.2 Ma, while the long-lived *Merycopotamus* lineage is recorded from 13.9 to 2.5 Ma (Badgley et al. 2008). The *Merycopotamus* lineage is well studied, and apart from a single occurrence in the late Miocene of Iraq, is known only from SSEA (Lihoreau et al. 2007a, b; Hanta et al. 2008). During the late Miocene, this lineage exhibited increased adaptation to aquatic habitats, including the evolution of elevated orbits (as in Hippopotamidae), increased body mass, and development of less cursorial limbs; its presence in the Tigris-Euphrates basin of Iraq suggests dispersal there from southern Asia during a sea level regression at about 10 Ma (Lihoreau et al. 2007a, b).

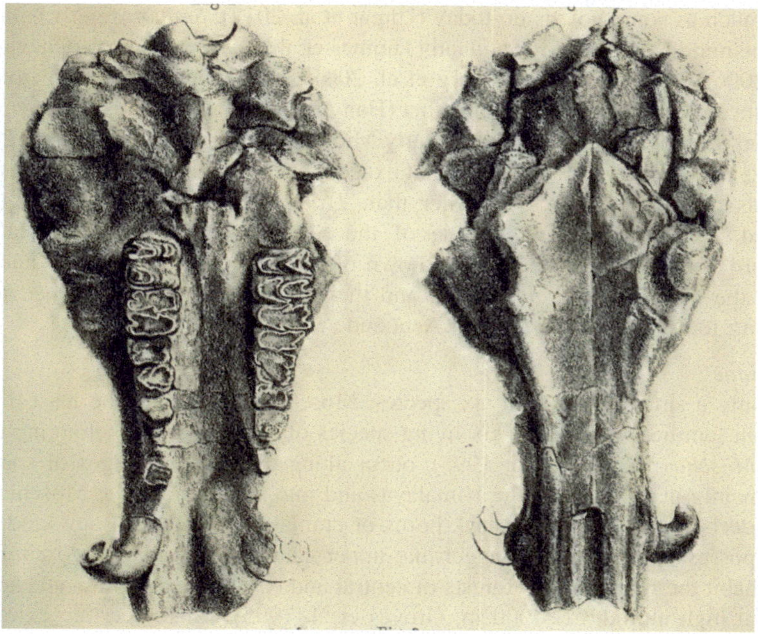

Fig. 2.11 Skull of the hippopotamus-like anthracothere, *Merycopotamus dissimilis*, recorded from the late Miocene to the terminal Pliocene in Pakistan, Nepal, Myanmar, and Indonesia (from Murchison 1868, p. 13)

Tragulidae

Tragulids comprise a clade of ruminants that is today relict in terms of its diversity and distribution, and that in morphology bears resemblances to early Paleogene artiodactyls, including primitive ruminants. Tragulids are the only extant survivors of an earliest Eocene diversification of hornless ruminants but their Paleogene fossil record is scanty. Interestingly, tragulids are physiologically conservative in that they only have a three-chambered stomach, rather than the more derived four-chambered stomach of most other ruminants. Today, tragulids are represented by several species of chevrotain or mouse deer (*Tragulus* spp.) from SE Asia including the Philippines, *Moschiola* spp. from India and Sri Lanka, and the larger water chevrotain *Hyemoschus aquaticus* from central Africa. The earliest fossil tragulid is *Aracheotragulus krabiensis* from the late Eocene of Thailand (Métais et al. 2001), and another enigmatic bunodont form (Métais et al. 2007), and they are otherwise documented in Europe, Turkey, and Pakistan. Early tragulids seem to have been restricted to densely forested and humid environments where they probably fed on fruits, seeds, soft leaves, and occasionally on invertebrates, as living tragulids do (Dubost 1978). There is evidence of varied mixed feeding strategies along a graze-browse spectrum among several species of *Dorcatherium* from the early Miocene of Kenya, suggesting that tragulids partitioned dietary niche space much as some bovids do today (Ungar et al. 2012). *Siamotragulus* is known from the middle Miocene of Thailand (Thomas et al. 1990), Myanmar (Chevasseau et al. 2006), and Pakistan (Ginsburg et al. 2001), and *Yunnannotherium* from the late Miocene of Lufeng, southern China (Han 1986). Another genus, *Afrotragulus*, has been recently described from the early Miocene of Africa (Sánchez et al. 2010). The size of these fossil tragulids did not exceed that of the extant *Hyemoschus*, and tragulids of size equal to or smaller than *Tragulus javanicus* (i.e., <2 kg) are reported from the middle Miocene of the Siwaliks (West 1980) and Kenya (Pickford 2001). Almost nothing is known of the evolution of tragulids since the end of the Miocene (Rössner 2007), and their fate today is probably tied to the shrinking forest habitats of tropical Asia and Africa.

Moschidae

With only a single genus and six species, Moschidae is among the least diverse ruminant families. Five of the six living species of musk deer, all belonging to the genus *Moschus* (Groves et al. 1995), occur along the northern edges of southern Asia, mainly in and around the Himalayas and into southern China. Moschids are small, deer-like ungulates that lack horns or cranial appendages of any kind, with males possessing prominent dagger-like upper canines. Nowadays, *Moschus* species inhabit the mountainous forests of central and eastern Asia, with some species living at high altitudes (≥4000 m, Groves et al. 1995). Musk deer are extremely territorial and feed on leaves, flowers, and various other soft items. Males are known to mark their territory with musk secretions (hence their name), and though endangered, their scent glands are still in high demand in China for medicinal or aromatic purposes.

Evolutionarily, moschids have long posed a problem because, unlike other advanced (pecoran) ruminants, they lack cranial appendages and possess large upper canines. These are primitive morphological features, similar to the condition in tragulids. Although lacking the double lacrimal orifice characteristic of cervids, morphological analyses have supported relationships to cervids and antilocaprids (e.g., Janis and Scott 1987). A supertree analysis combining behavioral, molecular, and morphological data also supported affinities with cervids (Hernández Fernández and Vrba 2005). In total contradiction, however, molecular analyses have consistently indicated that moschids are more closely related to bovids (Hassanin and Douzery 2003; Meredith et al. 2011; Hassanin et al. 2012).

Several putative moschids first appeared in the late Oligocene of Europe and Asia with late Oligocene and early Miocene taxa such as *Dremotherium* and *Amphitragulus*, and the enigmatic *Prodremotherium* from the Oligocene of Asia (Métais and Vislobokova 2007 and references therein). These potential stem moschids went extinct by the end of Agenian European Land Mammal Age (ca. 20 Ma), at the same time as the first appearance of cervids in Europe. From the middle Miocene, the moschids *Micromeryx* and *Hispanomeryx* are known from Europe and China (Sánchez et al. 2011). These two genera extended into the late Miocene (and probably into the mid-Pliocene for *Micromeryx*) and are thought to be closely related to the genus *Moschus*, which first appeared during the late

Miocene in China and eastern Europe (Vislobokova and Lavrov 2009). During the early Miocene, moschids also spread to North America, where they experienced an endemic radiation with six genera in the subfamily Blastomerycinae, all of which went extinct near the end of the Miocene (Prothero 2007).

Lophiomerycidae
This extinct family, probably closely related to tragulids, is known from Eocene to Oligocene fossil sites across Eurasia. The oldest lophiomerycids come from the middle Eocene of China (Guo et al. 2000) and are found in western Europe shortly after the Eocene–Oligocene transition (Stehlin 1909; Brunet and Sudre 1987a, b). Asian lophiomerycids diversified in the Oligocene with *Lophiomeryx,* reported from China, Pakistan, Central Asia, the Caucasus, and western Europe. In *Lophiomeryx*, the central metacarpals and metatarsals are unfused, while the lateral metatarsals are reduced (at least in western European species) (Fig. 2.12). These characters indicate that *Lophiomeryx* may have inhabited forested or swampy habitats, as in the living tragulid *Hyemoschus aquaticus*.

Bovidae
With over 130 living species, Bovidae is the most diverse family of extant large mammals. Thirty species are found in SSEA, most of these endemic to the region. Among these is the saola, *Pseudoryx nghetinhensis,* endemic to the forests of Laos and Vietnam, and which was only scientifically discovered in 1992 (Dung et al. 1993). Although somewhat goat-like in appearance, the saola is in fact the closest living relative of the clade comprising buffaloes and oxen (Bovini). The discovery of the saola was totally unexpected, since more than 50 years had passed since the last scientific discovery of a new living large mammal (the kouprey, from Cambodia) (Urbain 1937). This offered a reminder of the sheer richness of Southeast Asia's biodiversity, and of the potential for future discoveries to be made there.

Diets of bovid species may generally be characterized as relying either on browse (leaves) or graze (grass), or a combination of both. They inhabit a wide range of habitats from humid rainforests to semi-arid plains and deserts, and from coastal areas to the high altitudes of the Himalayas. The earliest fossil bovids are recorded at around 18 Ma from both SSEA (Pakistan) and Europe (France)

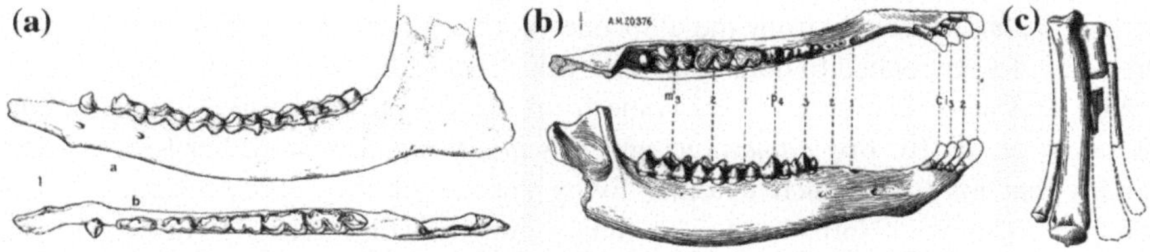

Fig. 2.12 **a** Lower jaw (holotype) of *Lophiomeryx mouchelini* from the early Oligocene of Villebramar (SW France), MP 22. (from Brunet and Sudre 1987a, b). Notice the retention of a p1 in this species, which is significantly smaller than in *L. chalaniati*, the type species of the genus. **b** Lower jaw (holotype) and **c** partial metacarpals of *Lophiomeryx angarae* from the late Eocene of Ardyn Obo, Mongolia (from Matthew and Granger 1925). Specimens not to scale

(Ginsburg and Heintz 1968; Solounias et al. 1995; Ginsburg et al. 2001; Antoine et al. 2013) and the crown clade probably originated around this time or shortly thereafter (Bibi 2013). SSEA has been an important arena of bovid evolution, especially since this region's faunas were partly isolated from those of Central Asia and Africa throughout much of the Neogene (Brunet et al. 1984; Bibi 2007; Lihoreau et al. 2007a, b).

Most extant SSEA bovids can be classified among four subclades (tribes): Bovini, comprising buffaloes and oxen; Antilopini, gazelles and relatives; Caprini, wild goats, sheep, and relatives; and Boselaphini, represented only by the nilgai and chousingha. With the exception of Boselaphini, which are today endemic to the Indian Subcontinent, these tribes are all widespread. Bovini today range across Eurasia, Africa, and North America, but both the fossil record and phylogenetic reconstructions indicate the origins of the Bovini are probably in the Middle Miocene of SSEA (Pilgrim 1939; Hassanin and Ropiquet 2004; Bibi 2007; Hassanin et al. 2013). Bovini developed many of their characteristic dental features in the late Miocene (around 9 Ma), and may have first spread out of SSEA and into Africa coincident with major environmental changes and the spread of tropical (C_4) grassland habitats (Bibi 2007). The lineage leading to the Asian buffalo (*Bubalus* spp.) is first recognized in the late Pliocene *Proamphibos* in the Siwaliks (Tatrot Formation) (Pilgrim 1939; Bibi 2009), and continues with *Hemibos* in the late Pliocene and Pleistocene (Pilgrim 1939; Gentry 1978; Barry et al. 1982; Bibi 2009). *Bubalus* populated Southeast Asian islands and differentiated into several 'dwarfed' species. These are represented today by the tamaraw (*Bubalus mindorensis*) of Mindoro Island, and the mountain and lowland anoas (*B. quarlesi* and *B. depressicornis*) of Sulawesi and Buton (Groves 1969). Records of fossil buffalo from the modern SE Asian archipelago include *Bubalus cebuensis* of Cebu (Croft et al. 2006), and the oldest records may be of *Bubalus palaeokerabau* at 1.2 Ma at Ci Saat in Java (Louys et al. 2007). This latter date matches the early Pleistocene age estimates for the dispersal of *Bubalus* from the mainland onto the islands (Tanaka et al. 1996; Bibi 2013; Hassanin et al. 2013).

Little is known about the evolution of wild oxen (*Bos* spp.), but their present ranges along with molecular phylogenies also suggest an SSEA origin for this clade, which includes the progenitor of domesticated cows (*Bos primigenius*) and the cold-adapted yak and bison clade (Hassanin and Ropiquet 2004; Hassanin et al. 2013). *Bison sivalensis* from the Pliocene or Pleistocene Siwaliks has been proposed as closely related to both bison and yak (Lydekker 1878; Pilgrim 1939; Olsen 1990), and its oldest records may come from the Tatrot formation at around 3 Ma (Khan et al. 2010). *Bos namadicus* and *Bos acutifrons* are two other SSEA fossil bovins that may be related to other living species of *Bos*. *Bos namadicus,* also referred to as the 'Narbada Ox' or Indian aurochs (Fig. 2.13) (Lydekker 1898; Gentry et al. 2004), had a horn span of about 1.25 m (as estimated from Lydekker 1878, p. 11) and was probably conspecific with the aurochs of Europe, *Bos primigenius* (Rütimeyer 1878; Lydekker 1898, p. 19; Gentry 1967). Molecular evidence suggests two or three separate domestication events for cattle around

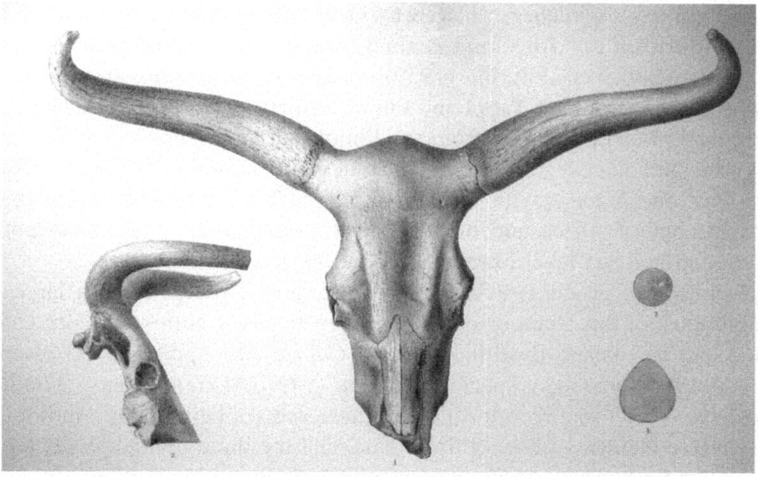

Fig. 2.13 The Narbada Ox or Indian aurochs, *Bos namadicus* or *B. primigenius namadicus*, may be the Pleistocene progenitor of domesticated cattle, and perhaps specifically of the zebu (*Bos taurus indicus*) (Gentry et al. 2004) (figure from Lydekker 1878, p. 11)

10,000 years ago, one of which occurred in the vicinity of northern India or Pakistan (Loftus et al. 1994; McTavish et al. 2013).

Gaur (*Bos gaurus*), banteng (*Bos javanicus*), and kouprey (*Bos sauveli*) are SSEA's wild oxen. Gaur and banteng have both been domesticated, but little is known about the kouprey. First described in 1937, the kouprey was only rarely sighted up through the 1960s, and a lack of convincing evidence of any viable populations since that time suggests that the kouprey may be extinct (MacKinnon and Stuart 1988; www.iucnredlist.org). Suggestions that the kouprey may be a feral hybrid of banteng and zebu cattle (Galbreath et al. 2006) have been refuted (Hassanin and Ropiquet 2004, 2007), and molecular age estimates suggest a divergence from the gaur and banteng in the late Pliocene or early Pleistocene (Hassanin and Ropiquet 2004).

Molecular evidence has shown the saola (*Pseudoryx nghetinhensis*) to be the living sister taxon to Bovini (Hassanin and Douzery 1999; Gatesy and Arctander 2000; Hassanin et al. 2013), and its geographic range and morphologically conserved features provide further support for a SSEA origin of Bovini. No fossils of the saola are known, but presumably its lineage has inhabited the forests of southeast Asia at least since its divergence from the Bovini, which molecular estimates place at between about 14 and 10 Ma (Hassanin et al. 2012; Bibi 2013).

While totally restricted to Africa today, Reduncini (e.g., waterbuck, reedbuck, lechwe) have an ancient fossil record in SSEA. Reduncins are documented in the late Miocene Siwaliks as early as 9.3 Ma (Pilgrim 1939; Barry et al. 2002; Badgley et al. 2008). Reduncin species are also recorded in the younger (late Pliocene–Pleistocene) formations of the Siwaliks (Pilgrim 1939; Khan et al. 2008), up into the Late Pleistocene (Vrba et al. 2015).

Alcelaphini (e.g., wildebeest, hartebeest) and Hippotragini (e.g., roan, oryx) are also today restricted to Africa (and Arabia), but show up as fossils in the younger levels of the Siwaliks. Pilgrim (1939) described *Sivatragus brevicornis* and *Sivatragus bohlini* from the Tatrot and Pinjor formations, respectively, and *Sivoryx sivalensis* and *Sivoryx cautleyi*, from the Pinjor and an unknown horizon, respectively. Vrba and Gatesy (1994) demonstrated the affinities of *Sivatragus* with *Hippotragus*, and *Sivoryx* with *Oryx*, and argued for a migration event of these hippotragins out of Africa and into SSEA at around 3–2.5 Ma, coincident with global cooling and sea level lowering (Vrba 1995).

Antilopini (e.g., gazelles) and Caprini (e.g., goats, sheep) have a large fossil record, but most of this is not resolved phylogenetically. Caprini are represented by eastern Asian and island dwelling serows (*Capricornis* spp.) and by essentially central Asian montane species (*Capra, Ovis, Hemitragus, Budorcas, Naemorhedus*), the fossil record for all of these species being poor. Antilopini are far less diverse elements of the SSEA fauna, and the three living species (*Gazella bennetti, G. subgutturosa, Antilope cervicapra*) are found in relatively dry habitats of the Indian Subcontinent. The blackbuck (*A. cervicapra*) is the only living antilopin with spiraling horns and may be the sole survivor of a late Miocene radiation of diverse spiral-horned antilopins (e.g., *Prostrepsiceros* spp.) that ranged from Greece through to Afghanistan (Gentry 2003, 2010).

Giraffidae

Living giraffes (*Giraffa, Okapia*) are also restricted to Africa, but the family has a fossil record spanning much of southern Europe and southern Asia. Early Miocene giraffids or giraffid-like taxa appear about the same time in Africa, the Mediterranean, and Pakistan (Solounias 2007). They persist in all these regions into the Pleistocene, and attain a peak diversity in the late Miocene and Pliocene, when it is not uncommon to find several species at a single locality (Solounias 2007). Across this geographic range, and since the late Miocene, giraffids have been represented by different forms including some similar to *Giraffa* (with a long neck and long limbs), others more like *Okapia* in form, and the extinct sivatheres (Sivatheriini) which reached large sizes and had short necks and short stocky limbs (Fig. 2.14).

SSEA fossil giraffes are known mainly from the Siwaliks and were described by Falconer and Cautley (1844), Lydekker (1882a, b), Pilgrim (1911), and Colbert (1935a, b) the latter author recognizing no less than 13 different species in the Siwaliks sequence. More recent accounts of Siwalik faunas (Barry et al. 2002; Badgley et al. 2008) recognize less diversity. *Giraffa punjabiensis* is recorded as ranging from 8.9 to 7.2 Ma (Badgley et al. 2008), followed by *Giraffa sivalensis* in the upper Siwaliks (late Pliocene to Pleistocene). Falconer and Cautley (1844, reprinted in Murchison 1868, p. 16), illustrate an elongated neck vertebra of *G. sivalensis*, demonstrating it had a long neck as in modern *Giraffa camelopardalis*.

Sivatheres are known from the Miocene to Pleistocene, from SSEA to Greece and into Arabia and across Africa. In the Siwaliks, they are represented by at least a single long-lived lineage comprising *Giraffokeryx punjabiensis* (13.6–10.3 Ma) and

Fig. 2.14 The head of *Sivatherium giganteum*, a massive giraffid with palmate antler-like ossicones that inhabited SSEA in the late Pliocene and Pleistocene. *Left* sketch from Colbert (1936), *right* sketches from Falconer (in Murchison 1868, p. 19 and 21)

Bramatherium megacephalum (10.3–7.1 Ma) (Barry et al. 2002; Badgley et al. 2008). Pliocene and Pleistocene sivatheres were massive, and are typified by *Sivatherium giganteum,* which had palmate branching ossicones (Fig. 2.14) (Falconer and Cautley 1836a, b, c). With a lower molar row length of around 170 mm (metrics in Khan et al. 2011), *S. giganteum* would have weighed around 3,000 kg (our estimate using the prediction equations in Janis 1990). *Sivatherium giganteum* persists in the Indian Subcontinent up to about 1.7 Ma (Dennell et al. 2006, 2008), or perhaps as late as 0.6 Ma (Rao et al. 1995; Nanda 2002).

Cervidae

The cervid (deer) family today spans arctic to tropical latitudes, and tundra to rainforest habitats in both the new and old worlds. They reach very high diversity in SSEA, with some 20 species (genera *Cervus, Axis, Muntjacus, Rucervus, Rusa*). Cervids are divided into two subclades, the Cervinae, including most Eurasian deer species (=Plesiometacarpalia), and the Capreolinae (or Odocoileinae), comprising primarily American deer (=Telemetacarpalia) (Groves 2007; Hassanin et al. 2012). All SSEA deer belong to Cervinae, which is distinguished by the retention of the proximal ends of the lateral metacarpals (distal in Capreolinae).

The earliest stem cervids are known from the early Miocene of Europe (Germany, France, Spain) (Azanza 1993; Gentry et al. 1999). The middle Miocene *Euprox*, also from Europe, may be close to the origin of the crown clade (Gentry 1994; Gentry et al. 1999). A mid-Miocene or earliest late Miocene origin for the crown clade is also supported by molecular age estimates (Gilbert et al. 2006; Hassanin et al. 2012; Bibi 2013).

If there is a single main ecological difference between deer (cervids) and antelopes (bovids) it is the fact that deer never truly adapted to and diversified in xeric habitats. Compared to bovids, which have evolved numerous independent radiations into a diversity of open habitats (in Bovini, Alcelaphini, Hippotragini, Caprini, Antilopini), cervids are by contrast still strongly tied to the presence of water and tree cover. One of the main features distinguishing the modern large mammal fauna of southern Asia from that of Africa is the presence of deer in southern Asia, but this may be the result of a relatively recent (late Pliocene) immigration southwards from Central Asia. Cervids are known in Asia from the middle Miocene (*Dicrocerus*, *Lagomeryx*, *Stephanocemas*), but, with the exception of a middle Miocene occurrence in northern Thailand (Ginsburg and Ukkakimapan 1983; Suraprasit et al. 2014), Miocene cervids are documented only from areas situated north of the Alpine-Himalayan Range. There are no fossil cervids from the rich Neogene sedimentary record of Pakistan (Sulaiman Range and Potwar Plateau) older than the first appearence of *Cervus* in the Pliocene and Pleistocene Tatrot and Pinjor beds (Fig. 2.15) (Barry et al. 1982, 2002). (Reported cervids from Miocene and Pliocene Siwaliks [Ghaffar et al. 2006, 2010, 2011a, b; Samiullah and Akhtar 2007] are fragmentary and appear to be bovids). Upper Siwaliks cervids were described by Lydekker (1876a, b) and Brown (1926), and revised by Azzaroli (1954).

Late Miocene to late Pliocene mammal localities in Iraq (Thomas et al. 1980), United Arab Emirates (Bibi et al. 2013), and Saudi Arabia (Thomas et al. 1982), all situated south of the Alpine-Himalayan chain, were, like SSEA at this time, also devoid of cervids. In contrast, the late Miocene locality of Maragheh in Iran, located north of the Taurus-Zagros Mountains, has yielded cervid remains (Tobien 1980; Bernor 1986; Ataabadi et al. 2013). Cervids are also documented in the late Miocene of Lufeng and Yuanmou in southern China, which is interesting considering that southern Chinese faunas were otherwise more similar to those of southern

Fig. 2.15 Cranium with antlers of *'Cervus' punjabiensis* from the upper Pinjor (Middle Pleistocene) of the Siwaliks (from Brown 1926). Views not to scale

Asia than to central or northern China (Dong and Qi 2013). Otherwise, it appears that cervids spread to Indonesia and the Philippines only later, in Pliocene to Pleistocene times (Reis and Garong 2001), never reaching Australia (until recent introduction by humans). Throughout most of their evolutionary history since the early Miocene, therefore, cervids were absent from SSEA, and their modern presence there can be traced no farther back than the late Pliocene (ca. 3 Ma). Extant cervids generally live in lowlands and it has been proposed that that the ~10,000 km-long mountain belt resulting from the closure of the Neotethys Ocean impeded their southward dispersal (Brunet et al. 1984). High cervid species richness in modern SSEA implies high rates of diversification or immigration into the area since the latest Pliocene and early Pleistocene.

Muntiacs (*Muntjacus, Elaphodus*) are small deer that possess massive upper canines, superficially similar to those in tragulids, and distinguishing them from all other pecorans. With their reduced antlers and large upper canines, muntiacs have been proposed to represent primitive deer (Azanza 1993), but their molecular affinities with Cervinae (Gilbert et al. 2006; Hassanin et al. 2012) suggest morphological reversions that accompanied adaptation to a forest environment (Groves 2007). The oldest fossil muntjacs are known from the late Miocene (9–7 Ma) of southwestern China (Dong et al. 2004), suggesting the presence there of forested landscapes, in contrast to the desert climate that prevails in that area today and which probably developed as a result of increased uplift in the Tibetan Plateau in the Plio-Pleistocene (An et al. 2001).

Camelidae

Although the natural range of camelids today is restricted to parts of Asia, Africa, and South America, camelids first evolved in the Eocene of North America, where they remained endemic until their first dispersal into Eurasia and Africa in the late Miocene (Honey et al. 1998). Camelids entered South America in the early Pleistocene and went extinct in North America in the end Pleistocene megafaunal extinctions (Honey et al. 1998). Their North American fossil record is abundant and diverse, but they are otherwise rather rare components of fossil sites in Eurasia and Africa.

Falconer and Cautley (1836a, b, c, reprinted in Murchison 1868) described a fossil camel from the Siwaliks, which they named *Camelus sivalensis* (Fig. 2.16). These authors, along with Colbert (1935a, b), believed the Siwalik fossil camel was very similar to both the bactrian and dromedary camels alive today, which is surprising considering there are major differences between the two living species. Colbert (1935a, b) records *C. sivalensis* in the Pleistocene Pinjor Formation, while Nanda (2002, 2008) notes its presence in the preceding late Pliocene Tatrot Formation (ca. 3.6–2.6 Ma) as well. *Camelus antiquus*, first mentioned by Falconer and Cautley (1836a, b, c) and later described by Lydekker (1885), was taken by Matthew (1929) and Colbert (1935a, b) to be a synonym of *C. sivalensis*.

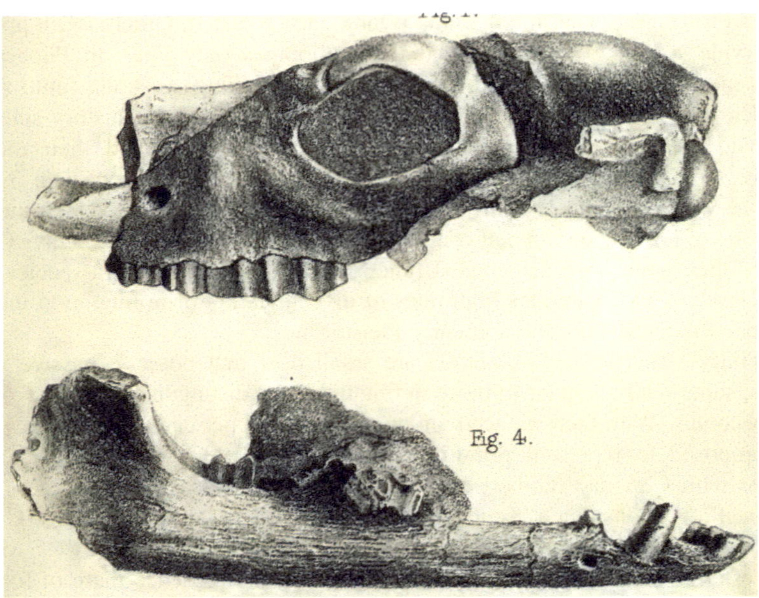

Fig. 2.16 *Camelus sivalensis*, a Pliocene-Pleistocene fossil camel from the Siwaliks (Falconer and Cautley 1836a, b, c; Murchison 1868, p. 18)

2.6.2 Perissodactyla

Living perissodactyls are limited to a few species of equids, rhinos, and tapirs, all of which today have restricted distributions. However, in the past, and especially during the Eocene, perissodactyls were much more diverse and widely distributed. Early perissodactyls may have originated from a taxon similar to *Radinskya* from the Paleocene of China (Froehlich 1999; Prothero and Schoch 2002). The earliest fossil perissodactyls (*Hyracotherium* spp.) appear at the beginning of the Eocene, almost simultaneously in Europe, Asia, and North America (Rose 2006). By the end of the early Eocene, perissodactyls had differentiated into lineages that would lead to equids, rhinocerotids, tapirids, and other extinct families. Perissodactyl diversity dropped steeply after the Eocene and, while competition with newly evolved ruminant artiodactyls has been proposed, the limited ability of perissodactyls to cope with increasingly dry and seasonal environments is more likely (Cifelli 1981a, b; Janis 1989).

Tapiridiae

A single tapir species (*Tapirus indicus*) inhabits the Old World, and is restricted to Southeast Asia, namely the Malay Peninsula and Sumatra. The closest living relatives of the Asian tapir are to be found in Central and South America, and this hints at the broader geographic distribution that tapirs once enjoyed. Living tapirs are remarkably conserved in their morphological features and most classifications include all four species into a single genus. This is remarkable given that molecular clock estimates put the divergence between them at about 30–20 Ma (Steiner and Ryder 2011). The earliest member of the tapir lineage (Tapiroidea) from the early Eocene of North America and Asia had many of the basic morphological features of this clade, and tapirs appear to have changed very little morphologically in the past 50 million years (Rose 2006). The main morphological changes since the early Eocene have primarily involved the development of the nasal trunk as recorded through the reduction in size of the nasal bones (Rose 2006). Their preferences for humid and forested habitats (less so *T. pinchaque*) may also be relatively unchanged since the Eocene, and perhaps their shrinking geographic distribution may have simply tracked the shrinking of tropical forest habitats in the later Cenozoic and especially during the late Neogene.

The earliest fossils assigned to Tapiridae (crown or stem) are from the middle Eocene of North America (Colbert and Schoch 1998; Rose 2006), while fossils assigned to *Tapirus* are as old as the middle Miocene of Europe (Heissig 1999) and the late Miocene of North America (Colbert and Schoch 1998). Tapiroids have been low in diversity since the Oligocene, and have generally declined in diversity since then (Schoch 1989; Colbert and Schoch 1998). They are relatively rare in the Neogene fossil record.

Fossil remains assigned to *Tapirus indicus*, though possibly representing mutliple *Tapirus* species, are recorded from as old as the early Pleistocene, from southern China to Java (Hooijer 1947; Tong 2005). The extant species appears to have coexisted with at least one other, the giant tapir *Megatapirus augustus*, which is known from the late middle Pleistocene of southern China and SE Asia (Hooijer 1947; Tougard 2001). This species differed from *Tapirus indicus* in its much larger size, being about 25 % larger in dental dimensions (Matthew and Granger 1923), which translates to about a doubling of body mass (our estimate based on the prediction equations in Janis 1990).

Rhinocerotoidea

Rhinocerotoids (amynodontids, hyracodontids, and rhinocerotids) were taxonomically and ecologically diverse during the Paleogene and early Neogene of Eurasia and North America. Since that time, hyracodontids and amynodontids have gone extinct and rhinocerotids have dispersed into Africa. The ancestry of rhinocerotoids is Holarctic and represented by *Hyrachyus*, an early to middle Eocene form that was widely distributed in Eurasia and North America (Prothero et al. 1986). In the Paleogene localities of Asia, rhinos outnumber all other ungulate mammals perhaps with the exception of tapiroids, another group of perissodactyls (generally smaller in size than rhinos) that flourished in Asia. Most Paleogene rhinos were hornless and the earliest horned rhinos possessed paired horns located at the tip of the nose. In rhinocerotids and hyracodontids, the upper third molar metastyle is reduced, resulting in a sub-triangular (to truly triangular in rhino) shape.

Amynodontidae

Amynodontids were ecologically diversified rhinocerotoids, with some resembling the hippopotamus (*Metamynodon*, with a body suggesting amphibious habits) and others more tapir-like (*Cadurcodon*, probably with a tapir-like proboscis) (Wall 1980). Their mode of life and feeding adaptations remain poorly understood, as are the origins and affinities of the basalmost forms. Amynodontids are represented by around a dozen genera that extend from the middle Eocene to the late Oligocene. Several Asiatic forms are poorly diagnosed, but they all seem to have been hornless. They probably originated in Asia during the Eocene (Wall and Manning 1986), though their earliest representatives are from the early middle Eocene of North America, and reached their maximum diversity in Asia during the Eocene–Oligocene transition (Wall 1989). *Cadurcotherium* is one of the 'Grande Coupure' immigrants into Europe, and the latest representative of the family in Asia (Métais et al. 2009).

Hyracodontidae

Hyracodontids flourished in the Eocene and Oligocene of Asia (Prothero 1998a, b). Most of these were medium-sized and fairly cursorial ungulates with relatively elongate and gracile limbs. Across Asia, *Hyracodon* and *Forstercooperia* are the most common hyracodontids of the Eocene, while *Eggysodon* is the common taxon of the Oligocene. Among hyracodontids, the paraceratherines, including the emblematic *Paraceratherium* (=*Indricotherium*, *Baluchitherium*, from Pakistan, Central Asia, China, and Turkey) attained gigantic sizes, reaching about 7 m in height and maximum weights of 10–20 tons (Lucas and Sobus 1989; Fortelius and Kappelman 1993), making them, along with some fossil proboscideans, among the largest terrestrial mammals ever known (Fig. 2.17). Besides their large size, paraceratherines can be differentiated from other hyracodontids by a derived anterior dentition with conical incisors, such as *Paraceratherium*, which was characterized by enlarged, procumbent first upper and lower incisors and a loss of the remaining two incisors, the canine, and the first deciduous premolars. The earliest paraceratherines were small hyracodontids known from the middle Eocene of North America (*Forstercooperia*) and China (*Juxia*). Large paraceratherines appeared during the early Oligocene in Asia, reaching maximum sizes during the

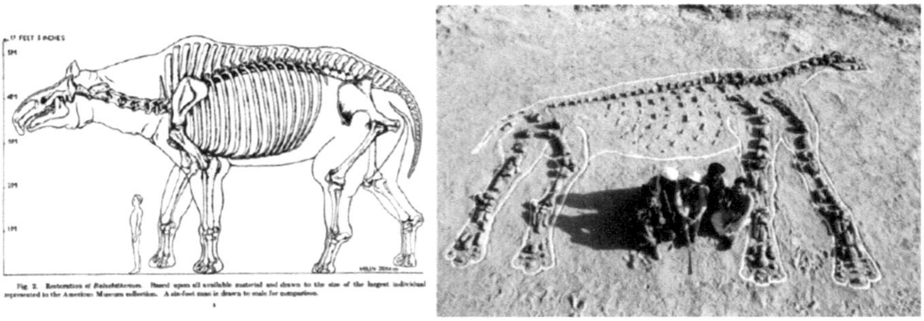

Fig. 2.17 The massive hyracodontid *Paraceratherium* as envisioned by Granger and Gregory (1935) (*left*), and a skeleton reconstructed in the field in the Bugti Hills showing the correct position and length of the neck (*photo* Mission Paléontologique Française au Balouchistan, MPFB)

late Oligocene and possibly persisting into the early Miocene. Paraceratherines appear to have possessed no sexual dimorphism, and were browsers in vast alluvial plains, dispersing often between Eurasia and North America (Radinsky 1967).

The sheer size of the gigantic paraceratherines (=indricotheres) raises functional and paleobiological problems (Fortelius and Kappelman 1993). For example, the cursorial ancestry of paraceratherines produced particular morphological features of the limbs (distinct from those of elephants) in response to the need to support such great body weight (over 15 tons in some individuals). Additionally, the quantity of plant matter (leaves or fruits) such a large animal would have to ingest daily to survive is a matter of question. Further investigation is required to better understand the extinction of these giants in the early Miocene, perhaps a result of climatic changes that took place in the Asian interior due to uplift of the Tibetan Plateau and shrinkage of the Paratethys Sea.

Rhinocerotidae

The five rhinoceros species living today are all in danger of extinction, but rhinocerotids comprised a large, diverse, and widespread clade throughout most of the Neogene. An important aspect of the evolutionary history of this clade involves the evolution of nasal horns and the transition from an older mode of combat using enlarged incisor tusks. The earliest fossil rhinocerotids (stem taxa) are known from the late Middle to Late Eocene of North America, Central Asia, and SSEA (Thailand and Vietnam) (Prothero 1998a, b; Antoine et al. 2003a, b; Böhme et al. 2013). These earliest rhinos did not have nasal horns, but instead had enlarged incisors forming a shearing complex with a chisel-like upper first upper incisor and a tusk-like lower second incisor. In contrast, all living rhinos have well-developed horns and incisors that are reduced (Asian species) or entirely lost (African species). The earliest horned rhino is the North American early Oligocene to early Miocene *Diceratherium*, which shows a morphological balance between the development of the horn and the incisor tusk complex (Heissig 1989; Prothero 1998a, b). Heissig (1989) reported that Asian rhinos use their tusk-like incisors to attack, and their horns only for interspecific combat, suggesting that rhino horns first evolved for sexual competition and display rather than defense.

Rhinos were diverse during the late Oligocene and Miocene of SSEA. All living rhinos belong to a single tribe, Rhinocerotini, but fossil rhinos represent a far greater diversity, with extinct clades such as Elasmotherini, Aceratheriini, and Teleoceratini. *Brachypotherium* (Teleoceratini), for example, was a long-lived lineage, recorded from the early to the late Miocene of SSEA, from Pakistan to Thailand (Chaimanee et al. 2004; Chavasseau et al. 2006; Bhandari et al. 2010; Antoine et al. 2013) In the earliest Miocene (23–20 Ma) of the Bugti Hills, Pakistan, no less than nine separate rhino species have been found to co-occur in a single level (Antoine et al. 2010). Neogene Siwalik rhinocerotids were first described by Falconer and Cautley (1847), and subsequently by Matthew (1929), Colbert (1934, 1935a, b, 1942), E. Khan (1971), and A.M. Khan (2010), among others. Antoine et al. (2003a, b) reviewed the Paleogene record, and Antoine (2012) reviewed the Quaternary rhinocerotids of Southeast Asia.

Though only found in SSEA today, *Dicerorhinus* appears to have been a latecomer to the region, having perhaps first evolved in Central Asia and entering SSEA only sometime during the Pliocene, as evidenced by *Dicerorhinus gwebinensis*, a fossil species from the early Pleistocene of the Upper Irrawady in Myanmar (Zin-Maung-Maung-Thein et al. 2008, 2010). Fossils of the living Sumatran rhino (*Dicerorhinus sumatrensis*) are as old as the early Pleistocene or late Pliocene of southern China (Tong and Guérin 2009; Zin-Maung-Maung-Thein et al. 2010). Apparently, the Sumatran rhino's dentition has changed little from Oligocene taxa like *Lartetotherium* (Heissig 1989), reflecting the conservation of morphological features in this lineage. *Rhinoceros* on the other hand has a much older Neogene presence in SSEA (Zin-Maung-Maung-Thein et al. 2010). The oldest records of this genus may be *Rhinoceros* aff. *sivalensis* from the Siwaliks at 8.5 Ma (Fig. 2.18) (Badgley et al. 2008). *Rhinoceros* is believed to have descended from *Gaindatherium* (Colbert 1934, 1935a, b, 1942; Prothero et al. 1989), which is known from the Chitarwata Formation (23–20 Ma) in the Bugti region of Pakistan (Antoine et al. 2010) as well as in the Siwaliks from 16.0 to 7.3 Ma (Badgley et al. 2008). In general agreement with the fossil record, molecular estimates put the origin of the crown clade Rhinocerotidae at between 30 and 20 Ma, with

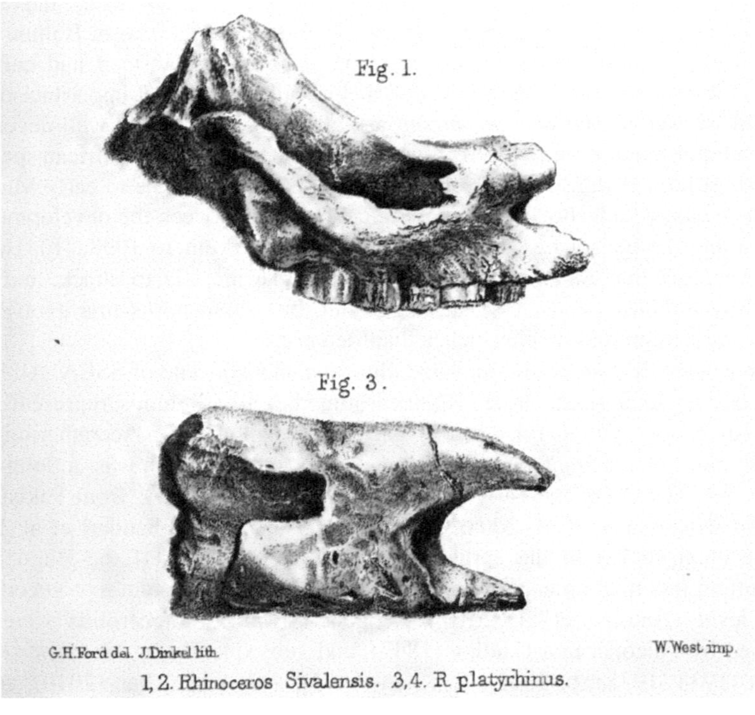

Fig. 2.18 Fossil skulls of *Rhinoceros sivalensis* and *Punjabitherium platyrhinus* from the Pleistocene (Pinjor) Siwaliks, both in lateral view (from Murchison 1868, p.14)

Rhinoceros and *Dicerorhinus* having been separated for just as long (Steiner and Ryder 2011).

Equidae

Equids are represented today by horses, zebras, and donkeys. Equids hold a special place of importance among humans, as the horse and donkey have supported agrarian and nomadic civilizations since the mid-Holocene. Today, nondomesticated equids in SSEA are basically limited to the endangered Asiatic wild ass, *Equus hemionus* (the kiang—*Equus kiang*—can be found in northernmost India, but its range is essentially in the Tibetan highlands). In SSEA, the IUCN (www.iucnredlist.org) records only a single wild population of the Asiatic wild ass, located in northwestern India.

The oldest recognized fossil (stem) equid is *Sifrihippus sandrae* (previously *Hyracotherium sandrae*) from the earliest Eocene (ca. 55.5 Ma) of North America (Froehlich 2002). Most of the evolutionary history of equids took place on the North American continent, so much so that the first appearances of progressive equid clades in the Old World are recognized as important biochronological events. Among these was the appearance of *Anchitherium* in the early Miocene (MN 3) in Europe and Asia, with anchitheriin equids persisting in the Old World until the early late Miocene (Bernor and Armour-Chelu 1999). Anchitheriins are not, however, recorded in SSEA (Bernor and Armour-Chelu 1999).

A second major dispersal event was the '*Hipparion* Datum,' marking the spread of hipparionin equids from North America and into Eurasia and Africa at around 11 Ma. One of the earliest Old World occurrences of hipparionin equids is from the Siwaliks at 10.7 Ma (Barry et al. 2002; Badgley et al. 2008). A number of fossil equid species from the Siwaliks were named by Falconer and Cautley (1849), Lydekker (1882a, b), and Pilgrim (1910a, b), with revisions by Matthew (1929), Colbert (1935a, b), Hussain (1971), MacFadden and Woodburne (1982), and Bernor and Hussain (1985). *Hipparion* makes its last appearance in SSEA in the Tatrot fauna, dating to about 3.3–2.7 Ma (Barry et al. 1982; Dennell et al. 2006), which is significantly earlier than its last appearances in Africa and North America, where it survived into the middle and late Pleistocene, respectively (MacFadden 1998; Bernor et al. 2010).

The third major dispersal out of North America was that of *Equus*, the genus to which all modern equids belong. The evolution of *Equus* also took place in North America, tracing back to an ancestor like *Dinohippus* of the North American late Miocene (Bernor and Armour-Chelu 1999; Prothero and Schoch 2002). The first appearance datum of *Equus* in Eurasia is at 2.6 Ma, and it appears simultaneously in Europe and the Siwaliks (Lindsay et al. 1980; Barry et al. 1982; Bernor and Armour-Chelu 1999), and shortly thereafter in Africa (Bernor et al. 2010). *Equus* differs from its numerous fossil relatives in several respects, but most easily observed among these may be the extreme reduction of the lateral metapodials, leaving only a single large and central element (metapodial III), and the great increase in tooth crown height, permitting extremely high rates of tooth wear as might occur with grass-feeding, especially in dry and dusty environments.

Pleistocene fossil *Equus* is recorded from the Pinjor fauna of the Siwaliks as *Equus sivalensis* and *E. namadicus* (Falconer and Cautley 1849; Colbert 1935a, b; Barry et al. 1982; Nanda 2002; Dennell et al. 2006; Nanda 2008), and in Myanmar as *E. yunnanensis* (Nanda 2008). There are overlapping records of both *E. namadicus* and the living Asiatic wild ass (*E. hemionus*) in the late Pleistocene of India (Chauhan 2008; Nanda 2008). The record of Indian fossil *Equus* is reviewed and discussed by Badam (1977) and Badam and Tewari (1974). Ironically, after over 50 million years of evolution in the North American continent, equids went extinct there in the latest Pleistocene megafaunal extinction, only to be reintroduced to the Americas by European colonialists in the fifteenth and sixteenth centuries.

Molecular phylogenetic analyses indicate a clade uniting onagers, asses, and zebras (Steiner et al. 2012). The fossil *Equus sivalensis* has been may be closely related to the living Asiatic ass (Prothero and Schoch 2002). If correct, this would indicate the separation between Asian and African asses may have already been in place by the early Pleistocene.

Chalicotheriidae

Chalicotheres are an extinct group of large and strange perissodactyls with no living analog. They were characterized by the possession of claws rather than hooves and low-crowned teeth that indicate a leaf-eating diet (Fig. 2.19). The earliest chalicotheres probably appeared in the Early Eocene of Asia (Mongolia) and North America, and were well represented in both continents by the Middle Eocene (Rose 2006). Chalicotheres are quite rare in fossil assemblages, so their evolutionary history and interrelationships are difficult to piece together, including also the story of their Pleistocene extinction.

The schizotheriine chalicothere *Phyllotillon naricus* is recorded in the Oligocene and early Miocene Chitarwata Formation in Pakistan (Forster-Cooper 1920; Métais

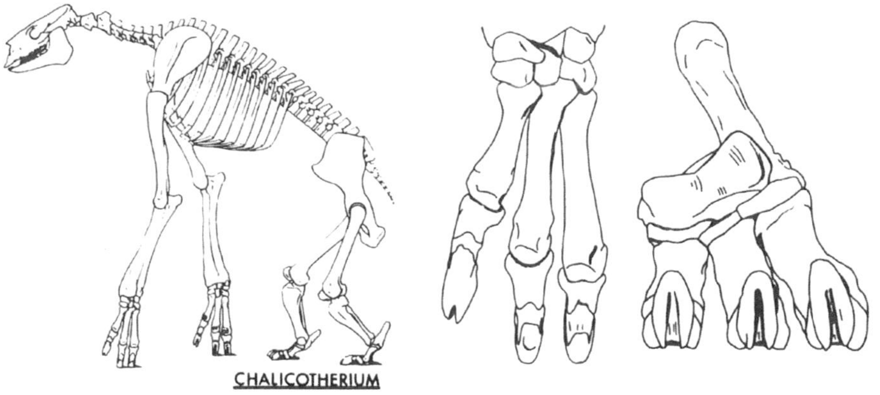

Fig. 2.19 Skeleton, hand, and foot of *Chalicotherium grande*, closely related to *Chalicotherium salinum* from the late Miocene of Pakistan. Species of *Chalicotherium* had long forelimbs, short hindlimbs, and large claws (from Zapfe, 1979 in Coombs 1989)

et al. 2009; Antoine et al. 2013). *Chalicotherium* is first recorded in the early Miocene of Africa, Europe, and Asia, and survives into the late Miocene. These were large animals marked by long forelimbs and shortened hindlimbs in gorilla-like proportions, and may have lived in forested habitats and adopted bipedal postures for browsing from trees, with the claw being used to hook onto branches (Fig. 2.19) (Coombs 1989). In SSEA, '*Chalicotherium*' *pilgrimi* is recorded in the early Miocene Chitarwata Formation in Pakistan (Métais et al. 2009; Antoine et al. 2013), and *Chalicotherium salinum* is recorded in the Siwaliks at 12.9–8.0 Ma (Forster-Cooper 1922a, b; Pickford 1982; Barry et al. 2002; Badgley et al. 2008; Khan et al. 2009). Further reports of Miocene chalicotheres from SSEA come from Thailand, southern China, and Myanmar (Chavasseau et al. 2009a, b and references therein).

Young SSEA chalicotheres are represented by *Anisodon* (=*Nestoritherium*) *sivalense* from the Pleistocene Pinjor (Falconer and Cautley 1843; Colbert 1935a, b) and *Nestoritherium/Anisodon* sp. from the Pleistocene of the Irrawady Formation in Myanmar (Takai et al. 2006). These records may represent the last appearance of chalicotheres (see Coombs 1989, Fig. 24.4).

2.6.3 *Proboscidea*

The order Proboscidea is named for the characteristic nasal proboscis (trunk), made up of a fused nose and upper lip. The earliest fossil Proboscideans are recognized from the earliest Eocene (around 55 Ma) of Africa (Gheerbrant and Tassy 2009; Sanders et al. 2010). Though the first appearance of proboscideans in Eurasia has been given important biochronological significance (Madden and Van Couvering 1976; Van Couvering and Berggren 1976), the timing of their dispersal from Africa remains poorly constrained. The early recognition of Proboscidea almost simultaneously in Europe, Asia, and North America has been termed as the 'Gomphotherium Datum,' or the Proboscidean Datum Event (PDE), and was long considered a reliable biochronologic marker dated at 17.5 Ma, with the implication that any non-African assemblage bearing proboscidean remains could not be older than the late early Miocene. Tassy (1990), however, cast serious doubt on the consistency of such a datum, and Antoine et al. (2003a, b) reported the occurrence of an elephantoid incisor from the late Oligocene Bugti Member. Lindsay et al. (2005) also pointed out remains of an indeterminate elephantoid in the lower Chitarwata Formation at Zinda Pir. The presence of elephantoids in the Oligocene of Pakistan confirms intermittent trans-Tethyan dispersals during the Paleogene as hypothesized by several authors (e.g., Ducrocq 1997; Gheerbrant et al. 2001; Gheerbrant and Rage 2006).

Proboscideans were of high diversity and wide geographic range, often with numerous sympatric species co-occurring at sites in Eurasia, Africa, and North America from the middle Miocene up to the latest Pleistocene. The end Pleistocene extinction of mammoths, mastodons, and other species left behind only the tropical species of *Loxodonta* and *Elephas*. Proboscideans have a relatively well-studied

Neogene fossil record, and the African record is well investigated with respect to ecological evolution and the evolution of grazing and browsing behaviors (Cerling et al. 1999; Uno et al. 2011; Lister 2013).

Elephantidae

Elephants are unique and peculiar animals. Elephants replace their teeth in a 'conveyor-belt' type replacement, with serial eruption, wear, and ejection of teeth one by one from the front of the tooth row, a version of the delayed dental eruption common to Afrotheres (Asher and Lehmann 2008). Living elephants are distinguished from many of their fossil relatives in the advanced development of their teeth, particularly the height of the tooth (hypsodonty) which presumably increased in order to sustain the higher rates of tooth wear demanded by a rougher diet. Strangely enough, living elephants subsist mainly on browse or mixed browse-graze diets, while most of their fossil relatives over the past six million years consumed much more grass than would have been expected from their dental evolution alone (Cerling et al. 1999).

Elephants are also important vectors of local ecological change. African elephants are known to destroy trees, making large clearings in forests, and often seriously changing the balance of trees to grassland habitats in local environments. While effects on habitats are also noted in Asian elephants, these are generally much less severe (Sukumar 2003). An important aspect of elephant ecology is their behavior, which is socially organized. In both African and Asian elephants, sexual size dimorphism is marked, and male and female social roles differ. Herds of varying size (five to 20 or more individuals—termed family units), are made up of closely related females and their young, and led by an older female (the matriarch). Upon reaching sexual maturity, males leave their family unit and lead basically solitary lives, meeting up with the female-dominated groups occasionally for mating. While proboscideans have a long and rich fossil record, almost nothing is known about their social behavior from the bone fossils themselves. Rather, one of the few clues to the antiquity of elephant social structure comes from a unique late Miocene trackway site in the United Arab Emirates (Bibi et al. 2012). The trackway site preserves footprints of a herd of proboscideans of varying sizes that walked together across what was a muddy and relatively featureless landscape. The site also fortuitously preserves the trackway of a lone individual that, by its large size, was probably made by a male. At around seven million years in age, these fossil trackways provide direct evidence for social herding and possible sexual segregation in late Miocene proboscideans.

Like rhinos and tragulids, the Asian elephant (*Elephas maximus*) has its closest living relatives in Africa, and is a surviving remnant of a once much more widespread and diverse clade of mammals. In fact, the earliest records of *Elephas* spp. are in the early Pliocene of Africa (Sanders et al. 2010). The lineage had spread to Eurasia by the late Pliocene, making its first appearance in Siwalik faunas at around 3.6 Ma with *Elephas planifrons* (Nanda 2002). This is followed by *E. hysudricus and E. platycephalus* alongside several species of *Stegodon, Stegolophodon,* and *Anancus* after 2.6 Ma (Nanda 2002, 2008). Kundal and Kundal

(2011) report a possible middle Pleistocene occurrence of *Elephas maximus* from Jammu Province, India. '*Elephas' celebensis* (=*Archidiskodon celebensis*) is a dwarf elephant recorded from the early Pleistocene (<2.5 Ma) Walanae Fauna of the island of Sulawesi (Hooijer 1949, 1964; van den Bergh et al. 2001a, b). If belonging to *Elephas*, then *'E.' celebensis* would represent at least a single dispersal event of this lineage as far southeast as Sulawesi, with subsequent insular dwarfing.

Extant wild Asian elephants were estimated to number between 40,000 and 50,000 individuals in 2006 (Sukumar 2006) though these estimates may not be very accurate (Blake and Hedges 2004). Besides occupying the southern and south-eastern Asian mainland, *Elephas maximus* is also known from the islands of Sri Lanka, Sumatra, and the northeastern tip of Borneo. Genomic analyses indicate that the Bornean, Sumatran, and Peninsular Malaysian populations all form distinct haplotypes indicating repeated incursions of the Sunda Shelf in the middle and late Pleistocene (Fernando et al. 2003).

Non-Elephantid Proboscideans

Extant African and Asian elephants are relatively closely related members of a single family, Elephantidae, to which also the famous mammoth (*Mammuthus*) belongs. The fossil record, however, documents the presence of a much greater diversity of proboscideans until the relatively recent past. Several non-elephantid proboscideans are recorded from the fossil record of SSEA. Deinotheres (Deinotheriidae) attained massive sizes and were characterized by a peculiar pair of down-curving lower tusks and relatively simple lophodont teeth suited to a browsing diet (Göhlich 1999; Sanders et al. 2010). Deinotheres are known from as early as the late Oligocene of Africa (Sanders et al. 2004), spreading to Eurasia in the early Miocene. In SSEA, deinotheres are represented by *Prodeinotherium* spp. from the early Miocene of Pakistan, and *Deinotherium* spp. from the middle and late Miocene of Pakistan and India (Forster-Cooper 1922a, b; Barry et al. 2002; Bhandari et al. 2010; Antoine et al. 2013).

Mammutids, represented by *Zygolophodon*, are recorded in SSEA from the late early Miocene to the early late Miocene of Pakistan (Tassy 1983a, b, c; Chavasseau et al. 2009a, b; Antoine et al. 2013). Gomphotheriids, represented by diverse taxa including *Gomphotherium, Protanancus, Tetralophodon*, and *Choerolophodon,* are recorded from the Miocene of Pakistan, Thailand, and southern China (Tassy 1983a, b, c; Chavasseau et al. 2009a, b; Antoine et al. 2013).

Stegodontid proboscideans, as represented by *Stegolophodon,* are recorded from the middle Miocene of Thailand and from numerous late Miocene localities in Thailand, the Siwaliks, and southern China (Tassy 1983a, b, c; Tassy et al. 1992; Saegusa et al. 2005; Chavasseau et al. 2009a, b; Thasod et al. 2012). In the Pleistocene, stegodontids are represented in Southeast Asia by many different insular dwarfed species of *Stegodon*, as recorded from Sulawesi, Flores, Timor, Mindanao, and Luzon among others (Hooijer 1953, 1964; van den Bergh et al. 2001a, b, 2008; Meijer et al. 2010). The smallest among these, *Stegodon sondaari* from Flores at 900 ka, weighed around 300 kg (Van den Bergh 1999; van den Bergh et al. 2008). In early Pleistocene of Sulawesi, *Stegodon* co-existed with another 'pygmy elephant',

'Elephas' celebensis (Hooijer 1949, 1964; van den Bergh et al. 2001a, b). *Stegodon* survived into the late Pleistocene in Southeast Asia, and into the Holocene in southern China (Saegusa 2001, van den Bergh et al. 2008).

Proboscidean diversity in SSEA, as for so many other herbivorous groups reviewed here, was much higher throughout the Neogene than it is today. The dynamic Southeast Asian island archipelago in particular appears to have been a landscape of proboscidean diversification.

2.7 Modern Large Herbivore Faunas: Continental Comparisons

With around 85 living large herbivore species (83 of which are found in SSEA, Table 1.1), the Indomalayan Realm is about 25 % less species-rich than the Afrotropics (~106 species), but more than three times as rich than the Neotropics (~25 species). These numbers becomes astounding, however, when one considers that the land area of the Indomalayan Realm is only a third the size of sub-Saharan Africa, and less than half the size of the Neotropics. This extraordinary richness of herbivores in SSEA relies greatly on Southeast Asia and the southeastern island archipelago, a hotspot of speciation and endemism. Other reasons are the effects of the end Pleistocene extinctions, which took out a much larger proportion of South America's large mammals than it did Africa's or southern Asia's.

Sub-Saharan African large herbivore fauna may be diverse (106 species in 8 families), but is overwhelmingly dominated by bovids (88 species). The Neotropical large herbivore fauna is species-poor (25 species in 4 families) but is overwhelmingly dominated by cervids (16 species). An obvious major difference between these two continental areas is that the Afrotropics have no cervids, while the Neotropics have no bovids (Fig. 2.20). Like the Afrotropics, the Indomalayan Realm is very diverse (85 species in 9 families) but in contrast to both the Afro- and Neotropics, species richness in Indomalaya is much more evenly distributed among mammalian families, with the richest family (Bovidae) comprising only a third of the total (30 spp). Also in contrast to both the Afrotropics and the Neotropics, the Indomalayan Realm is home to both bovids and cervids. Taken together, these clades comprise 60 % of SSEA species richness, which approaches the African and South American ratios for bovids and cervids, respectively.

However, it should be kept in mind that the modern taxonomic diversity of the Indomalayan Realm is a relatively recent, Pleistocene or Holocene, configuration. For example, while absent there today, we know that species of hippopotamids and giraffids were to be found in SSEA through most of this area's Neogene history. In contrast, cervids only dispersed into the region in the early Pleistocene (discussed above).

From the late Miocene Siwaliks of northern Pakistan, some 55 species of large herbivores in 11 families are recorded between 11 and 5 Ma (Badgley et al. 2008).

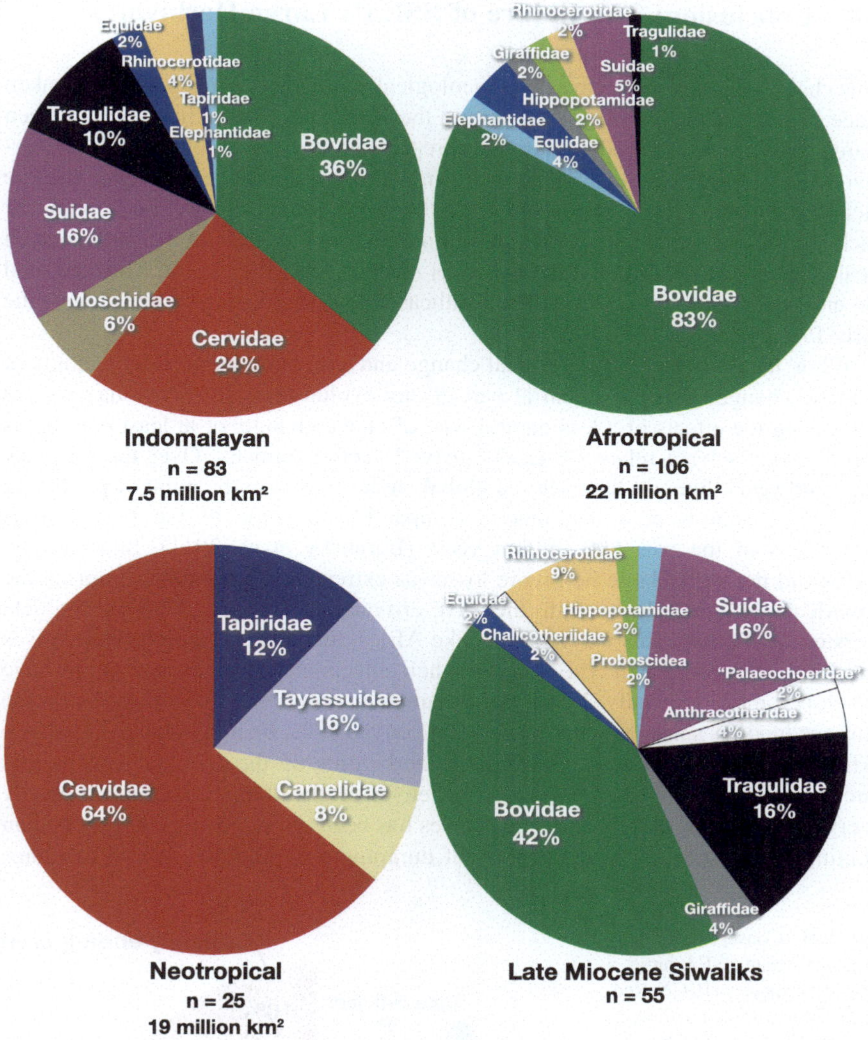

Fig. 2.20 Comparison of taxonomic diversity among the large herbivore faunas of the Indomalayan, Afrotropical, and Neotropical Realms, and to the late Miocene fauna (11–5 Ma) of the Siwaliks of northern Pakistan (faunal list of Badgley et al. 2008). Similarities between Indomalaya today and in the late Miocene include high proportions of bovids, suids, and tragulids

This rough estimate spanning 6 million years of evolution from a small geographic area allows only a superficial comparison with the modern day tropical diversity (Fig. 2.20). Even so, the late Miocene of northern Pakistan still shows some similarities to the modern Indomalayan Realm, namely in the high proportions of bovids, tragulids, and suids. The lack of cervids, and the presence of hippopotamids, giraffids, and several entirely extinct families are obvious and major differences.

2.8 Conclusion: The Future of SSEA's Large Herbivores

This chapter has covered many of the geological and faunal changes that have taken place in South and Southeast Asia during the Cenozoic. These changes have been many, and in many respects the modern representation and diversity of large herbivores is very different today than it was in the Miocene, Pliocene, or even the late Pleistocene. However, many elements of this evolutionary history are familiar far back in time, including the relative isolation and distinctiveness of the Indomalayan Realm from surrounding continental areas since at least the middle Miocene, and the endemic species 'factory' of the Southeast island archipelago since at least the early Pleistocene.

While naturally reoccurring faunal change and extinction induced by tectonic or climatic changes may be a central part of any evolutionary history, today we are witnessing the effects of a less natural type of faunal change, or at least one that is being wrought by a relatively newly arrived vector: humans. Over the past few thousand years, humans have altered global environments to the point of producing species extinctions at a rate that is comparable to some of the largest mass extinctions of the past 600 million years (Barnosky et al. 2011). Interestingly, SSEA did not suffer large herbivore losses as extreme as in mainland Eurasia, the Americas, or Australasia with the first arrivals of humans there in the late Pleistocene. Perhaps this was because, like Africa, humans and their fossil relatives had long been present in SSEA and so their effects on the landscape might have developed more gradually. Many large mammals, however, have suffered extreme range reductions since the late Pleistocene (Louys 2012a, b), and today over half the extraordinarily rich assemblage of large herbivores of the Indomalayan area is threatened or in danger of extinction (Grenyer et al. 2006; Fig. 2.21). Probably every single one of these threatened species has been put in this position by human hunting, deforestation, and the effects of dumping and pollution. These pressures

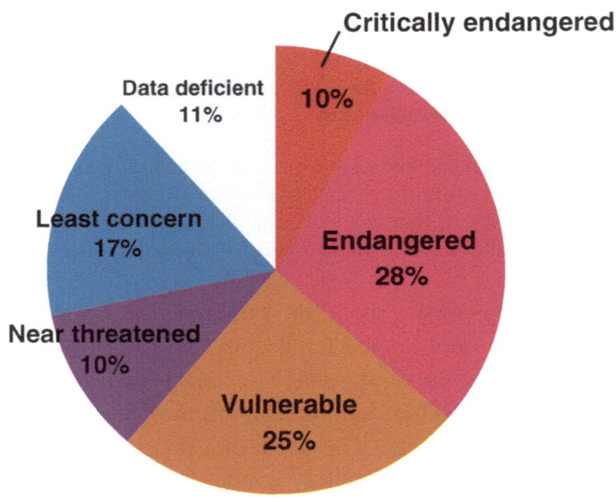

Fig. 2.21 Conservation status of extant SSEA large herbivore species (IUCN Red List). Of the 85 Indomalayan species, 52 are vulnerable or in danger of extinction

will only be compounded by the effects of ongoing global warming, such as global sea level rise and the inundation of low-lying islands. Southeast Asia, a particularly fragile region on account of its small islands and large number of endemics, stands to lose some 85 % of its mammal species by the end of the twenty-first century (Sodhi et al. 2010).

The evolutionary history of SSEA's large herbivores has been shaped by extinction probably just as much as it has been shaped by speciation. The fossil record informs us that local or global biotas have always rebounded following major extinction events, though always in a permanently altered state. With the ongoing human-induced mass extinction and with current rates of habitat and species loss, we might ask whether SSEA's faunas might ever be able to recuperate to preindustrial levels, and if so, in what kind of altered state.

Acknowledgments We would like to thank Farshid S. Ahrestani and Mahesh Sankaran for inviting us to contribute to this chapter and for their editorial work. Mikael Fortelius, Tang Hui, Ross MacPhee, Rajeev Patnaik, and Zhijie J. Tseng read and provided helpful comments that much improved the text. F.B. was supported at the AMNH by a Gerstner Fellowship. G.M. was supported by the National Geographic Society, and the CNRS-INSU-InterrVie and ATM-MNHN Programs.

Bibliography

Agustí J, Cabrera L, Garces M et al (1997) The Vallesian mammal succession in the Vallès-Penedès basin (northeast Spain): paleomagnetic calibration and correlation with global events. Palaeogeogr Palaeoclimatol Palaeoecol 133:149–180

Agustí J, Sanz De Siria A, Garcés M (2003) Explaining the end of the hominoid experiment in Europe. J Hum Evol 45:145–153

Allen MB, Armstrong HA (2008) Arabia-Eurasia collision and the forcing of mid-Cenozoic global cooling. Palaeogeogr Palaeoclimatol Palaeoecol 265:52–58

Alroy J (2001) A multispecies overkill simulation of the end-Pleistocene megafaunal mass extinction. Science 292:1893–1896

An Z, Kutzbach JE, Prell WL et al (2001) Evolution of Asian monsoons and phased uplift of the Himalaya-Tibetan plateau since Late Miocene times. Nature 411:62–66

Antoine P-O (2012) Pleistocene and Holocene rhinocerotids (Mammalia, Perissodactyla) from the Indochinese Peninsula. CR Palevol 11:159–168

Antoine P-O, Ducrocq S, Marivaux L et al (2003a) Early rhinocerotids (Mammalia: Perissodactyla) from South Asia and a review of the Holarctic Paleogene rhinocerotid record. Can J Earth Sci 40:365–374

Antoine P-O, Welcomme J-L, Marivaux L et al (2003b) First record of Paleogene Elephantoidea (Mammalia, Proboscidea) from the Bugti Hills of Pakistan. J Vertebr Paleontol 23:977–980

Antoine P-O, Downing KF, Crochet JY et al (2010) A revision of Aceratherium blanfordi Lydekker, 1884 (Mammalia: Rhinocerotidae) from the early Miocene of Pakistan: postcranials as a key. Zoo J Linnean Soc 160:139–194

Antoine P-O, Métais G, Orliac M et al (2013) Mammalian Neogene biostratigraphy of the Sulaiman Province, Pakistan. In: Wang X, Flynn LJ, Fortelius M (eds) Fossil mammals of Asia: Neogene biostratigraphy and chronology. Columbia University Press, New York, pp 400–422

Asher R, Lehmann T (2008) Dental eruption in afrotherian mammals. BMC Biol 6:14

Ataabadi MM, Bernor Raymond L, Kostopoulos Dimitris S et al (2013) Recent advances in paleobiological research of the late Miocene Marageh fauna, northwest Iran. In: Wang X, Flynn LJ, Fortelius M (eds) Fossil mammals of Asia: Neogene biostratigraphy and chronology. Columbia University Press, New York, pp 546–565

Azanza B (1993) Systématique et évolution du genre Procervulus, cervidé (Artiodactyla, Mammalia) du Miocène inférieur d'Europe. Comptes rendus de l'Académie des Sciences Paris, II 316:717–723

Azzaroli A (1954) Critical observations upon Siwalik deer. Proc Linnean Soc Lond 165:75–83

Badam G (1977) First record of a Middle Palaeolithic fossil from Gujarat, India. J Palaeontol Soc India 20:314–319

Badam G, Tewari B (1974) On the Zebrine affinities of the Pleistocene horse Equus sivalensis Falconer and Cautley. Bull Deccan College Res Inst 34:7–11

Badgley C, Barry JC, Morgan ME et al (2008) Ecological changes in Miocene mammalian record show impact of prolonged climatic forcing. Proc Natl Acad Sci 105:12145–12149

Bande MB, Prakash U (1986) The tertiary flora of Southeast Asia with remarks on its palaeoenvironment and phytogeography of the Indo-Malayan region. Rev Palaeobot Palynol 49:203–233

Barnosky AD, Koch PL, Feranec RS et al (2004) Assessing the causes of Late Pleistocene extinctions on the continents. Science 306:70–75

Barnosky AD, Matzke N, Tomiya S et al (2011) Has the Earth's sixth mass extinction already arrived? Nature 471:51–57

Barry JC, Lindsay EH, Jacobs LL (1982) A biostratigraphic zonation of the middle and upper Siwaliks of the Potwar Plateau of northern Pakistan. Palaeogeogr Palaeoclimatol Palaeoecol 37:95–130

Barry JC, Morgan ME, Flynn LJ et al (2002) Faunal and environmental change in the late Miocene Siwaliks of northern Pakistan. Paleobiology 28(S2):1–71

Barry JC, Cote S, Maclatchy L et al (2005) Oligocene and Early Miocene Ruminants (Mammalia, Artiodactyla) from Pakistan and Uganda. Palaeontol Electron 8(22A):29p

Barry JC, Behrensmeyer AK, Badgley CE et al (2013) The Neogene Siwaliks of the Potwar Plateau, Pakistan. In: Wang X, Flynn LJ, Fortelius M (eds) Fossil mammals of Asia: Neogene Biostratigraphy and Chronology. Columbia University Press, New York, pp 373–399

Beck RA, Burbank DW, Sercombe WJ et al (1995) Stratigraphic evidence for an early collision between northwest India and Asia. Nature 373:55–58

Beden M, Brunet M (1986) Faunes de mammifères et paléobiogéographie des domaines indiens et péri-indiens au Néogène. Sci Terre 47:61–87

Benammi M, Chaimanee Y, Jaeger J-J et al (2001) Eocene Krabi Basin (southern Thailand): paleontology and magnetostratigraphy. Bull Geol Soc Am 113:265–273

Benammi M, Urrutia-Fucugauchi J, Alva-Valdivia LM et al (2002) Magnetostratigraphy of the Middle Miocene continental sedimentary sequences of the Mae Moh Basin in northern Thailand: evidence for counterclockwise block rotation. Earth Planet Sci Lett 204:373–383

Bender F (1983) Geology of Burma. Gebrüder Borntraeger, Berlin

Bender F, Raza HA, (1995) Geology of Pakistan. Gebruder Borntraeger Verlagsbuchhandlung, Science Publishers, Stuttgart, Germany

Bernor RL (1986) Mammalian biostratigraphy, geochronology, and zoogeographic relationships of the late Miocene Maragheh fauna, Iran. J Vertebr Paleontol 6:76–95

Bernor RL, Armour-Chelu M (1999) Family Equidae. In: Rössner GE, Heissig K, Fahlbusch V (eds) Pfeil, Munich

Bernor RL, Hussain ST (1985) An assessment of the systematic, phylogenetic and biogeographic relationships of Siwalik hipparionine horses. J Vertebr Paleontol 5:32–87

Bernor RL, Brunet M, Ginsburg L et al (1987) A consideration of some major topics concerning Old World Miocene mammalian chronology, migrations and paleogeography. Géobios 20:431–439

Bernor R, Armour-Chelu M, Gilbert H et al (2010) Equidae. In: Werdelin L, Sanders WJ (eds) Cenozoic Mammals of Africa. University of California Press, Berkeley, pp 691–727

Bhandari A, Mohabey DM, Bajpai S et al (2010) Early Miocene mammals from central Kutch (Gujarat), Western India: implications for geochronology, biogeography, eustacy and intercontinental dispersals. Neues Jahrbuch für Geologie und Paläontologie-Abhandlungen 256:69–97

Bibi F (2007) Origin, paleoecology, and paleobiogeography of early Bovini. Palaeogeogr Palaeoclimatol Palaeoecol 248:60–72

Bibi F (2009) Evolution, systematics, and paleoecology of Bovinae (Mammalia: Artiodactyla) from the Late Miocene to the recent. Dissertation, Yale University

Bibi F (2011) Mio-Pliocene faunal exchanges and African biogeography: the record of fossil bovids. PLoS ONE 6:e16688

Bibi F (2013) A multi-calibrated mitochondrial phylogeny of extant Bovidae (Artiodactyla, Ruminantia) and the importance of the fossil record to systematics. BMC Evol Biol 13:166. http://www.biomedcentral.com/1471–2148/13/166

Bibi F, Kraatz B, Craig N et al (2012) Early evidence for complex social structure in Proboscidea from a late Miocene trackway site in the United Arab Emirates. Biol Lett 8:670–673

Bibi F, Hill A, Beech M et al (2013) Late Miocene fossils from the Baynunah Formation, United Arab Emirates: summary of a decade of new work. In: Wang X, Flynn LJ, Fortelius M (eds) Fossil Mammals of Asia: Neogene biostratigraphy and chronology. Columbia University Press, New York, pp 583–594

Bishop LC, Hill A (1999) Fossil Suidae from the Baynunah Formation, Emirate of Abu Dhabi, United Arab Emirates. In: Whybrow PJ, Hill A (eds) Fossil Vertebrates of Arabia, with Emphasis on the Late Miocene Faunas, Geology, and Palaeoenvironments of the Emirate of Abu Dhabi, United Arab Emirates. Yale University Press, New Haven, pp 254–270

Blake S, Hedges S (2004) Sinking the flagship: the case of forest elephants in Asia and Africa. Conserv Biol 18:1191–1202

Blakey RC (2008) Gondwana paleogeography from assembly to breakup—a 500 my odyssey. In: Fielding CR, Frank TD, Isbell JL (eds) Resolving the late paleozoic ice age in time and space: geological society of America Special Paper 441, pp 1–28

Blanford WT (1883) Geological notes on the hills in the neighbourhood of the Sind and Punjab Frontier between Quetta and Dera Ghazi Khan. Mem Geol Surv India 20:1–136

Böhme M, Prieto J, Schneider S, Hung NV, Quang DD, Tran DN (2011) The Cenozoic on-shore basins of Northern Vietnam: Biostratigraphy, vertebrate and invertebrate faunas. J. Asian Earth Sci 40:672–687

Böhme M, Aiglstorfer M, Antoine P-O et al (2013) Na Duong (northern Vietnam)–an exceptional window into Eocene ecosystems from Southeast Asia. Zitteliana A 53:121–167

Boisserie J-R (2005) The phylogeny and taxonomy of Hippopotamidae (Mammalia: Artiodactyla): a review based on morphology and cladistic analysis. Zoo J Linnean Soc 143:1–26

Boisserie J-R, White TD (2004) A new species of pliocene hippopotamidae from the Middle Awash, Ethiopia. J Vertebr Paleontol 24:464–473

Boisserie J-R, Brunet F, Lihoreau M (2005) The position of Hippopotamidae within Cetartiodactyla. Proc Natl Acad Sci 102:1537–1541

Boisserie J-R, Lihoreau F, Orliac M et al (2010) Morphology and phylogenetic relationships of the earliest known hippopotamids (Cetartiodactyla, Hippopotamidae, Kenyapotaminae). Zoo J Linnean Soc 158:325–366

Brown B (1926) A new deer from the Siwaliks. Amer Mus Novit 242:1–6

Brunet M, Sudre J (1987a) Evolution et systématique du genre Lophiomeryx Pomel 1853 (Mammalia, Artiodactyla). Münchner Geowissenschaftliche Abhandlungen (A) 10:225–242

Brunet M, Sudre J (1987b) Evolution et systématique du genreLophiomeryxPomel 1853 (Mammalia, Artiodactyla). Münchner Geowissenschaftliche Abhandlungen (A) 10:225–242

Brunet M, Heintz E, Battail B (1984) Molayan (Afghanistan) and the Khaur Siwaliks of Pakistan; an example of biogeographic isolation of late Miocene mammalian faunas. Geol Mijnbouw 63:31–38

Campos PF, Willerslev E, Sher A et al (2010) Ancient DNA analyses exclude humans as the driving force behind late Pleistocene musk ox (Ovibos moschatus) population dynamics. Proc Natl Acad Sci 107:5675
Cerling TE, Harris JM, Macfadden BJ et al (1997) Global vegetation change through the Miocene/Pliocene boundary. Nature 389:153–158
Cerling TE, Harris JM, Leakey MG (1999) Browsing and grazing in elephants: the isotope record of modern and fossil proboscideans. Oecologia 120:364–374
Chaimanee Y, Jolly D, Benammi M et al (2003) A Middle Miocene hominoid from Thailand and orangutan origins. Nature 422:61–65
Chaimanee Y, Suteethorn V, Jintasakul P et al (2004) A new orangutan relative from the Late Miocene of Thailand. Nature 427:439–441
Chatterjee S, Goswami A, Scotese CR (2013) The longest voyage: tectonic, magmatic, and paleoclimatic evolution of the Indian plate during its northward flight from Gondwana to Asia. Gondwana Res 23:238–267
Chauhan PR (2008) Large mammal fossil occurrences and associated archaeological evidence in the Pleistocene contexts of peninsular India and Sri Lanka. Quatern Int 192:20–42
Chavasseau O, Chaimanee Y, Tun ST et al (2006) Chaungtha, a new Middle Miocene mammal locality from the Irrawaddy Formation, Myanmar. J Asian Earth Sci 28:354–362
Chavasseau O, Chaimanee Y, Coster P et al (2009a) First record of a chalicothere from the Miocene of Myanmar. Acta Palaeontol Pol 55:13–22
Chavasseau O, Chaimanee Y, Yamee C et al (2009b) New Proboscideans (Mammalia) from the middle Miocene of Thailand. Zoo J Linnean Soc 155:703–721
Cifelli RL (1981a) Patterns of evolution among the Artiodacyla and Perissodactyla (Mammalia). Evolution 35:433–440
Cifelli RL (1981b) Patterns of evolution among the Artiodacyla and Perissodactyla (Mammalia). Evolution 433–440
Claude J, Naksri W, Boonchai N et al (2011) Neogene reptiles of northeastern Thailand and their paleogeographical significance. Annales de Paléontologie 97:113–131
Clift PD, Plumb RA (2008) The Asian monsoon: causes, history and effects. Cambridge University Press, Cambridge
Colbert EH (1933) An upper Tertiary peccary from India. Amer Mus Novit 635:1–9
Colbert EH (1934) A new rhinoceros from the Siwalik beds of India. Amer Mus Novit 749:1–13
Colbert EH (1935a) Distributional and phylogenetic studies on Indian fossil mammals, V; the classification and the phylogeny of the Giraffidae American Museum of Natural History, New York, NY, United States
Colbert EH (1935b) Siwalik mammals in the American Museum of Natural History. Trans Amer Philos Soc 26:278–294
Colbert EH (1936) Was the extinct giraffe (Sivatherium) known to the early Sumerians? Am Anthropol 38:605–608
Colbert EH (1938) Fossil mammals from Burma in the American Museum of Natural History. Bull Amer Mus Nat Hist 74:255–434
Colbert EH (1942) Notes on the lesser one-horned rhinoceros, Rhinoceros sondaicus. Amer Mus Novit 1207:1–6
Colbert EH (1943) Pleistocene vertebrates collected in Burma by the American Southeast Asiatic expedition. Trans Amer Philos Soc 32:395–429
Colbert M, Schoch R (1998) Tapiroidea and other moropomorphs. In: Janis C, Scott KM, Jacobs LL (eds) Evolution of tertiary mammals of North America. Cambridge University Press, Cambridge, pp 569–582
Coombs MC (1989) Interrelationships and diversity in the Chalicotheriidae. In: Prothero DR, Schoch RM (eds) The evolution of Perissodactyls. Oxford University Press, Oxford, pp 438–457
Copley A, Avouac JP, Royer JY (2010) India-Asia collision and the Cenozoic slowdown of the Indian plate: implications for the forces driving plate motions. J Geophys Res 115:B03410. doi:10.1029/2009JB006634(1-14)

Coster P, Benammi M, Chaimanee Y et al (2010) A complete magnetic-polarity stratigraphy of the Miocene continental deposits of Mae Moh Basin, northern Thailand, and a reassessment of the age of hominoid-bearing localities in northern Thailand. Geol Soc Am Bull 122:1180–1191
Covert HH, Hamrick MW, Dzanh T et al (2001) Fossil mammals from the late Miocene of Vietnam. J Vertebr Paleontol 21:633–636
Croft DA, Heaney LR, Flynn JJ et al (2006) Fossil remains of a new, diminutive Bubalus (Artiodactyla: Bovidae: Bovini) from Cebu Island, Philippines. J Mammal 87:1037–1051
Dawson MR, Marivaux L, Li C-K, Beard KC, Métais G (2006) Laonastes and the "Lazarus Effect" in recent mammals. Science 311:1456–1458
De Bonis L, Brunet M, Heintz E et al (1992) La province greco-irano-afghane et la répartition des faunes mammaliennes au Miocène supérieur. Paleontologia y Evolució 24–25:103–112
Dennell R, Coard R, Turner A (2006) The biostratigraphy and magnetic polarity zonation of the Pabbi Hills, northern Pakistan: an Upper Siwalik (Pinjor Stage) Upper Pliocene-Lower Pleistocene fluvial sequence. Palaeogeogr Palaeoclimatol Palaeoecol 234:168–185
Dennell R, Coard R, Turner A (2008) Predators and scavengers in Early Pleistocene southern Asia. Quatern Int 192:78–88
Dewey JF, Pitman WC, Ryan WBF et al (1973) Plate tectonics and the evolution of the Alpine system. Geol Soc Am Bull 84:3137–3180
Dong W, Qi G-Q (2013) Hominoid-producing localities and biostratigraphy in Yunnan. In: Wang X, Flynn LJ, Fortelius M (eds) Fossil mammals of Asia: Neogene biostratigraphy and chronology. Columbia University Press, New York, pp 293–313
Dong W, Pan Y, Liu J (2004) The earliest Muntiacus (Artiodactyla, Mammalia) from the Late Miocene of Yuanmou, southwestern China. CR Palevol 3:379–386
Downing KF, Lindsay EH, Downs WR et al (1993) Lithostratigraphy and vertebrate biostratigraphy of the early Miocene Himalayan Foreland, Zinda Pir Dome, Pakistan. Sed Geol 87:25–37
Dubost G (1978) Un aperçu sur l'écologie du chevrotain africain Hyemoschus aquaticus Ogilby, Artiodactyle Tragulidé. Mammalia 42:1–62
Ducrocq S (1994) An Eocene peccary from Thailand and the biogeographical origins of the artiodactyl family Tayassuidae. Palaeontology 37:765–780
Ducrocq S (1996) The Eocene terrestrial mammal from Timor, Indonesia. Geol Mag 133:763–766
Ducrocq S (1997) The anthracotheriid genus Bothriogenys (Mammalia, Artiodactyla) in Africa and Asia during the Paleogene: phylogenetical and paleobiogeographical relationships. Stuttgarter Beiträge zur Naturkunde Serie B (Geologie und Paläontologie) 250:1–44
Ducrocq S, Lihoreau F (2006a) The occurrence of bothriodontines (Artiodactyla, Mammalia) in the Paleogene of Asia with special reference to Elomeryx: Paleobiogeographical implications. J Asian Earth Sci 27:885–891
Ducrocq S, Benammi M, Chavasseau O, Chaimanee Y, Suraprasit K, Pha PD, Phuong Vl, Phach PV, Jaeger J-J (2015) New anthracotheres (Cetartiodactyla, Mammalia) from the Paleogene of northeastern Vietnam: biochronological implications. J Vert Paleont 35:e929139
Ducrocq S, Chaimanee Y, Suteethorn V et al (1994) Ages and paleoenvironment of Miocene mammalian faunas from Thailand. Palaeogeogr Palaeoclimatol Palaeoecol 108:149–163
Ducrocq S, Chaimanee Y, Suteethorn V et al (1995) Mammalian faunas and the ages of the continental Tertiary fossiliferous localities from Thailand. J SE Asian Earth Sci 12:65–78
Ducrocq S, Chaimanee Y, Suteethorn V et al (1998) The earliest known pig from the upper Eocene of Thailand. Palaeontology 41:147–156
Ducrocq S, Naing Soe A, Bo B et al (2000) First record of an Anthracobunidae (Mammalia,? Tethytheria) from the Eocene of the Pondaung Formation, Myanmar. Comptes Rendus de l'Académie des Sciences-Series IIA-Earth and Planetary Science 330:725–730
Ducrocq S, Chaimanee Y, Suteethorn V et al (2003) Occurrence of the anthracotheriid Brachyodus (Artiodactyla, Mammalia) in the early Middle Miocene of Thailand. CR Palevol 2:261–268
Ducrocq S, Chaimanee Y, Jaeger J-J et al (2006) A new Ceratomorph (Perissodactyla, Mammalia) from the Late Eocene of Southeast Asia. J Vertebr Paleontol 26:1024–1027

Dung VV, Giao PM, Chinh NN et al (1993) A new species of living bovid from Vietnam. Nature 363:443–445
Ducrocq S, Soe AN, Sein C, Lazzari V, Chaimanee Y, Valentin X, Jaeger JJ (in press) First record of a diacodexeid artiodactyl in the middle Eocene Pondaung Formation (Myanmar)
Edwards MA, Grasemann B (2009) Mediterranean snapshots of accelerated slab retreat: subduction instability in stalled continental collision. In: Van Hinsbergen DJJ, Edwards MA, Govers R (eds) Collision and collapse at the Africa-Arabia-Eurasia subduction zone. Geological Society Special Publication, London, pp 155–192
Falconer H, Cautley P (1836a) Note on the fossil hippopotamus of the Siwalik Hills. Asiatic Res Calcutta 19:39–53
Falconer H, Cautley PT (1836b) On the fossil camel of the Sewalik Hills. Asiatic Researches 19:115–134
Falconer H, Cautley PT (1836c) Sivatherium giganteum, a new fossil ruminant genus, from the valley of the Markanda, in the Sivalik branch of the Sub-Himalayan Mountains. Asiatic Res 19:1–24
Falconer H, Cautley PT (1843) On some fossil remains of Anoplotherium and giraffe, from the Sewalik Hills, in the north of India. Proc Geol Soc Lond 4:235–249
Falconer H, Cautley PT (1844) On some fossil remains of Anoplotherium and giraffe, from the Sewalik Hills, in the north of India. Proc Geol Soc Lond 4:235–249
Falconer H, Cautley P (1847) Fauna Antiqua Sivalensis, Part VIII, Suidae and Rhinocerotidae Smith, Elder, and Co, London
Falconer H, Cautley P (1849) Equidae, Camelidae and Sivatherium. Fauna Antiqua Sivalensis Part 9
Fernando P, Vidya TC, Payne J et al (2003) DNA analysis indicates that Asian elephants are native to Borneo and are therefore a high priority for conservation. PLoS Biol 1:110–115
Flower BP, Kennett JP (1994) The middle Miocene climatic transition: East Antarctic ice sheet development, deep ocean circulation and global carbon cycling. Palaeogeogr Palaeoclimatol Palaeoecol 108:537–555
Fluteau F, Ramstein G, Besse J (1999) Simulating the evolution of the Asian and African monsoons during the past 30 Myr using an atmospheric general circulation model. J Geophys Res 104:11995–12018
Flynn LJ (1983) Sur l'âge de la faune de vertébrés du Bassin de Bamian, Afghanistan. Comptes Rendus de l'Académie des Sciences, Paris, séries II 297:687–690
Flynn LJ, Qi GQ (1982) Age of the Lufeng, China, hominoid locality. Nature 298:746–747
Flynn LJ, Lindsay EH, Pilbeam D et al (2013) The Siwaliks and Neogene evolutionary biology in South Asia. In: Wang X, Flynn LJ, Fortelius M (eds) Fossil mammals of Asia: Neogene biostratigraphy and chronology. Columbia University Press, New York, pp 353–372
Forster-Cooper C (1913) New anthracotheres and allied forms from Baluchistan. Ann Mag Nat Hist 19:514–522
Forster-Cooper C (1915) New genera and species of mammals from the Miocene deposits of Baluchistan. Ann Mag Nat Hist 16:404–410
Forster-Cooper C (1920) Chalicotheroidea from Baluchistan. Proc Zool Soc Lond 90:357–366
Forster-Cooper C (1922) Macrotherium salinum, sp. n., a new Chalicothere from India. Ann Mag Nat Hist (Series 9) 10:542–544
Forster-Cooper C (1922b) Miocene Proboscidia from Baluchistan. Proc Zoo Soc Lond 92:609–626
Forster-Cooper C (1924) The Anthracotheriidae of the Dera Bugti deposits in Baluchistan. Mem Geol Surv India 4:1–59
Fortelius M, Kappelman J (1993) The largest land mammal ever imagined. Zoo J Linnean Soc 107:85–101
Friedman R, Gee J, Tauxe L et al (1992) The magnetostratigraphy of the Chitarwata and lower Vihowa formations of the Dera Ghazi Khan area, Pakistan. Sed Geol 81:253–268
Friend PF, Raza SM, Geehan G et al (2001) Intermediate-scale architectural features of the fluvial Chinji Formation (Miocene), Siwalik Group, northern Pakistan. J Geol Soc Lond 158:163–177

Froehlich DJ (1999) Phylogenetic systematics of basal perissodactyls. J Vertebr Paleontol 19:140–159
Froehlich DJ (2002) Quo vadis eohippus? The systematics and taxonomy of the early Eocene equids (Perissodactyla). Zoo J Linnean Soc 134:141–256
Galbreath G, Mordacq J, Weiler F (2006) Genetically solving a zoological mystery: was the kouprey (Bos sauveli) a feral hybrid? J Zool 270:561–564
Gatesy J, Arctander P (2000) Hidden morphological support for the phylogenetic placement of Pseudoryx nghetinhensis with bovine bovids: a combined analysis of gross anatomical evidence and DNA sequences from five genes. Syst Biol 49:515–538
Geisler JH, Uhen MD (2005) Phylogenetic relationships of extinct cetartiodactyls: results of simultaneous analyses of molecular, morphological, and stratigraphic data. J Mammal Evol 12:145–160
Gentry AW (1967) Pelorovis oldowayensis Reck, an extinct bovid from East Africa. Bull Br Mus (Nat Hist) Geol Ser 14:245–299
Gentry AW (1978) Bovidae. In: Maglio VJ, Cooke HBS (eds) Evolution of African Mammals. Harvard University Press, Cambridge, Massachusetts, pp 540–572
Gentry AW (1994) The Miocene differentiation of Old World Pecora (Mammalia). Hist Biol 7:115–158
Gentry AW (1999) Fossil pecorans from the Baynunah Formation, Emirate of Abu Dhabi, United Arab Emirates. In: Whybrow PJ, Hill A (eds) Fossil Vertebrates of Arabia. Yale University Press, New Haven, pp 290–316
Gentry AW (2003) Ruminantia (Artiodactyla). In: Fortelius M, Kappelman J, Sen S, Bernor RL (eds) Geology and paleontology of the Miocene Sinap formation, Turkey. Columbia University Press, New York
Gentry AW (2010) Bovidae. In: Werdelin L, Sanders WJ (eds) Cenozoic mammals of Africa. University of California Press, Berkeley, pp 747–803
Gentry AW, Rössner GE, Heizmann EPJ (1999) Suborder Ruminantia. In: Rössner GE, Heissig K (eds) The Miocene land mammals of Europe. Verlag Friedrich Pfeil, Munich, pp 225–258
Gentry A, Clutton-Brock J, Groves CP (2004) The naming of wild species and their domestic derivatives. J Archaeol Sci 31:645–651
Ghaffar A, Khan MA, Akhtar M et al (2006) The oldest cervid from the Siwalik Hills of Pakistan. J Appl Sci 6:127–130
Ghaffar A, Khan M, Akhtar M (2010) Early Pliocene Cervids (Artiodactyla-Mammalia) from the Siwaliks of Pakistan. Yerbilimleri 31:217–231
Ghaffar A, Akhtar M, Khan MA et al (2011a) Cervus cf. rewati (Cervidae, Mammalia) from the Pliocene Dhok Pathan Formation (Middle Siwaliks), Pakistan. Aust J Earth Sci 104:107–113
Ghaffar A, Akhtar M, Nayyer A (2011b) Evidences of early Pliocene fossil remains of tribe Cervini (Mammalia, Artiodactyla, Cervidae) from the Siwaliks of Pakistan. J Anim Plant Sci 21:830–835
Gheerbrant E, Rage J-C (2006) Paleobiogeography of Africa: how distinct from Gondwana and Laurasia? Palaeogeogr Palaeoclimatol Palaeoecol 241:224–246
Gheerbrant E, Tassy P (2009) L'origine et l'évolution des éléphants. CR Palevol 8:281–294
Gheerbrant E, Sudre J, Iarochene M et al (2001) First ascertained African "Condylarth" mammals (primitive ungulates: cf. Bulbulodentata and cf. Phenacodonta) from the earliest Ypresian of the Ouled Abdoun Basin, Morocco. J Vertebrat Paleontol 21:107–118
Gilbert C, Ropiquet A, Hassanin A (2006) Mitochondrial and nuclear phylogenies of Cervidae (Mammalia, Ruminantia): systematics, morphology, and biogeography. Mol Phylogenet Evol 40:101–117
Gingerich PD (2008) Early evolution of whales: a century of research in Egypt. In: Fleagle JG, Gilbert CC (eds) Elwyn Simons: a search for origins. Springer, New York, pp 107–124
Gingerich PD, Haq MU, Zalmout IS et al (2001) Origin of whales from early artiodactyls: hands and feet of Eocene Protocetidae from Pakistan. Science 293:2239–2242
Ginsburg L (1984) Les faunes tertiaires du Nord de la Thaïlande. Mémoire de la Société Geologique de France 147:67–69

Ginsburg L, Heintz E (1968) La plus ancienne antilope d'Europe, Eotragus artenensis du Burdigalien d'Artenay. Bulletin du Muséum National d'Histoire Naturelle 40:837–842

Ginsburg L, Ukkakimapan Y (1983) Un Cervidé nouveau du Miocène d'Asie et l'âge des lignites des bassins intramontagneux du nord de la Thaïlande. Comptes Rendus des Séances de l'Académie des Sciences Sér 2(297):297–300

Ginsburg L, Van Minh L, Qui Nam K et al (1992) Premières découvertes de vertébrés continentaux dans le Néogène du Nord du Vietnam. Comptes Rendus de l'Académie des Sciences de Paris, série IIa 314:627–630

Ginsburg L, Morales J, Soria D (2001) Les Ruminantia (Artiodactyla, Mammalia) du Miocène des Bugti (Balouchistan, Pakistan). Estudios Geol 57:155–170

Göhlich UB (1999) Order Proboscidea. In: Rössner GE, Heissig K (eds) The Miocene land mammals of Europe. Verlag Friedrich Pfeil, Munich, pp 157–168

Gongora J, Cuddahee RE, Nascimento FFd et al (2011) Rethinking the evolution of extant sub-Saharan African suids (Suidae, Artiodactyla). Zoolog Scr 40:327–335

Granger W, Gregory WK (1935) A revised restoration of the skeleton of Baluchitherium, gigantic fossil rhinoceros of Central Asia. Amer Mus Novit 787:1–3

Grenyer R, Orme CDL, Jackson SF, Thomas GH, Davies RG, Davies TJ, Jones KE, Olson VA, Ridgely RS, Rasmussen PC, Ding T-S, Bennett PM, Blackburn TM, Gaston KJ, Gittleman JL, Owens IPF (2006) Global distribution and conservation of rare and threatened vertebrates. Nature 444:93–96

Groves CP (1969) Systematics of the Anoa (Mammalia, Bovidae). Beaufortia 17:1–12

Groves C (2001) Mammals in Sulawesi: where did they come from and when, and what happened to them when they got there. In: Metcalfe I, Smith JMB, Morwood MJ, Davidson I (eds) Faunal and floral migrations and evolution in SE Asia-Australia, Balkema, Lisse, pp 333–342

Groves CP (2001b) Mammals in Sulawesi: where did they come from and when, and what happened to them when they got there. In: Metcalfe I, Smith JMB, Morwood MJ, Davidson I (eds) Faunal and floral migrations and evolution in SE Asia-Australia. Balkema, Lisse, pp 333–342

Groves C (2007) Family Cervidae. In: Dr P, Se F (eds) The evolution of Artiodactyls. Johns Hopkins University Press, Baltimore, pp 249–256

Groves CP, Wang Y, Grubb P (1995) Taxonomy of musk-deer, genus Moschus (Moschidae, Mammalia). Acta Theriol Sinica 15:181–197

Guillot S, Garzanti E, Baratoux D, Marquer D, Mahéo G, De Sigoyer J (2003) Reconstructing the total shortening history of the NW Himalaya. Geochemistry, Geophysics, Geosystems 1064, doi:10.1029/2002GC000484, 1–22

Guo J, Dawson MR, Beard KC (2000) Zhailimeryx, a new lophiomerycid artiodactyl (Mammalia) from the late middle Eocene of Central China and the early evolution of ruminants. J Mammal Evol 7:239–258

Guo ZT, Sun B, Zhang ZS et al (2008) A major reorganization of Asian climate by the early Miocene. Clim Past 4:153–174

Han D (1986) Fossils of tragulidae from Lufeng, Yunnan. Acta Anthropologica Sinica 5:73–78.

Hanta R, Ratanasthien B, Kunimatsu Y et al (2008) A new species of Bothriodontinae, Merycopotamus thachangensis (Cetartiodactyla, Anthracotheriidae) from the late Miocene of Nakhon Ratchasima, northeastern Thailand. J Vertebr Paleontol 28:1182–1188

Hardjasasmita H (1987) Taxonomy and phylogeny of the Suidae (Mammalia) in Indonesia. Scripta Geol 85:1–68

Harris JM, Liu L (2007) Superfamily Suoidea. In: Prothero DR, Foss SE (eds) The evolution of Artiodactyls. Johns Hopkins University Press, Baltimore, pp 130–150

Harzhauser M, Piller WE, Steininger FF (2002) Circum-Mediterranean Oligo-Miocene biogeographic evolution–the gastropods' point of view. Palaeogeogr Palaeoclimatol Palaeoecol 183:103–133

Harzhauser M, Kroh A, Mandic O et al (2007) Biogeographic responses to geodynamics: a key study all around the Oligo-Miocene Tethyan Seaway. Zoologischer Anzeiger—J Compar Zoo 246:241–256

Hassanin A, Douzery EJP (1999) Evolutionary affinities of the enigmatic saola (Pseudoryx nghetinhensis) in the context of the molecular phylogeny of Bovidae. Proc R Soc Lond Ser B-Biol Sci 266:893–900

Hassanin A, Douzery EJP (2003) Molecular and morphological phylogenies of Ruminantia and the alternative position of the Moschidae. Syst Biol 52:206–228

Hassanin A, Ropiquet A (2004) Molecular phylogeny of the tribe Bovini (Bovidae, Bovinae) and the taxonomic status of the Kouprey, Bos sauveli Urbain 1937. Mol Phylogenet Evol 33:896–907

Hassanin A, Ropiquet A (2007) Resolving a zoological mystery: the kouprey is a real species. Proc R Soc Lond Sers B-Biol Sci 274:2849–2855

Hassanin A, Delsuc F, Ropiquet A et al (2012) Pattern and timing of diversification of Cetartiodactyla (Mammalia, Laurasiatheria), as revealed by a comprehensive analysis of mitochondrial genomes. CR Biol 335:32–50

Hassanin A, An J, Ropiquet A et al (2013) Combining multiple autosomal introns for studying shallow phylogeny and taxonomy of Laurasiatherian mammals: application to the tribe Bovini (Cetartiodactyla, Bovidae). Mol Phylogenet Evol 66:766–775

Hay WW, Soeding E, Deconto RM et al (2002) The Late Cenozoic uplift-climate paradox. Int J Earth Sci 91:746–774

Heintz E, Brunet M (1982) Rôle de la Téthys et de la chaîne alpine asiatique dans la distribution spatio-temporelle des cervidés. Comptes Rendus de l'Académie des Sciences de Paris, série IIa 294:1391–1394

Heintz E, Brunet M, Carbonnel J-P (1978a) Découverte du premier grand gisement à Mammifères fossiles d'Afghanistan. Comptes Rendus de l'Académie des Sciences de Paris, série IIa 286:945–947

Heintz E, Ginsburg L, Hartenberger J-L (1978b) Mammifères fossiles en Afghanistan: état des connaissances et résultats d'une prospection. Bulletin du Museum d'Histoire naturelle, Paris 69:101–119

Heissig K (1989) The Rhinocerotidae. In: Prothero DR, Schoch R (eds) The evolution of Perissodactyls. Oxford University Press, Oxford, pp 399–417

Heissig K (1999) Family Tapiridae. In: Roessner G, Heissig K (eds) The Miocene land mammals of Europe. K. Verlag Dr. Friedrich Pfeil, Munich, pp 171–175

Hernández Fernández M, Vrba ES (2005) A complete estimate of the phylogenetic relationships in Ruminantia: a dated species-level supertree of the extant ruminants. Biol Rev Camb Philos Soc 80:269–302

Hill A, Bibi F, Beech M et al (2012) Before archaeology: life and environments in the Miocene of Abu Dhabi. In: Potts D, Hellyer P (eds) Fifty years of emirates archaeology. Ministry of Culture, Youth and Community Development, Abu Dhabi, pp 20–23

Holt BG, Lessard J-P, Borregaard MK et al (2013) An update of Wallace's zoogeographic regions of the world. Science 339:74–78

Honey JG, Harrison JA, Prothero DR et al (1998) Camelidae. In: Janis C, Scott KM, Jacobs LL (eds) Evolution of tertiary mammals of North America. Cambridge University Press, Cambridge, pp 439–462

Hooijer DA (1947) On fossil and prehistoric remains of Tapirus from Java, Sumatra and China. Zoologische Mededelingen 27:253–299

Hooijer DA (1948a) Pleistocene vertebrates from Celebes I. Celebochoerus heekereni nov. gen. nov. spec. Proceedings Koninklijke Nederlandse Akademie Van Wetenschappen Series B-Phys Sci 51:1024–1032

Hooijer DA (1948b) Pleistocene vertebrates from Celebes I.Celebochoerus heekereninov. gen. nov. spec. Proceedings Koninklijke Nederlandse Akademie Van Wetenschappen Series B-Phys Sci 51:1024–1032

Hooijer DA (1949) Pleistocene vertebrates from Celebes IV. Archidiskodon celebensis nov. spec. Zoologische Mededelingen 30:205–226
Hooijer DA (1950) The fossil Hippopotamidae of Asia, with notes on the recent species. Zoologische Verhandelingen 8:1–123
Hooijer DA (1953) Pleistocene vertebrates from Celebes: VI. Stegodon spec. Zoologische Mededelingen 32:107–112
Hooijer DA (1964) Pleistocene vertebrates from Celebes XII–Notes on pygmy Stegodonts. Zoologische Mededelingen 40:37–44
Hooker JJ, Sánchez-Villagra MR, Goin FJ et al (2008) The origin of Afro-arabian 'didelphimorph' marsupials. Palaeontology 51:635–648
Hoorn C, Ohja T, Quade J (2000) Palynological evidence for vegetation development and climatic change in the Sub-Himalayan Zone (Neogene, Central Nepal). Palaeogeogr Palaeoclimatol Palaeoecol 163:133–161
Huang Z, Zhang W (2006) The Quaternary faunas and climatic fluctuation in tropical China. Trop Geogr 26:6–11
Huber M, Goldner A (2012) Eocene monsoons. J Asian Earth Sci 44:3–23
Hussain ST (1971) Revision of Hipparion (Equidae, Mammalia) from the Siwalik Hills of Pakistan and India. Bayerische Akademie der Wissenschaften, Abhandlungen 147:1–68
Jaeger J-J, Naing Soe A, Chavasseau O et al (2012) First Hominoid from the late Miocene of the Irrawaddy formation (Myanmar). PLoS ONE 6:e17065
Janis CM (1989) A climatic explanation for patterns of evolutionary diversity in ungulate mammals. Palaeontology 32:463–481
Janis CM (1990) Correlation of cranial and dental variables with body size in ungulates and macropodoids. In: Damuth J, Macfadden BJ (eds) Body size in mammalian paleobiology. Cambridge University Press, Cambridge, pp 255–300
Janis CM, Scott KM (1987) The interrelationships of higher ruminant families with special emphasis on the members of the Cervoidea. Amer Mus Novit 2893:1–85
Ji H (1982) The living environment of the Quaternary mammalian faunas of South China. Vertebrata PalAsiatica 20:148–154
Kappelman JW, Rasmussen DL, Sanders WJ et al (2003) Oligocene mammals from Ethiopia and faunal exchange between Afro-Arabia and Eurasia. Nature 426:549–552
Khan E (1971) Punjabitherium gen. nov., an extinct rhinocerotid of the Siwaliks, Punjab, India. Proc Ind Nat Sci Acad 37:105–109
Khan AM (2010) Taxonomy and distribution of Rhinoceroses from the Siwalik Hills, Pakistan. Dissertation, University of the Punjab
Khan M, Khan A, Iqbal M et al (2008) Reduncine fossils from the Upper Siwaliks of Tatrot. J Anim Plant Sci 18:50–52
Khan MA, Iqbal M, Akhtar M et al (2009) Chalicotheres in the Siwaliks of Pakistan. Pakistan J Zool 41:429–435
Khan MA, Kostopoulos DS, Akhtar M et al (2010) Bison remains from the Upper Siwaliks of Pakistan. Neues Jahrbuch für Geologie und Paläontologie-Abhandlungen 258:121–128
Khan A, Khan M, Iqbal M et al (2011) Sivatherium (Artiodactyla, Ruminantia, Giraffidae) from the Upper Siwaliks, Pakistan. J Anim Plant Sci 21:202–206
Kundal YP, Kundal SN (2011) Elephas cf. E. maximus indicus (Elephantidae, Mammalia) from the post Siwalik deposits of Jammu Province, Jammu and Kashmir, India. Vertebrata Palasiatica 49:348–361
Kunimatsu Y, Ratanasthien B, Nakaya H et al (2004) Hominoid fossils discovered from Chiang Muan, northern Thailand: the first step towards understanding hominoid evolution in Neogene Southeast Asia. Anthropological Science Published online in J-STAGE (www.jstage.jst.go.jp): 1–9
Lang J, Lavocat R (1968) Première découverte d'une faune de vertébrés dans le tertiaire d'Afghanistan et datation de la série de Bamian. Comptes Rendus de l'Académie des Sciences, Paris 266:79–82

Legendre S (1989) Les communautés de mammifères du Paléogène (Eocène supérieur et Oligocène) d'Europe Occidentale: structures, milieux et évolution. Münchner Geowissenschaftliche Abhandlungen (A) 16:1–100
Li Y (1981) On the subdivisions and evolution of the Quaternary mammalian faunas of South China. Vertebrata Palasiatica 19:67–76
Licht A, Van Cappelle M, Abels HA et al (2014) Asian monsoons in a late Eocene greenhouse world. Nature 513:501
Licht A, Boura A, De Franceschi D, Utescher T, Sein C, Jaeger J-J (2015) Late middle eocene fossil wood of Myanmar: Implications for the landscape and the climate of the Eocene Bengal Bay. Rev Palaeobot Palynol 216:44–54
Lihoreau F, Barry J, Blondel C et al (2007a) Anatomical revision of the genus Merycopotamus (Artiodactyla; Anthracotheriidae): its significance for Late Miocene mammal dispersal in Asia. Palaeontology 50:503–524
Lihoreau F, Ducrocq S, Prothero DR et al (2007b) Family Anthracotheriidae. In: Prothero DR, Foss SE (eds) The evolution of Artiodactyls. Johns Hopkins University Press, Baltimore, pp 89–105
Lindsay EH, Opdyke ND, Johnson NM (1980) Pliocene dispersal of the horse Equus and late Cenozoic mammalian dispersal events. Nature 287:135–138
Lindsay EH, Flynn LJ, Cheema IU et al (2005) Will Downs and the Zinda Pir Dome. Palaeontol Electron 8:1–19
Lister AM (2013) The role of behaviour in adaptive morphological evolution of African proboscideans. Nature 331–334
Liu L-P (2001) Eocene suoids (Artiodactyla, Mammalia) from Bose and Yongle basins, China, and the classification and evolution of the Paleogene suoids. Vert. PalAsia 39:115–128
Loftus RT, Machugh E, Bradley DG et al (1994) Evidence for two independent domestications of cattle. Proc Natl Acad Sci USA 91:2757–2761
Louys J (2007) Limited effect of the Quaternary's largest super-eruption (Toba) on land mammals from Southeast Asia. Quatern Sci Rev 26:3108–3117
Louys J (2012a) The future of mammals in Southeast Asia: conservation insights from the fossil record. In: Louys J (ed) Paleontol Ecol Conserv. Springer, Berlin, pp 227–238
Louys J (2012b) Mammal community structure of Sundanese fossil assemblages from the Late Pleistocene, and a discussion on the ecological effects of the Toba eruption. Quatern Int 258:80–87
Louys J, Meijaard E (2010) Palaeoecology of Southeast Asian megafauna-bearing sites from the Pleistocene and a review of environmental changes in the region. J Biogeogr 37:1432–1449
Louys J, Curnoe D, Tong H (2007) Characteristics of Pleistocene megafauna extinctions in Southeast Asia. Palaeogeogr Palaeoclimatol Palaeoecol 243:152–173
Lucas SG, Sobus JC (1989) The systematics of indricotheres. In: Prothero DR, Schoch RM (eds) The evolution of perissodactyls. Oxford University Press, New York, pp 358–378
Lydekker R (1876) Molar teeth and other remains of mammalia. Palaeontologia Indica, Memoirs of the Geological Survey of India Ser 10, V.1:19–87
Lydekker R (1876b) Notes on the fossil mammalian faunae of India and Burma. Rec Geol Surv Ind 9:86–106
Lydekker R (1878) Crania of ruminants. Palaeontologia Indica. Mem Geol Surv Ind Ser 10, 1:88–171
Lydekker R (1882a) Indian Tertiary and post-Tertiary vertebrata, Siwalik and Narbada Equidae. Palaeontologia Indica 2:67–98
Lydekker R (1882b) Siwalik Camelopardalidae. Mem Geol Surv Ind 2:99–142
Lydekker R (1885) Catalogue of the fossil mammalia in the British Museum. Part II Taylor and Francis, London
Lydekker R (1898) Wild oxen, sheep & goats of all lands, living and extinct R. Ward, London
Lyell C (1830) Principles of geology. John Murray, London
Macfadden B (1998) Equidae. In: Janis C, Scott KM, Jacobs LL (eds) Evolution of tertiary mammals of North America. Cambridge University Press, Cambridge, pp 537–559

Macfadden BJ (2006) Extinct mammalian biodiversity of the ancient New World tropics. Trends Ecol Evol 21:157–165
Macfadden BJ, Woodburne MO (1982) Systematics of the Neogene Siwalik hipparions (Mammalia, Equidae) based on cranial and dental morphology. J Vertebr Paleontol 2:185–218
Mackinnon JR, Stuart SN (1988) The kouprey: an action plan for its conservation IUCN. Glan, Switzerland
Macphee RD (1999) Extinctions in near time: causes, contexts, and consequences. Kluwer Academic/ Plenum Publishers, New York
Macphee RD, Marx PA (1997) The 40,000-year plague: humans, hyperdisease, and first-contact extinctions. In: Goodman SM, Patterson BD (eds) Natural change and human impact in Madagascar. Smithsonian Institution Press, Washington, DC, pp 169–217
Madsen O, Scally M, Douady CJ et al (2001) Parallel adaptive radiations in two major clades of placental mammals. Nature 409:610–614
Martin PS (1966) Africa and Pleistocene overkill. Nature 212:339–342
Martin PS (1967) Prehistoric overkill. In: Martin PS, Wright Jr HE (eds) Pleistocene extinctions: the search for a cause. Yale University Press, New Haven, pp 75–120
Mattauer M, Matte P, Olivet J-L (1999) A 3D model of the India-Asia collision at plate scale. Comptes Rendus de l'Académie des Sciences-Series IIA-Earth and Planetary Science 328:499–508
Matthew WD (1929) Critical observations upon Siwalik mammals. Bull Amer Mus Nat Hist 56:437–560
Matthew WD, Granger W (1923) New fossil mammals from the Pliocene of Sze-Chuan, China. Bull Amer Mus Nat Hist 48:563–598
Matthew WD, Granger W (1925) New ungulates from the Ardyn Obo Formation of Mongolia. Amer Mus Novit 195:1–12
Mcquarrie N, Stock JM, Verdel C et al (2003) Cenozoic evolution of Neotethys and implications for the causes of plate motions. Geophys Res Lett 30. doi:10.1029/2003GL017992 art. no. 2036
Mctavish EJ, Decker JE, Schnabel RD et al (2013) New World cattle show ancestry from multiple independent domestication events. Proc Natl Acad Sci 110:E1398–E1406
Meijer HJ, Van Den Hoek Ostende LW, Van Den Bergh GD et al (2010) The fellowship of the hobbit: the fauna surrounding Homo floresiensis. J Biogeogr 37:995–1006
Mein P, Ginsburg L (1997) Les mammifères du gisement miocène inférieur de Li Mae Long, Thaïlande: systématique, biostratigraphie et paléoenvironnement. Geodiversitas 19:783–844
Mennessier G (1961) Les caractères structuraux des montagnes de la région de Kaboul (Afghanistan). Bulletin de la Société Géologique de France, Séries 3(7):40–49
Merceron G, Blondel C, Brunet M et al (2004) The Late Miocene paleoenvironment of Afghanistan as inferred from dental microwear in artiodactyls. Palaeogeogr Palaeoclimatol Palaeoecol 207:143–163
Meredith RW, Janečka JE, Gatesy J et al (2011) Impacts of the Cretaceous Terrestrial Revolution and KPg Extinction on Mammal Diversification. Science 334:521–524
Métais G, Antoine P-O, Marivaux L, Welcomme J-L, Ducrocq S (2003) New artiodactyl ruminant (Mammalia) from the late Oligocene of Pakistan. Acta Palaeontol Pol 48:375–382
Métais G (2006) New basal selenodont artiodactyls from the Pondaung Formation (late middle Eocene, Myanmar) and the phylogenetic relationships of early ruminants. Ann Carnegie Mus Nat Hist 75:51–67
Métais G, Chaimanee Y, Jaeger J-J et al (2001) New remains of primitive ruminants from Thailand: evidence of the early evolution of the Ruminantia in Asia. Zoolog Scr 30:231–249
Métais G, Soe AN, Marivaux L et al (2007) Artiodactyls from the Pondaung Formation (Myanmar): new data, and new interpretations on the South Asian faunal Province during the middle Eocene. Naturwissenschaften 94:759–768
Métais G, Chaimanee Y, Jaeger J-J, Ducrocq S (2007) Eocene bunoselenodont Artiodactyla from southern Thailand, and the early evolution of Ruminantia in South Asia. Naturwissenschaften 94:493–498

Métais G, Vislobokova IA (2007) Basal Ruminants. In: Prothero DR, Foss SE (eds) The evolution of artiodactyls. The Johns Hopkins University Press, Baltimore, pp 189–212
Métais G, Antoine P-O, Baqri SRH et al (2009a) Lithofacies, depositional environments, regional biostratigraphy and age of the Chitarwata formation in the Bugti Hills, Balochistan, Pakistan. J Asian Earth Sci 34:154–167
Métais G, Welcomme J-L, Ducrocq S (2009b) New lophiomerycid ruminants from the Oligocene of the Bugti Hills (Balochistan, Pakistan). J Vertebr Paleontol 29:231–241
Molnar P, Tapponnier P (1977) The collision between India and Asia. Sci Am 236:30–41
Molnar P, Boos WR, Battisti DS (2010) Orographic controls on climate and paleoclimate of Asia: thermal and mechanical roles for the Tibetan Plateau. Annu Rev Earth Planet Sci 38:77–102
Montgelard C, Catzeflis F, Douzery E (1997) Phylogenetic relationships of artiodactyls and cetaceans as deduced from the comparison of cytochrome *b* and 12S rRNA mitochondrial sequences. Mol Biol Evol 14:550–559
Morley CK (2002) A tectonic model for the Tertiary evolution of strike-slip faults and rift basins in SE Asia. Tectonophysics 347:189–215
Murchison C (1868) Palaeontological memoirs and notes of the late hugh falconer robert hardwicke, London
Najman Y, Pringle M, Godin L et al (2001) Dating of the oldest continental sediments from the Himalayan foreland basin. Nature 410:194–197
Nanda AC (2002) Upper Siwalik mammalian faunas of India and associated events. J Asian Earth Sci 21:47–58
Nanda AC (2008) Comments on the Pinjor mammalian fauna of the Siwalik Group in relation to the post-Siwalik faunas of Peninsular India and Indo-Gangetic Plain. Quatern Int 192:6–13
Nelson SV (2007) Isotopic reconstructions of habitat change surrounding the extinction of Sivapithecus, a Miocene hominoid, in the Siwalik Group of Pakistan. Palaeogeogr Palaeoclimatol Palaeoecol 243:204–222
Ni X, Qiu Z (2002) The micromammalian fauna from the Leilao, Yuanmou hominoid locality: Implications for biochronology and paleoecology. J Hum Evol 42:535–546
Nicholson KN, Khan M, Mahmood K (2010) Geochemistry of the Chagai–Raskoh arc, Pakistan: Complex arc dynamics spanning the Cretaceous to the Quaternary. Lithos 118:338–348
Ogino S, Egi N, Takai M (2011) New species of Agriotherium (Mammalia, Carnivora) from the late Miocene to early Pliocene of central Myanmar. J Asian Earth Sci 42:408–414
O'leary MA, Bloch JI, Flynn JJ et al (2013) The placental mammal ancestor and the post-K-Pg radiation of placentals. Science 339:662–667
Olsen SJ (1990) Fossil ancestry of the yak, its cultural significance and domestication in Tibet. Proc Acad Nat Sci Philadelphia 142:73–100
Orliac MJ, Antoine P-O, Roohi G et al (2010a) Suoidea (Mammalia, Cetartiodactyla) from the Early Oligocene of the Bugti Hills, Balochistan, Pakistan. J Vertebr Paleontol 30:1300–1305
Orliac MJ, Pierre-Olivier A, Ducrocq S (2010b) Phylogenetic relationships of the Suidae (Mammalia, Cetartiodactyla): new insights on the relationships within Suoidea. Zoolog Scr 39:315–330
Orliac M, Guy F, Chaimanee Y et al (2011) New remains of Egatochoerus jaegeri (Mammalia, Suoidea) from the late Eocene of peninsular Thailand. Palaeontology 54:1323–1335
Patnaik R (2013) Indian Neogene Siwalik mammalian biostratigraphy: an overview. In: Wang X, Flynn LJ, Fortelius M (eds) Fossil mammals of Asia: Neogene biostratigraphy and chronology. Columbia University Press, New York, pp 423–444
Patnaik R (2015) Diet and habitat changes among Siwalik herbivorous mammals in response to Neogene and Quaternary climate changes: an appraisal in the light of new data. Quatern Int. doi:10.1016/j.quaint.2014.11.025
Patnaik R, Chauhan PR, Rao M et al (2009) New geochronological, paleoclimatological, and archaeological data from the Narmada Valley hominin locality, central India. J Hum Evol 56:114–133
Patnaik R, Gupta AK, Naidu PD et al (2012) Indian monsoon variability at different time scales: marine and terrestrial proxy records. Proc Ind Nat Sci Acad 78:535–547

Patriat P, Achache J (1984) India-Eurasia collision chronology has implications for crustal shortening and driving mechanism of plates. Nature 311:615–621
Pickford M (1982) Miocene Chalicotheriidae of the Potwar Plateau, Pakistan. Tertiary Res 1:13–29
Pickford M (2001) Africa's smallest ruminant: a new tragulid from the Miocene of Kenya and the biostratigraphy of East African Tragulidae. Géobios 34:437–447
Pickford M, Nakaya H, Kunimatsu Y et al (2004) Age and taxonomic status of the Chiang Muan (Thailand) hominoids. CR Palevol 3
Pilgrim GE (1907) Description of some new Suidae from the Bugti Hills, Baluchistan. Rec Geol Surv Ind 36:45–56 (1 plate)
Pilgrim GE (1908) The Tertiary and Post-Tertiary freshwater deposits of Baluchistan and Sind with notices of new vertebrates. Rec Geol Surv Ind 37:139–167, 4 plates
Pilgrim GE (1910a) Notices of new mammalian genera and species from the Tertiaries of India. Rec Geol Surv Ind 40:63–71
Pilgrim GE (1910b) Preliminary note on a revised classification of the Tertiary freshwater deposits of India. Rec Geol Surv Ind 40:185–205
Pilgrim GE (1911) The fossil Giraffidae of India. Mem Geol Surv Ind 4:1–29
Pilgrim GE (1912) The vertebrate fauna of the Gaj series in the Bugti hills and the Punjab. Mem Geol Surv Ind 4:1–83
Pilgrim GE (1928) The Artiodactyla of the Eocene of Burma. Palaeontologia Indica 13:1–39
Pilgrim GE (1939) The fossil Bovidae of India. Palaeontologia Indica NS 26:1–356
Pilgrim GE, Cotter GDP (1916) Some newly discovered Eocene mammals from Burma. Records of the Geological Survey of India 47:42–77
Pitra C, Fickel J, Meijaard E et al (2004) Evolution and phylogeny of old world deer. Mol Phylogenet Evol 33:880–895
Popov SV, Rögl F, Rozanov AY et al (2004) Lithological-paleogeographic maps of Paratethys. Courier Forschungsinstitut Senckenberg 250:1–46
Prell WL, Kutzbach JE (1992) Sensitivity of the Indian monsoon to forcing parameters and implications for its evolution. Nature 360:647–652
Prothero DR (1998a) Hyracodontidae. In: Janis CM, Scott KM, Jacobs LL (eds) Evolution of tertiary mammals of North America. Cambridge University Press, Cambridge, pp 589–593
Prothero DR (1998b) Rhinocerotidae. In: Janis C, Scott KM, Jacobs LL (eds) Evolution of tertiary mammals of North America. Cambridge Univesity Press, Cambridge, pp 595–605
Prothero DR (2007) Family Moschidae. In: Prothero D, Foss S (eds) The evolution of artiodactyls. Johns Hopkins University Press, Baltimore, pp 221–226
Prothero DR, Schoch RM (2002) Horns, tusks, and flippers: the evolution of hoofed mammals. Johns Hopkins University Press, Baltimore
Prothero DR, Manning E, Hanson CB (1986) The phylogeny of the Rhinocerotoidea (Mammalia, Parissodactyla). Zoo J Linnean Soc 87:341–366
Prothero DR, Guérin C, Manning E (1989) The history of the Rhinocerotoidea. In: Prothero DR, Schoch RM (eds) The evolution of perissodactyls. Oxford University Press, Oxford, pp 321–340
Pubellier M, Rangin C, Phach PV, Que BC, Sang CL (2008) The Cao Bang-Tien Yen Fault: Implications on the relationships between the Red River Fault and the South China Coastal Belt. Adv Nat Sci 4:347–361
Qiu ZX, Qiu ZD (1995) Chronological sequence and subdivision of Chinese Neogene mammalian faunas. Palaeogeogr Palaeoclimatol Palaeoecol 116:41–70
Qiu Z-X, Qiu Z-D, Deng T et al (2013) Neogene land mammal stages/ages of China. In: Wang X, Flynn LJ, Fortelius M (eds) Fossil mammals of Asia: Neogene biostratigraphy and chronology. Columbia University Press, New York, pp 29–90
Quade J, Cerling TE (1995) Expansion of C-4 grasses in the late Miocene of northern Pakistan—evidence from stable isotopes in paleosols. Palaeogeogr Palaeoclimatol Palaeoecol 115:91–116
Quade J, Cerling TE, Bowman JR (1989) Development of Asian monsoon revealed by marked ecological shift during the latest Miocene in northern Pakistan. Nature 342:163–166

Quade J, Cerlinga TE, Barry JC et al (1992) A 16-Ma record of paleodiet using carbon and oxygen isotopes in fossil teeth from Pakistan. Chem Geol Isotope Geosci Sect 94:183–192
Radinsky LB (1967) Hyrachyus, Chasmotherium, and the early evolution of helaletid tapiroids. Amer Mus Novit 2313:1–23
Ramstein G, Fluteau F, Besse J et al (1997) Effect of orogeny, plate motion and land-sea distribution on Eurasian climate change over the past 30 million years. Nature 386:788–794
Rao AR, Nanda AC, Sharma UN et al (1995) Magnetic polarity stratigraphy of the Pinjor Formation (Upper Siwalik) near Pinjore, Haryana. Curr Sci 68:1231–1236
Raufi F, Sickenberg O (1973) Zur Geologie der weiteren Umgebung von Tirin und Misan im Südosten des Hochlandes von Zentral-Afghanistan. Beih Geol Jb 96:63–99
Reis KR, Garong AM (2001) Late Quaternary terrestrial vertebrates from Palawan Island, Philippines. Palaeogeogr Palaeoclimatol Palaeoecol 171:409–421
Reuter M, Kern AK, Harzhauser M et al (2013) Global warming and South Indian monsoon rainfall? Lessons from the Mid-Miocene. Gondwana Res 23:1172–1177
Rose KD (2006) The beginning of the age of mammals. Johns Hopkins University Press, Baltimore
Rössner GE (2007) Family Tragulidae. In: Prothero DR, Foss SE (eds) The evolution of artiodactyls. Johns Hopkins University Press, Baltimore, pp 213–220
Ruddiman WF (1997) Tectonic uplift and climate change. Plenum Press, New York
Russell DE, Zhai RJ (1987) The Paleogene of Asia: mammals and stratigraphy. Mémoire du Museum Nationale d'Histoire Naturelle 53:1–488
Rütimeyer L (1878) Die Rinder der Tertiär-Epoche. Abhandlungen der Schweizerischen Paläontologischen Gesellschaft 5:3–205
Saegusa H (2001) Comparisons of stegodon and elephantid abundances in the Late Pleistocene of southern China. In: Cavarretta G, Gioia P, Mussi M, Palombo MR (eds) The world of elephants. Proceedings of the 1st international congress, Consiglio Nazionale delle Ricerche, Rome, pp 345–349
Saegusa H, Thasod Y, Ratanasthien B (2005) Notes on Asian stegodontids. Quatern Int 126:31–48
Samiullah K, Akhtar M (2007) An evidence of Cervus punjabiensis from the Lower Siwaliks of the Punjab, Pakistan. Punjab Univ J Zoo 22:63–68
Sánchez IM, Quiralte V, Morales J et al (2010) A new genus of tragulid ruminant from the early Miocene of Kenya. Acta Palaeontol Pol 55:177–187
Sánchez IM, Demiguel D, Quiralte V et al (2011) The first known Asian Hispanomeryx (Mammalia, Ruminantia, Moschidae). J Vertebr Paleontol 31:1397–1403
Sanders WJ, Kappelman J, Rasmussen DT (2004) New large-bodied mammals from the late Oligocene site of Chilga, Ethiopia. Acta Palaeontol Pol 49:365–392
Sanders WJ, Gheerbrant E, Harris JM et al (2010) Proboscidea. In: Werdelin L, Sanders WJ (eds) Cenozoic Mammals of Africa. University of California Press, Berkeley, pp 161–251
Sandom C, Faurby S, Sandel B et al (2014) Global late Quaternary megafauna extinctions linked to humans, not climate change. Proc R Soc B: Biol Sci 281:20133254
Schoch RM (1989) A review of the tapiroids. In: Prothero DR, Schoch RM (eds) The evolution of Perissodactyls. Oxford University Press, Oxford, pp 298–320
Schweinfurth GA (1886) Reise in das Depressionsgebiet im Umkreise des Fajum im Januar 1886. Z Gesellschafte Erdkurde Berlin 21:96–149
Sclater PL (1858) On the general geographic distribution of the members of the class Aves. J Proc Linnean Soc Lond Zoo 2:130–136
Seiffert ER (2006) Revised age estimates for the later Paleogene mammal faunas of Egypt and Oman. Proc Natl Acad Sci 103:5000–5005
Seiffert ER (2012) Early primate evolution in Afro-Arabia. Evol Anthropol 21:239–253
Sen S (1998) The age of the Molayan mammal locality, Afghanistan. Géobios 31:385–391
Sen S (2001) Rodents and insectivores from the Upper Miocene of Molayan, Afghanistan. Palaeontology 44:913–932
Sen S, Blieck A, Bouvrain G et al (1997) Late Miocene mammals from Taghar, Khurdkabul Basin, Afghanistan. Annales de Paléontologie 83:233–266

Şengör AMC, Kidd WSF (1979) Post-collisional tectonics of the Turkish-Iranian plateau and a comparison with Tibet. Tectonophysics 55:361–376

Sepulchre P, Jolly D, Ducrocq S et al (2010) Mid-Tertiary paleoenvironments in Thailand: pollen evidence. Climate Past 6:461–473

Shoshani J, Walter RC, Abraha M et al (2006) A proboscidean from the late Oligocene of Eritrea, a "missing link" between early Elephantiformes and Elephantimorpha, and biogeographic implications. Proc Natl Acad Sci USA 103:17296–17301

Kappelman J, Simons EL Swisher CC, III (1992) New age determinations for the Eocene-Oligocene boundary sediments in the Fayum Depression, Northern Egypt. J Geol 100:647–668

Simpson GG (1977) Too many lines: the limits of the Oriental and Australian zoogeographic regions. Proc Am Philos Soc 121:107–120

Sodhi NS, Posa MRC, Lee TM et al (2010) The state and conservation of Southeast Asian biodiversity. Biodivers Conserv 19:317–328

Soe AN, Myitta Tun ST et al (2002) Sedimentary facies of the late Middle Eocene Pondaung Formation (central Myanmar) and the palaeoenvironments of its Anthropoid Primates. CR Palevol 1:153–160

Solounias N (2007) Family Giraffidae. In: Prothero DR, Foss SE (eds) The evolution of artiodactyls. Johns Hopkins University Press, Baltimore, pp 257–277

Solounias N, Barry JC, Bernor RL et al (1995) The oldest bovid from the Siwaliks, Pakistan. J Vertebr Paleontol 15:806–814

Spaulding M, O'leary MA, Gatesy J (2009) Relationships of Cetacea (Artiodactyla) among mammals: increased taxon sampling alters interpretations of key fossils and character evolution. PLoS ONE 4:e7062

Steensma K, Hussain S (1992) Merycopotamus dissimilis [Artiodactyla, Mammalia] from the Upper Siwalik Subgroup and its affinities with Asian and African forms. Proceedings of the Koninklijke Nederlandse Akademie van Wetenschappen 95:97–108

Stehlin HG (1909) Remarques sur les faunules de mammifères de l'Eocène et de l'Oligocène du Bassin de Paris. Bulletin de la Société Géologique de France 4(9):488–520

Steiner CC, Ryder OA (2011) Molecular phylogeny and evolution of the Perissodactyla. Zoo J Linnean Soc 163:1289–1303

Steiner CC, Mitelberg A, Tursi R et al (2012) Molecular phylogeny of extant equids and effects of ancestral polymorphism in resolving species-level phylogenies. Mol Phylogenet Evol 65:573–581

Stille P, Steinmann M, Riggs SR (1996) Nd isotope evidence for the evolution of the paleocurrents in the Atlantic and Tethys Oceans during the past 180 Ma. Earth Planet Sci Lett 144:9–19

Stuart A (1999) Late Pleistocene megafaunal extinctions. In: Macphee R (ed) Extinctions in near time: causes, contexts, and consequences. Kluwer Academic, New York, pp 257–269

Suganuma Y, Hamada T, Tanaka S et al (2006) Magnetostratigraphy of the Miocene Chiang Muan Formation, northern Thailand: implication for revised chronology of the earliest Miocene hominoid in Southeast Asia. Palaeogeogr Palaeoclimatol Palaeoecol 239:75–86

Sukumar R (2003) The living elephants: evolutionary ecology, behavior, and conservation. Oxford University Press, New York

Sukumar R (2006) A brief review of the status, distribution and biology of wild Asian elephants Elephas maximus. International Zoo Yearbook 40:1–8

Sun X, Wang P (2005) How old is the Asian monsoon system? Paleobotanical records from China. Paleogeogr Paleoclimatol Paleoecol 222:181–222

Suraprasit K, Chaimanee Y, Bocherens H et al (2014) Systematics and phylogeny of middle Miocene Cervidae (Mammalia) from Mae Moh Basin (Thailand) and a paleoenvironmental estimate using enamel isotopy of sympatric herbivore species. J Vertebr Paleontol 34:179–194

Suraprasit K, Chaimanee Y, Chavasseau O, Jaeger JJ (2015) Middle Miocene bovids from Mae Moh Basin, Northern Thailand: The first record of the genus Eotragus from Southeast Asia. Acta Palaeontol Pol 60:67–78

Swisher CC, Curtis GH, Jacob T et al (1994) Age of the earliest known hominids in Java, Indonesia. Science 263:1118–1121

Tabbutt K, Sheikh K Johnson N (1997) A fission track age for the Bugti bonebeds. Baluchistan, Pakistan. Extended abstract, Proceedings of the third workshop on Siwaliks of South Asia. Rec Geol Surv Pakistan 109:53

Takai M, Saegusa H, Thaung-Htike et al (2006) Neogene mammalian fauna in Myanmar. Asian Paleoprimatol 4:143–172

Tanaka K, Solis CD, Masangkay JS et al (1996) Phylogenetic relationship among all living species of the genus Bubalus based on DNA sequences of the cytochrome b gene. Biochem Genet 34:443–452

Tapponnier P, Peltzer G, Armijo R (1986) On the mecanism of collision between India and Asia. In: Tectonics Collision (ed) Coward M P Ries A C. Geological Society of London, Special, Publication, pp 115–157

Tassy P (1983a) Les Elephantoidea Miocènes du plateau du Potwar, Groupe de Siwalik, Pakistan. Deuxième partie: Choerolophodontes et Gomphothères. Annales de Paléontologie 69:235–297

Tassy P (1983b) Les Elephantoidea Miocènes du plateau du Potwar, Groupe de Siwalik, Pakistan. Premiere partie: introduction, cadre chronologique et géographie, mammutidés, amébélodontidés. Annales de Paléontologie 69:99–136

Tassy P (1983c) Les Elephantoidea Miocènes du plateau du Potwar, Groupe de Siwalik, Pakistan. Troisième partie: Stégodontidés, Eléphantoïdes indéterminés. Restes postcraniens. Conclusions. Annales de Paléontologie 69:317–354

Tassy P (1990) The "Proboscidean Datum Event": how many proboscideans and how many events? In: Lindsay EH, Fahlbusch V, Mein P (eds) European Neogene mammal chronology. Plenum Press, New York, pp 237–252

Tassy P, Anupandhanant P, Ginsburg L et al (1992) A new Stegolophodon (Proboscidea, Mammalia) from the early Miocene of northern Thailand. Geobios 25:511–523

Thasod Y, Jintasakul P, Ratanasthien B (2012) Proboscidean Fossil from the Tha Chang Sand Pits, Nakhon Ratchasima Province, Thailand. J Sci Technol MSU 31:33–44

Thenius E (1970) Zur evolution und verbreitungsgeschichte der Suidae (Artiodactyla, Mammalia). Zeitschrift für Säugetierkunde 35:321–342

Thewissen JGM, Williams EM, Hussain ST (2001a) Eocene mammal faunas from Northern Indo-Pakistan. J Vertebr Paleontol 21:347–366

Thewissen JGM, Williams EM, Roe LJ et al (2001b) Skeletons of terrestrial cetaceans and the relationship of whales to artiodactyls. Nature 413:277–281

Thewissen JGM, Cooper LN, Clementz MT et al (2007) Whales originated from aquatic artiodactyls in the Eocene epoch of India. Nature 450:1190–1194

Thomas H (1984) Les Bovidae (Artiodactyla; Mammalia) du Miocène du sous-continent indien, de la péninsule arabique et de l'Afrique: biostratigraphie, biogéographie et écologie. Palaeogeogr Palaeoclimatol Palaeoecol 45:251–299

Thomas H, Sen S, Ligabue G (1980) La faune Miocene de la Formation Agha Jari du Jebel Hamri (Irak). Koninklijke Nederlandse Akademie van Wetenschappen, Amsterdam, Series B 83:269–287

Thomas H, Sen S, Khan M et al (1982) The lower Miocene fauna of Al-Sarrar (Eastern province, Saudi Arabia). J Saudi Arabian Archaeol 5:109–136

Thomas H, Ginsburg L, Hintong C, Suteethorn V (1990) A new tragulid, Siamotragulus sanyathanai n.g., n.sp. (Artiodactyla, Mammalia) from the Miocene of Thailand (Amphoe Pong, Phayao Province). C R Acad Sci Paris, serie IIa 310:989–995

Tobien H (1980) Cerviden-Reste (Ruminantia, Mammalia) aus dem Obermiozän von Maragheh (Iran). Annalen des Naturhistorischen Museums in Wien 83:329–341

Tong H (2005) Dental characters of the Quaternary tapirs in China, their significance in classification and phylogenetic assessment. Geobios 38:139–150

Tong H-w Guérin C (2009) Early Pleistocene Dicerorhinus sumatrensis remains from the Liucheng Gigantopithecus Cave, Guangxi, China. Geobios 42:525–539

Tong Y, Wang J (2006) Fossil mammals from the early Eocene Wutu formation of Shandong province. Palaeontol Sinica 192:1–223

Tougard C (2001) Biogeography and migration routes of large mammal faunas in South-East Asia during the Late Middle Pleistocene: focus on the fossil and extant faunas from Thailand. Palaeogeogr Palaeoclimatol Palaeoecol 168:337–358

Tripathi C (1986) Siwaliks of the Indian subcontinent. J Palaeontol Soc Ind 31:1–8

Tsubamoto T (2010) Recognition of Microbunodon (Artiodactyla, Anthracotheriidae) from the Eocene of China. Paleontol Res 14:161–165

Tsubamoto T, Takai M, Egi N et al (2002) The Anthracotheriidae (Mammalia; Artiodactyla) from the Eocene Pondaung Formation (Myanmar) and comments on some other anthracotheres from the Eocene of Asia. Paleontol Res 6:363–384

Tsubamoto T, Egi N, Takai M et al (2005) Middle Eocene ungulate mammals from Myanmar: a review with description of new specimens. Acta Palaeontol Pol 50:117–138

Udvardy MDF (1975) A Classification of the Biogeographical Provinces of the World IUCN Occasional Paper No. 18. International Union for Conservation of Nature and Natural Resources, Morges, Switzerland

Ungar P, Scott J, Curran S et al (2012) Early Neogene environments in East Africa: evidence from dental microwear of tragulids. Palaeogeogr Palaeoclimatol Palaeoecol 342:84–96

Uno KT, Cerling TE, Harris JM et al (2011) Late Miocene to Pliocene carbon isotope record of differential diet change among East African herbivores. Proc Natl Acad Sci 108:6509–6514

Urbain A (1937) Le kou prey ou boeuf sauvage Cambodgien. Mammalia: Morphologie. Biologie, Systématique des Mammifères 1:257–258

Van Couvering JA, Berggren WA (1976) Biostratigraphical basis of the Neogene time scale. In: Hagel JE, Kauffman EH (eds) Concepts in biostratigraphy. The Paleontological Society, Lawrence, Kansas, pp 283–306

Van Den Bergh GD (1999) The late Neogene elephantoid-bearing faunas of Indonesia and their palaeozoogeographic implications: a study of the terrestrial faunal succession of Sulawesi, Flores and Java, including evidence for early hominid dispersal east of Wallace's Line. Scripta Geol 117:1–419

Van Den Bergh GD, De Vos J, Aziz F et al (2001) Elephantoidea in the Indonesian region: new Stegodon findings from Flores. In: Cavarretta G, Gioia P, Mussi M, Palombo MR (eds) The world of elephants. Proceedings of the 1st international congress., Consiglio Nazionale delle Ricerche, Rome, pp 623–627

Van Den Bergh GD, De Vos J, Sondaar PY (2001b) The late Quaternary palaeogeography of mammal evolution in the Indonesian Archipelago. Palaeogeogr Palaeoclimatol Palaeoecol 171:385–408

Van Den Bergh GD, Awe RD, Morwood MJ et al (2008) The youngest Stegodon remains in Southeast Asia from the Late Pleistocene archaeological site Liang Bua, Flores, Indonesia. Quatern Int 182:16–48

Van Der Geer A, Dermitzakis M De Vos J (2008) Fossil folklore from India: the Siwalik Hills and the Mahabharata. Folklore 119:71–92

Van Der Voo R, Spakman W, Bijwaard H (1999) Tethyan subducted slabs under India. Earth Planet Sci Lett 171:7–20

Van Hinsbergen DJ, Edwards MA Govers R (2009) Collision and collapse at the Africa-Arabia-Eurasia subduction zone Geological Society Special Publication, London

Vislobokova I, Lavrov A (2009) The earliest musk deer of the genus Moschus and their significance in clarifying of evolution and relationships of the family Moschidae. Paleontol J 43:326–338

Von Koenigswald GHR (1959) A mastodon and other fossil mammals from Thailand. Report Invest Roy Dept Mines 2:25–28

Von Koenigswald GHR (1967) An upper Eocene mammal of the family Anthracotheriidae from the island of Timor, Indonesia. Proceedings of the Koninklijke Nederlandse Akademie van Wetenschappen, Series B: Palaeontol Geol Phys Chem 70:529–533

Voris HK (2000) Maps of Pleistocene sea levels in Southeast Asia: shorelines, river systems and time durations. J Biogeogr 27:1153–1167

Vrba ES (1995) The fossil record of African antelopes (Mammalia, Bovidae) in relation to human evolution and paleoclimate. In: Vrba Elisabeth S, Denton George H, Partridge Timothy C, Burckle Lloyd H (ed) Paleoclimate and evolution, with emphasis on human origins. Yale University Press, New Haven, CT, United States, pp 385–424

Vrba ES, Bibi F Costa AG (2015) First Asian record of a late Pleistocene reduncine (Artiodactyla, Bovidae, Reduncini), *Sivacobus sankaliai*, sp. nov., from Gopnath (Miliolite Formation) Gujarat, India, and a revision of the Asian genus *Sivacobus* Pilgrim, 1939. J Vertebr Paleontol 35:e943399

Vrba ES, Gatesy J (1994) New antelope fossils from Awash, Ethiopia, and phylogenetic analysis of Hippotragini (Bovidae, Mammalia). Palaeontologia Africana 31:55–72

Wall WP (1980) Cranial evidence for a proboscis in Cadurcodon and a review of snout structure in the family Amynodontidae (Perissodactyla: Rhinocerotoidae). J Paleontol 54:968–977

Wall WP (1989) The phylogenetic history and adaptative radiation of the Amynodontidae. In: Prothero DR, Schoch RM (eds) The evolution of perissodactyls. Oxford University Press, New York, pp 341–354

Wall WP, Manning E (1986) Rostriamynodon grangeri n. gen., n. sp. of amynodontid (Perissodactyla, Rhinocerotoidae) with comments on the phylogenetic history of Eocene Amynodontidae. J Paleontol 60:911–919

Wallace AR (1876b) The geographical distribution of animals. Harper & Brothers, New York

Wang X, Flynn LJ, Fortelius M (2013) Toward a continental Asian biostratigraphic and geochronologic framework. In: Wang X, Flynn LJ, Fortelius M (ed) Fossil Mammals of Asia: Neogene biostratigraphy and chronology. Columbia University Press, New York, pp 1–28

Webb SD, Taylor BE (1980) The phylogeny of hornless ruminants and a description of the cranium of Archaeomeryx. Bull Amer Mus Nat Hist 167:117–158

Wei D (1987) Miocene mammalian fauna of Xiaolongtan, Kaiyuan, Yunnan province. Vertebrata PalAsiatica 25:116–123

Welcomme J-L, Benammi M, Crochet J-Y et al (2001) Himalayan Forelands: Palaeontological evidence for Oligocene detrital deposits in the Bugti Hills (Balochistan, Pakistan). Geol Mag 138:397–405

West RM (1980) A minute new species of Dorcatherium (Tragulidae: Mammalia) from the Chinji Formation near Daud Khel, Mianwali District, Pakistan. Contribut Biol Geol 33:1–6

Whybrow PJ, Hill A (1999) Fossil Vertebrates of Arabia, with emphasis on the late Miocene Faunas, geology, and palaeoenvironments of the Emirate of Abu Dhabi, United Arab Emirates. Yale University Press, New Haven

Woodruff DS (2010) Biogeography and conservation in Southeast Asia: how 2.7 million years of repeated environmental fluctuations affect today's patterns and the future of the remaining refugial-phase biodiversity. Biodivers Conserv 19:919–941

Xia K, Su T, Liu Y-SC et al (2009) Quantitative climate reconstructions of the late Miocene Xiaolongtan megaflora from Yunnan, southwest China. Palaeogeogr Palaeoclimatol Palaeoecol 276:80–86

Xie Y, Li DM, Mackinnon J (2002) Preliminary researches on bio-geographical divisions of China. Acta Ecologica Sinica 22:1599–1615

Yao Y-F, Bruch AA, Mosbrugger V et al (2011) Quantitative reconstruction of Miocene climate patterns and evolution in Southern China based on plant fossils. Palaeogeogr Palaeoclimatol Palaeoecol 304:291–307

Yin A (2010) Cenozoic tectonic evolution of Asia: a preliminary synthesis. Tectonophysics 488:293–325

You Y, Huber M, Müller RD et al (2009) Simulation of the Middle Miocene climate optimum. Geophys Res Lett 36:L04702. doi:10.1029/2008GL036571

Zachos J, Pagani M, Sloan L et al (2001) Trends, rhythms, and aberrations in global climate 65 Ma to present. Science 292:686–693

Zazzo A, Mariotti A, Lécuyer C et al (2002) Intra-tooth isotope variations in late Miocene bovid enamel from Afghanistan: paleobiological, taphonomic, and climatic implications. Palaeogeogr Palaeoclimatol Palaeoecol 186:145–161

Zhang Z, Wang H, Guo Z et al (2007) What triggers the transition of palaeoenvironmental patterns in China, the Tibetan Plateau uplift or the Paratethys Sea retreat? Palaeogeogr Palaeoclimatol Palaeoecol 245:317–331

Zheng H, Clift PD, Wang P et al (2013) Pre-Miocene birth of the Yangtze River. Proc Natl Acad Sci 110:7556–7561

Zhuang G, Najman Y, Guillot S, Roddaz M, Antoine P-O, Métais G, Carter A, Marivaux L, Solangi SH (2015) Constraints on the collision and the pre-collision tectonic configuration between India and Asia from detrital geochronology, thermochronology, and geochemistry studies in the lower Indus basin, Pakistan. Earth Planet Sci Lett 432:363–373

Zin-Maung-Maung-Thein Takai M, Tsubamoto T et al (2008) A new species of Dicerorhinus (Rhinocerotidae) from the Plio-Pleistocene of Myanmar. Palaeontology 51:1419–1433

Zin-Maung-Maung-Thein Takai M, Tsubamoto T et al (2010) A review of fossil rhinoceroses from the Neogene of Myanmar with description of new specimens from the Irrawaddy Sediments. J Asian Earth Sci 37:154–165

Zin-Maung-Maung-Thein Takai M, Uno H et al (2011) Stable isotope analysis of the tooth enamel of Chaingzauk mammalian fauna (late Neogene, Myanmar) and its implication to paleoenvironment and paleogeography. Palaeogeogr Palaeoclimatol Palaeoecol 300:11–22

Zu Dehm R, Oettingen-Spielberg T (1958) Paläontologische und geologische Untersuchungen im Tertiär von Pakistan. 2. Die mitteleocänen Saügetiere von Ganda Kas bei Basal in Nordwest-Pakistan. Abhandlungen der Bayerischen Akademie der Wissenschaften 91:1–54

Chapter 3
Species Richness and Size Distribution of Large Herbivores in the Himalaya

Charudutt Mishra, Yash Veer Bhatnagar and Kulbhushansingh R. Suryawanshi

Abstract Species diversity across several taxa ranging from plants to vertebrates is reported to decrease with altitude, or to show a mid-elevation peak in mountain systems. Plant biomass availability for herbivores is similarly expected to decline with altitude as temperature becomes limiting. However, the relationship between herbivore species richness and altitude has not been examined in detail. We show that while the overall regional pattern (gamma-richness) for 25 large-herbivore species (56 % grazers, 44 % browsers/mixed feeders) in the Western Himalayas shows a mid-elevation peak, the species richness of grazers increases nearly monotonically with altitude peaking at 4000–5000 m. Median body mass of herbivores decreased with altitude, suggesting greater suitability of higher elevations for smaller bodied herbivores. We propose that seasonal altitudinal migration patterns, biogeographic influences, increases in the abundance of graminoids, and an increase in plant nutrients with altitude might explain the unusual high grazer species richness at higher altitudes in the Himalayan Mountains.

Keywords Gamma-richness · Ungulate diversity patterns · Mid-domain effect · Herbivore species richness · Grazer diversity

Large-mammalian herbivores occupy half of the Earth's total land surface area and play significant ecological roles as modifiers of their landscapes, as dominant drivers of ecosystem function and as prey for large carnivores (Olff et al. 2002; Prins and Gordon 2008). Their economic importance is reflected in the fact that over

---

C. Mishra (✉) · Y.V. Bhatnagar · K.R. Suryawanshi
The Snow Leopard Trust, 4649 Sunnyside Ave N, Suite 325, Seattle, WA 98103, USA
e-mail: charu@ncf-india.org

Y.V. Bhatnagar
e-mail: yash@ncf-india.org

K.R. Suryawanshi
e-mail: kulbhushan@ncf-india.org

C. Mishra · Y.V. Bhatnagar · K.R. Suryawanshi
Nature Conservation Foundation, 3076/5, IV Cross, Gokulam Park, 570002 Mysore, India

© Springer Science+Business Media Dordrecht 2016
F.S. Ahrestani and M. Sankaran (eds.), *The Ecology of Large Herbivores in South and Southeast Asia*, Ecological Studies 225,
DOI 10.1007/978-94-017-7570-0_3

one-fourth of the Earth's total land surface area is currently under managed grazing (Asner et al. 2004). Yet, the diversity of large wild herbivores across the world is threatened by human activities including anthropogenic habitat modification, hunting, and the competitive and other negative impacts of domestic herbivores (Prins and Gordon 2008, Kinnaird and O'brien 2012; Berger et al. 2013, also see Chaps. 1 and 11). Understanding the patterns and mechanisms underlying the diversity of large wild herbivores, therefore, has been a preoccupation of ecologists for almost 50 years now.

Analyses of large-herbivore species richness patterns traditionally focused on resource partitioning mechanisms that allow species of different body masses and adaptive syndromes (ruminants and hindgut fermenters, grazers and browsers; Clauss et al. 2008) to exploit different plant species or plant parts, thereby enabling niche differentiation and higher species packing (Jarman 1974; Jarman and Sinclair 1979; Vesey-Fitzgerald 1960). Based on patterns of body-mass structuring of large-herbivore assemblages, species interactions (competition and facilitation) were proposed to be the important underlying mechanisms for such niche differentiation and species richness (Prins and Olff 1998; Du Toit and Olff 2014). Gradients in moisture and nutrients in the soil available to plants were proposed as being the fundamental determinants of large-herbivore richness, with areas of intermediate moisture and high soil nutrients predicted to hold the highest potential herbivore richness globally (Olff et al. 2002).

In mountain systems, across a variety of taxa from plants to vertebrates, species diversity is generally a linearly decreasing function of altitude, or approximates a quadratic function with a mid-elevation peak (MacArthur 1972; Rahbek 2005, 1995; Stevens 1992; Namgail et al. 2012; Kraft et al. 2011; Tang et al. 2012; Presley et al. 2012). Temperature acts as an important limiting factor for plant growth at high elevations by reducing growing season length (Körner 2000; Mishra 2001; Hoch and Körner 2012). Consequently, for herbivores, plant biomass availability is expected to decline with altitude, and indeed, some of the lowest reported estimates of peak-season plant biomass come from ecosystems at high elevations (Mishra 2001). This pattern of reduced plant availability due to low temperature is further intensified in tropical mountains, where aridity additionally increases with altitude, particularly at higher elevations, despite reduced temperatures (Leuschner 2000). A hump shaped relationship between plant biomass and altitude has also been reported recently in the Western Trans-Himalayas (Namgail et al. 2012). However, the relationship between large-herbivore richness and altitude has remained unexplored.

We explored patterns of gamma-diversity of large herbivores in the highest mountain chain on earth, the Himalaya. Within vertical elevation bands of 1000 m in a 2.5 million km^2 region of the Western Himalaya-Tibetan Plateau complex (Fig. 3.1), we specifically examined patterns in richness and body mass distribution of all terrestrial large-mammalian herbivores (>2 kg). We also explored the relationships between observed richness patterns and potential correlates including climatic (temperature, precipitation, potential evapotranspiration) and biotic (plant species richness) variables, area within each band, and the geometrical constraints

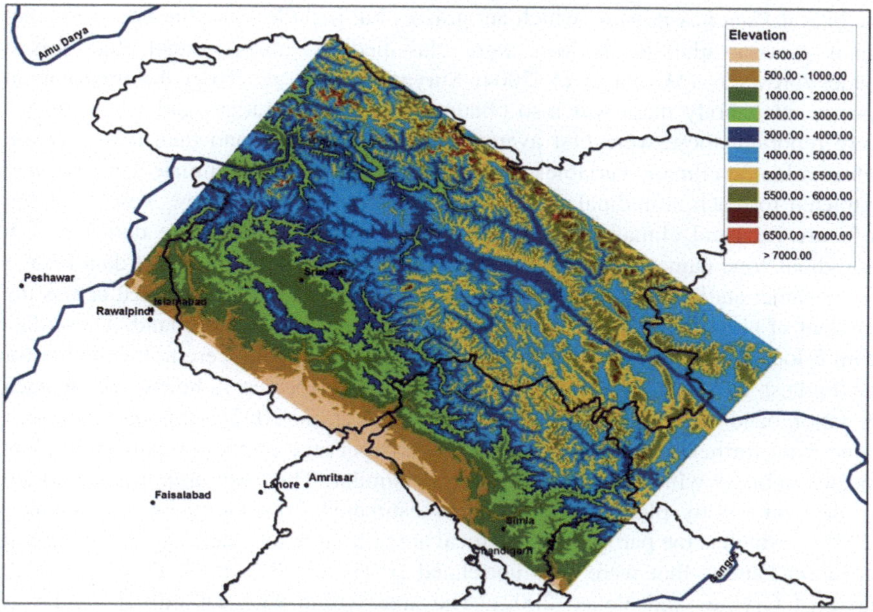

Fig. 3.1 Map of the Western Himalayan region showing the area under different elevation bands

of range-size distribution, commonly called the mid-domain effect (Colwell et al. 2004; Currie and Kerr 2008; Letten et al. 2013). We then separately examined the altitudinal richness patterns of grazers and browsers (including mixed feeders), since species belonging to these morpho-physiological types have different nutritional strategies (Clauss et al. 2008; Duncan and Poppi 2008), and can be expected to respond differently to elevation gradients. For instance, a greater occurrence of shrubs at relatively lower or mid-elevations could be expected to lead to greater browser richness compared to grazers.

We focused on the Western Trans-Himalayan region of India as the large-herbivore assemblage of this region is relatively well documented. From published records as well as our own observations and knowledge of the altitudinal range of the Western Himalayan large herbivores, we assigned presence/absence for each species within each of the 1000 m elevation bands. Information from only the Western Himalaya was used since the lower latitudes and higher precipitation of the Eastern Himalaya make the elevation bands incomparable. Further, the entire large-mammalian assemblage of the Eastern Himalayan region still remains poorly documented, and even within the last 10 years, several new ungulate records have been reported from the region (Datta et al. 2003; Mishra et al. 2006), and even a large primate species new to science has been described (Sinha et al. 2005).

The bands were deliberately chosen at a coarse resolution to minimize errors in this preliminary analysis. Each species was categorized as a grazer or a browser/mixed feeder based on information on species diets in the literature. Species such as the bharal *Pseudois nayaur*, which are grazers but include some browse in the diet when grass availability is low, were classified as grazers rather than mixed feeders/browsers (Mishra et al. 2004; Suryawanshi et al. 2010). Information on species adult body mass was also obtained from the literature, and where ranges were reported, these were first averaged within each sex, and then across sexes. Information on climatic variables was obtained from the FAO climate database, and extracted for each altitudinal zone of the study area using the new New_LocClim software for local climate estimation (http://www.fao.org/sd/). The data for each altitudinal zone came from 5 points (each point representing averaged data from 1 to 7 weather stations located within the study site or nearby) distributed across the gradient of altitude, latitude, and longitude within each elevation band. These data from 5 locations were averaged for each band. Climate data were not available for the highest elevation band (5000–6000 m). Plant species richness within each elevation band was estimated from Grytnes and Vetaas (2002). Although their data came from further east in the Himalayas, we expect the altitudinal pattern in plant species richness within our study area to be similar. This assumption is supported by data on woody plant species richness estimated from Oommen and Shanker (2005)—which came partly from our focal area (though not extending to the highest elevation band)—that were well correlated (Pearson's $R = 0.74$, $P = 0.07$) with those of Grytnes and Vetas (2002). The area within each altitudinal band was calculated based on a digital map of the region (Fig. 3.1). Predicted large-herbivore species richness along the elevation gradient due to geometrical constraints of range size was estimated through Monte Carlo simulations using the RangeModel software, with altitudinal ranges and mid-points for each species forming the input data (Colwell 2008; Colwell et al. 2004).

The regional species pool for the entire 2.5 million km^2 study area comprised 25 large-herbivore species (Table 3.1), with each elevation band holding a subset ranging from 28 to 56 % of the total pool. Fifty six percent of the species were grazers, the remaining 44 % being browsers/mixed feeders. Across this stretch of the Western Himalaya-Tibetan plateau complex (Fig. 3.1), the species richness of large herbivores was highest within the 3000–4000 m band (Fig. 3.2)—which also harbors the tree line—consistent with the commonly observed diversity pattern in mountains with a peak at mid-elevations (Rahbek 1995, 2005). Surprisingly, even within the 5000–6000 m band (5500 m being the limit to plant growth), where the growing season length is one–fifth of that at lower altitudes, species richness of large herbivores was still quite high (Fig. 3.2). In contrast, species richness was considerably lower at low elevations, i.e., 1000–2000 m band, where a significant number of species were lost when compared to the mid-elevation band of 3000–4000 m (Fig. 3.2). The observed altitudinal species richness pattern appeared to be correlated with climatic variables, particularly potential evapotranspiration and annual temperature, although the relationships were not statistically significant (Table 3.2). No relationship could be detected between overall herbivore species richness and plant species richness, the area within each elevation band, or the geometrical constraints imposed by range-size distribution (Table 3.2, Fig. 3.2).

Table 3.1 Large-herbivore species occurring in the 2.5 million km^2 Western Himalayan study site, their diet (G: grazer, B: browser or mixed feeder), and their occurrence (+ present, . absent) in 1000 m elevation bands across the altitudinal gradient (*A* <1000 m, *B* 1001–2000 m, *C* 2001–3000 m, *D* 3001–4000 m, *E* 4001–5000 m, and *F* 5001–6000 m)

Species	Mean body mass (kg)	Diet	Altitudinal zone in m above mean sea level					
			A	B	C	D	E	F
Elephas maximus	3000	B	+	.	.	.	.	.
Bos grunniens	413	G	.	.	.	+	+	+
Equus kiang	275	G	.	.	.	+	+	+
Cervus unicolor	250	B	+	+	+	+	.	.
Boselaphus tragocamelus	180	B	+	.	.	.	.	.
Cervus elaphus	154	B	.	+	+	+	.	.
Sus scrofa	125	B	+	+	+	.	.	.
Nemorhaedus sumatraensis	95	B	.	.	+	+	.	.
Ovis ammon	87	G	.	.	.	.	+	+
Capra sibirica	76	G	.	.	+	+	+	+
Hemitragus jemlahicus	75	G	.	.	+	+	.	.
Capra falconeri	71	B	.	.	+	+	.	.
Axis axis	70	G	+	+	.	.	.	.
Ovis vignei	60	G	.	.	+	+	+	.
Pseudois nayaur	55	G	.	.	+	+	+	+
Pantholops hodgsoni	32	G	.	.	.	.	+	+
Muntiacus muntjak	25	B	+	+	+	.	.	.
Nemorhaedus goral	25	G	+	+	+	.	.	.
Procapra picticaudata	14	B	.	.	.	.	+	+
Moschus chrysogaster	10	B	.	.	+	+	+	.
Marmota caudata	8.5	G	.	.	.	+	+	+
Marmota himalayana	6	G	.	.	.	.	+	+
Lepus capensis	4.5	G	.	.	.	+	+	+
Lepus oiostolus	3	B	.	.	.	+	+	+
Lepus nigricollis	2.1	G	+	+	+	.	.	.

The median body mass of the species assemblage decreased with altitude, the decline being particularly sharp between the 3000–4000 and 4000–5000 m band, suggesting greater suitability of higher elevations for smaller bodied herbivores (Fig. 3.3).

Further analyses revealed that the combined diversity of browsers and mixed feeders showed a mid-elevation peak in the 2000–3000 m band, with a steeper loss of species at higher altitudes when compared to grazers (Fig. 3.4). This pattern in browser diversity was positively correlated with plant species richness, but was also consistent with the predictions of the mid-domain effect (Table 3.2). Browser

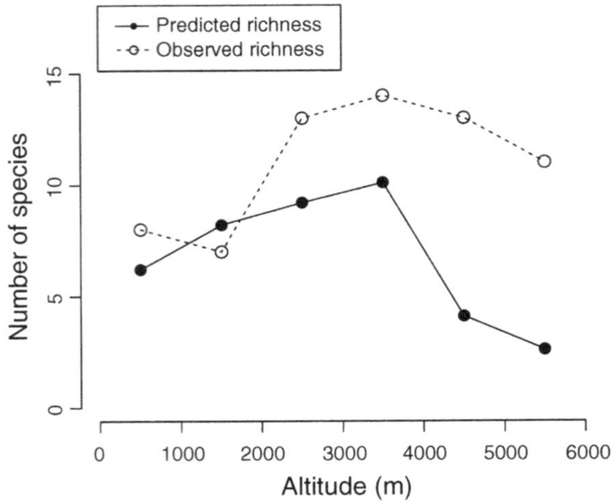

Fig. 3.2 Large-herbivore species richness within each altitudinal zone in the Western Himalaya (*open circles*). *Closed circles* depict the expected null distribution generated by mid-domain effects or geometrical constraints of range size

Table 3.2 Correlation coefficients (Pearson's R) of the relationship between large-herbivore species richness and independent climatic, geometric, and biotic variables across an elevation gradient in the Western Himalaya (see Fig. 3.1)

	All species	Grazers	Browsers
Mean temperature	−0.850	−0.975*	0.158
Precipitation	−0.561	−0.826	0.461
Potential evapotranspiration (PET)	−0.849	−0.819	−0.145
Mid-domain effect (MDE)	0.128	−0.425	0.884*
Log area	0.498	0.808	−0.532
Plant species richness	0	−0.554	0.893**

$^*P \leq 0.05$, $^{**}P \leq 0.01$

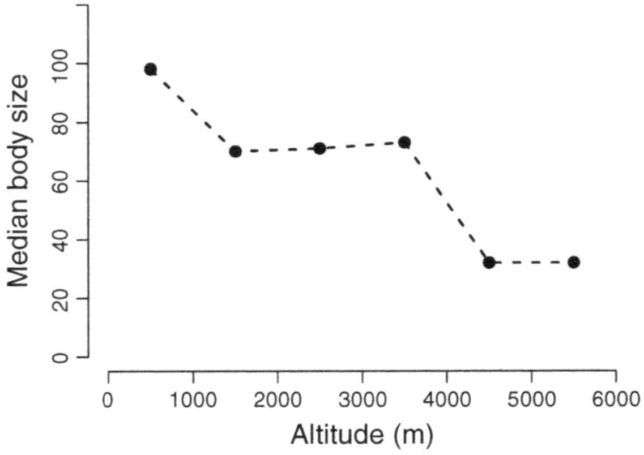

Fig. 3.3 Median body mass of the large-herbivore assemblage within 1000 m *vertical bands* across an elevation gradient in the Western Himalaya

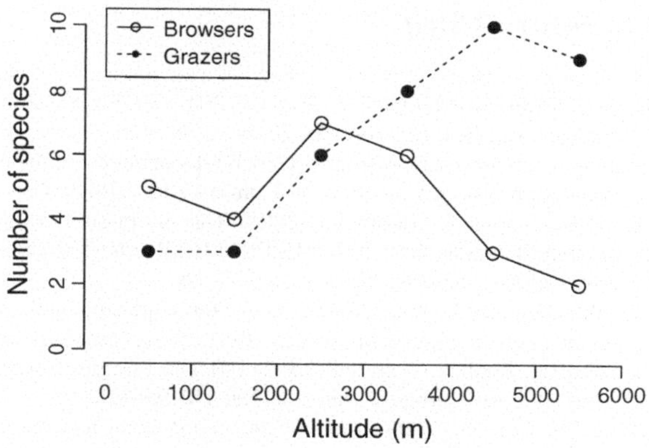

Fig. 3.4 Species richness of large-bodied grazers and browsers within 1000 m *vertical bands* across an elevation gradient in the Western Himalaya

species diversity was not correlated with climatic or biotic variables (Table 3.2). On the other hand, grazer species showed a sharp, near monotonic increase in diversity with elevation reaching their peak at 4000–5000 m (Fig. 3.4). The observed pattern in grazer diversity was negatively correlated with temperature (Table 3.2). Grazers did not show any relationship with plant species richness (Table 3.2). Even though we only compared grazer species diversity with overall plant species richness, we expect grass species richness to show a similar pattern with altitude as overall plant species richness (see Namgail et al. 2012).

The relatively high overall herbivore species richness at higher altitudes was surprising. Even more surprising and perhaps counterintuitive was the continuous increase in the grazer species richness with altitude. Temperature was the only significant correlate of grazer species richness. Seasonal altitudinal migration by herbivores of the high elevation bands could, presumably, be one mechanism allowing diversity at high altitudes. Variation in biogeographic histories, which we did not consider in our analysis, could be another explanation for the relatively high diversity at high altitudes.

We also propose that increase in plant nutrients with altitude may be an important mechanism of large-herbivore diversity in high mountain systems. It is empirically well established that the nutrient content of plants, especially the nitrogen content of leaves tends to increase with altitude independently of nutrient availability in the soil (Koerner 1989; Morecroft and Woodward 1996). This increase in nitrogen content has been linked to increased carboxylation efficiency at higher altitudes (Friend et al. 1989; Friend and Woodward 1990). This suggests that large-herbivore diversity in high mountains can be relatively greater at higher elevations than at mid-elevations, as long as there is sufficient plant biomass to support herbivore populations. Grazers and relatively smaller bodied herbivores are presumably better able to respond to increased plant nutrition at higher elevations as (i) climatic regimes may lead to a predominance of grassland systems at higher altitudes, especially above the tree line, and, (ii) the reduction in plant biomass with altitude limits species richness of larger bodied herbivores more than that of smaller bodied herbivores at higher elevations.

Literature Cited

Asner GP, Elmore AJ, Olander LP et al (2004) Grazing Systems, ecosystem eesponses, and global change. Ann Rev Environ Res 29:261–299

Berger J, Bayarbaatar B, Mishra C (2013) Globalization of the cashmere market and the decline of large mammals in Central Asia. Conserv Biol 27:679–689

Clauss M, Kaiser T, Hummel J (2008) The morphophysiological adaptations of browsing and grazing mammals. In: Gordon IJ, Prins HHT (eds) The ecology of large mammalian herbivore browsing and grazing. Springer, pp 47–88

Colwell RK (2008) RangeModel: tools for exploring and assessing geometric constraints on species richness (the mid-domain effect) along transects. Ecogr 31:4–7

Colwell RK, Rahbek C, Gotelli NJ (2004) The mid-domain effect and species richness patterns: what have we learned so far. Am Nat 163:E1–E23

Currie DJ, Kerr JT (2008) Tests of the mid-domain hypothesis: a review of the evidence. Ecol Monogr 78:3–18

Datta A, Pansa J, Madhusudan MD, Mishra C (2003) Discovery of the leaf deer (*Muntiacus putaoensis*) in Arunachal Pradesh: an addition to the large mammals of India. Curr Sci 84: 454–458

Duncan AJ, Poppi DP (2008) Nutritional ecology of grazing and browsing ruminants. Ecol Stu 195:89

Du Toit JT, Olff H (2014) Generalities in grazing and browsing ecology: using across-guild comparisons to control contingencies. Oecol 174:1–9

Friend AD, Woodward FI (1990) Evolutionary and ecophysiological responses of mountain plants to the growing season environment. Adv Ecol Res 20:59–124

Friend AD, Woodward FI, Switsur VR (1989) Field measurements of photosynthesis, stomatal conductance, leaf nitrogen and d 13 C along altitudinal gradients in Scotland. Funct Ecol 3:117–122

Grytnes JA, Vetaas OR (2002) Species richness and altitude: a comparison between null models and interpolated plant species richness along the Himalayan altitudinal gradient, Nepal. Am Nat 159:294–304

Günter H, Körner C (2012) Global patterns of mobile carbon stores in trees at the high-elevation tree line. Glob Ecol and Biogeogr 21:861–871

Jarman PJ (1974) The social organisation of antelope in relation to their ecology. Behav 58: 215–267

Jarman PJ, Sinclair ARE (1979) Feeding strategy and the pattern of resource partitioning in ungulates. In: Sinclair ARE, Norton-Griffiths N (eds) Serengeti: dynamics of an ecosystem. University of Chicago Press, Chicago, pp 130–163

Kinnaird MF, O'brien TG (2012) Effects of private land use, livestock management, and human tolerance on diversity, distribution, and abundance of large African mammals. Conserv Biol 26:1026–1039

Körner C (1989) The nutritional status of plants from high altitudes: a worldwide comparison. Oecol 81:379–391

Körner C (2000) Why are there global gradients in species richness? Mountains might hold the answer. TREE 15:513–514

Kraft NJB, Comita LS, Chase JM et al (2011) Disentangling the drivers of β diversity along latitudinal and elevational gradients. Science 333:1755–1758

Letten AD, Lyons SK, Moles AT (2013) The mid-domain effect: it's not just about space. J Biogeogr 40:2017–2019

Leuschner C (2000) Are high elevations in tropical mountains arid environments for plants? Ecol 81:1425–1436

MacArthur RH (1972) Geographical ecology: patterns in the distribution of species. Princeton University Press, Princeton

Mishra C (2001) High altitude survival: conflicts between pastoralism and wildlife in the Trans-Himalaya. Dissertation. Wageningen University, Wageningen, The Netherlands
Mishra C, Van Wieren SE, Ketner P et al (2004) Competition between domestic livestock and wild bharal *Pseudois nayaur* in the Indian Trans-Himalaya. J Appl Ecol 41:344–354
Mishra C, Datta A, Madhusudan MD (2006) Record of the Chinese goral Naemorhedus caudatus in Arunachal Pradesh. J Bom Nat Hist Soc 102:225–227
Morecroft MD, Woodward F (1996) Experiments on the causes of altitudinal differences in the leaf nutrient contents, size and d13C of Alchemilla alpina. New Phytol 134:471–479
Namgail T, Rawat GS, Mishra C, Van Wieren SE et al (2012) Biomass and diversity of dry alpine plant communities along altitudinal gradients in the Himalayas. J Plan Res 125:93–101
Olff H, Ritchie ME, Prins HHT (2002) Global environmental determinants of diversity in large herbivores. Nature 415:901–905
Oommen MA, Shanker K (2005) Elevational species richness patterns emerge from multiple local mechanisms in Himalayan woody plants. Ecol 86:3039–3047
Presley SJ, Cisneros LM, Patterson BD et al (2012) Vertebrate metacommunity structure along an extensive elevational gradient in the tropics: a comparison of bats, rodents and birds.". Glob Ecol Biogeogr 21:968–976
Prins HHT, Gordon IJ (2008) Introduction: grazers and browsers in a changing world. In: Gordon IJ, Prins HHT (eds) The ecology of large mammalian herbivore browsing and grazing. Springer: Ecological Studies 195:1
Prins HHT, Olff H (1998) Species richness of African grazer assemblages: towards a functional explanation. In: Newberry DM, Prins HHT, Brown ND (eds) Dynamics of tropical communities. Blackwell Science, London, UK, pp 449–490
Rahbek C (1995) The elevational gradient of species richness: a uniform pattern? Ecogr 18: 200–205
Rahbek C (2005) The role of and the perception of large-scale species-richness patterns. Ecol Lett 8:224–239
Sinha A, Datta A, Madhusudan MD et al (2005) *Macaca munzala*: a new species from western Arunachal Pradesh, northeastern India. Int J Primatol 26:977–989
Stevens GC (1992) The elevational gradient in altitudinal range: an extension of Rapoport's latitudinal rule to altitude. Am Nat 140:893
Suryawanshi KR, Bhatnagar YV, Mishra C (2010) Why should a grazer browse? Livestock impact on winter resource use by bharal *Pseudois nayaur*. Oecol 162:453–462
Tang Z, Fang J, Chi X et al (2012) Patterns of plant beta-diversity along elevational and latitudinal gradients in mountain forests of. Ecogr 35:1083–1091
Vesey-Fitzgerald DF (1960) Grazing succession among East African game animals. J Mammal 41:161–172

Chapter 4
Grazing and Browsing by Large Herbivores in South and Southeast Asia

Farshid S. Ahrestani, Ignas M.A. Heitkönig, Hisashi Matsubayashi and Herbert H.T. Prins

Abstract Understanding a species' foraging habits and preferences is fundamental to understanding its overall ecology and essential for its management and conservation. In general, large herbivores are classified as either grazers, browsers, or mixed feeders, and a species' diet preference is related to its body mass and digestive trait syndrome. Here, we analyze feeding strategies of large herbivores in South and Southeast Asia (SSEA) as related to their body mass and digestive trait syndromes. Overall, our results are similar to patterns observed on other continents. The majority of large herbivore species in SSEA are mixed feeders. Browsers and frugivores dominate the smallest body mass classes, while bulk feeders, predominantly grazers, dominate the largest body mass classes. There is an absence of hindgut fermenters in the lower body mass classes, and an absence of ruminants in the megaherbivore class (>1000 kg). Cervids in SSEA do not get as large as Bovids, and in both Cervids and Bovids the greatest number of species are found in the smaller body mass classes. Although large herbivores in SSEA occur across a wide range of different habitat types, there are discernible habitat associations with different groups of species. While this chapter sheds light on this important facet of large herbivore ecology in the region, there remains an acute lack of data on the foraging ecology of the majority of species in SSEA.

F.S. Ahrestani (✉)
Department of Ecology, Evolution and Environmental Biology,
Columbia University, New York, NY 10027, USA
e-mail: farshid.ahrestani@gmail.com

I.M.A. Heitkönig · H.H.T. Prins
Resource Ecology Group, Wageningen University,
6708 PB Wageningen, The Netherlands
e-mail: ignas.heitkonig@wur.nl

H.H.T. Prins
e-mail: herbert.prins@wur.nl

H. Matsubayashi
Department of Forest Science, Tokyo University of Agriculture, Setagaya, Japan
e-mail: matsubayashi.hisashi@gmail.com

© Springer Science+Business Media Dordrecht 2016

F.S. Ahrestani and M. Sankaran (eds.), *The Ecology of Large Herbivores in South and Southeast Asia*, Ecological Studies 225,
DOI 10.1007/978-94-017-7570-0_4

Keywords Body mass relationships · Digestive physiology · Forage preference · Hindgut fermenters · Ruminants · Vegetation quality

4.1 Introduction

Understanding a species' food requirements and resource acquisition strategies is fundamental to understanding its ecology. Based on their preference for the two major plant types, monocots (grasses and sedges) and dicots (forbs, shrubs, and trees), large herbivores have historically been classified as either (a) grazers (i.e., bulk and roughage feeders), (b) browsers (selectors of concentrated dicotyledonous herbage), or (c) intermediate or mixed feeders that consume either grass or browse at different seasons or in different areas (Hofmann and Stewart 1972; Hofmann 1973, 1989; Duncan and Poppi 2008). Grass and browse represent very different food resources and their consumption poses different constraints for herbivores (Duncan and Poppi 2008). Browse species, when compared to grasses, have higher levels of soluble cell content and nitrogen that are beneficial for large herbivores, but at the same time have higher levels of lignin and secondary metabolites that are detrimental to large herbivores (Demment and Van Soest 1985; Gordon and Illius 1994; Gordon 2003; Duncan and Poppi 2008). These fundamental differences between grasses and browse have led to different adaptations in animals specializing on one or the other plant type, with implications for all aspects of their ecology and life history (Hofmann and Stewart 1972; Hofmann 1973, 1989; Gordon 2003; Duncan and Poppi 2008).

Based on differences found in the relationships between stomach structure and feeding habits of East African ruminants, Hofmann et al. proposed that a species' digestive trait syndrome was the primary factor deciding forage selection and feeding habits of large herbivores (Hofmann and Stewart 1972; Hofmann 1973, 1989). However, subsequent analyses that statistically accounted for differences in body mass among species found limited evidence for morphological and anatomical differences between large herbivores belonging to different feeding categories (Gordon and Illius 1994; Gordon 2003). Thus, while additional studies have found evidence to support Hofmann's proposition (Clauss et al. 2003, 2008, 2010; Hofmann et al. 2008), it is now widely accepted that body mass plays a critical role in the resource ecology of large herbivores (Prins and Olff 1998).

Hofmann and Stewart (1972) originally proposed their foraging categories based on differences between only ruminant species, which was not surprising considering that 92 % of the ~260 large herbivore species worldwide are ruminants. All ruminants belong to the order Artiodactyla (even-toed ungulates) and the remaining 8 % of large herbivores are hindgut fermenters that are confined to the mammalian orders Perissodactyla (odd-toed ungulates) and Proboscidea (elephants). Ruminants are foregut fermenters, i.e., they derive the bulk of their energy from the forage they consume before passage through the gut, primarily by the action of symbiotic microbes on plant material during a fermentation process in the rumen, or a similar morphological structure as in hippos or camelids (Van Soest 1994). In hindgut fermentation, though energy is first derived when food passes through the stomach, additional energy is derived from fermentation of food in the hindgut, primarily the caecum.

The dominance of the large herbivore guild by ruminants is the result of a ruminant radiation that began ~10 Ma during the Miocene, which coincided with the expansion and diversification of grasslands worldwide (Perez-Barberia et al. 2001; Prins and Gordon 2008; Bouchenak-Khelladi et al. 2009). The first large herbivores date back to the early Cenozoic (~55 Ma), a time when grasslands were not common, which suggests that browsing was the more primitive of the two major large herbivore diets (Bodmer and Ward 2006). However, when grasslands began expanding across the globe around 10 Ma, especially in the tropics, the first specialized grazers—which were primarily ruminants—emerged (Cerling et al. 1993, 1997; Bouchenak-Khelladi et al. 2009). An increasing speciation of ruminants continued to match the expansion of grasslands worldwide, which in turn was being fuelled by the emergence of an alternate to the prevailing C_3 photosynthetic pathway in plants—the C_4 photosynthetic pathway, which enabled plants to function at lower CO_2 levels (Janis 2008; Damuth and Janis 2011).

Although ruminants dominate the large herbivore guild and many ruminants are grazers, the majority of the smallest ruminants primarily subsist as browsers and/or frugivores. Ruminants, in contrast to hindgut fermenters, are morpho-physiologically limited in their daily intake due to their longer digestive retention times and passage rates (Beekman and Prins 1989; Clauss et al. 2003, 2010). The longer retention times help maximize the energy extracted by the microbial community in the gut from low quality forage, often grasses. These longer retention times, however, are better tolerated by larger rather than smaller bodied species. The innate capacity for larger bodied animals to tolerate forage of lower quality is a function of the mismatch between the nonlinear relationship of metabolic requirements (MR) and body mass (BM), $MR = 70\ BM^{0.75}$, and the near-linear relationship of gut capacity (GC) and body mass, $\log GC = 1.03 \log BM - 0.94$ (Demment and Van Soest 1985). Therefore, compared to larger species, ruminants of smaller body mass are more dependent on higher quality forage, and to satisfy their nutritional requirements, smaller herbivores forage on browse and fruits that are of higher quality more than they forage on grasses (Bell 1970, 1971; Jarman 1974).

Large herbivores also have to contend with differences in the physical demands of feeding on grasses versus browse. Grasses in general are more fibrous, i.e., they have a greater proportion of cell wall (cellulose and hemicellulose) to cell content, and have higher levels of abrasive components like silica (Kaiser et al. 2009). Therefore when compared to browsers, grazers tend to have bigger masseter muscles, larger and deeper angles of the jaw, broader muzzles, and longer masseteric fossa on the skull (Gordon and Illius 1988; Clauss et al. 2008; Hofmann et al. 2008; Janis et al. 2010).

Building on what we know from studies of large herbivores in Europe, North America, and Africa, this chapter explores patterns in the feeding preferences of different large herbivore species in South and Southeast Asia (SSEA). We classified species as *grazers* if they consumed primarily graminoids year-round, *browsers* if they consumed primarily non-graminoids year-round, and *mixed feeders* if they either switched between grazing and browsing or consumed both browse and grasses. The wide range in body mass of species in the large herbivore assemblage in SSEA, which is similar to the body mass range observed in the large herbivore assemblage in Africa, allows us to analyze broad trends in foraging preferences as related to body mass and digestive syndromes.

One significant difference between the herbivore assemblages found in SSEA and Africa is that while there is only one Cervid (deer) species within the assemblage of 94 large herbivore species found in Africa (the Atlas deer *Cervus elaphas barbarus* found in the Atlas mountains in Africa's extreme Northwest), ~30 % of large herbivore species that inhabit SSEA are Cervid species. It is, however, unclear how Bovids and Cervids differ in their digestive physiology and therefore, species of both groups were grouped together for our analysis (Langvatn and Hanley 1993; Hobbs et al. 1996; Mysterud 2000; Seagle 2003; Hewison et al. 2009; Beguin et al. 2011; Redjadj et al. 2014). Although pig species (Suidae) are sometimes included in analyses related to large herbivore forage selection, they are indeed omnivores and were therefore excluded from this study (Rerat 1978). Excluding the 13 suid species, we were left with 70 large herbivore species to analyze: six belonging to the order Perissodactyla, one belonging to the order Proboscidea (the Asian elephant), and the remaining 63 species (~90 %) are ruminants and belong to the order Artiodactyla.

4.2 Forage Preference—Body Mass, Roughage, and Vegetation Quality

Given the paucity of data on forage preferences of many large herbivore species in SSEA, we pooled data and classified species in related groups, primarily by genus (Table 4.1). In general, forage selection by large herbivores is guided by the body mass of a species, its anatomical and morphological specializations, and the quantity, quality (mainly nitrogen and phosphorus), and the composition (level of roughage) of plant material (Bell 1971; Jarman 1974; Gordon and Illius 1996). In addition, the ratio of the abundance of grasses and browse at local scales influences the composition of a large herbivore assemblage found in a given area (Gordon 2003; Prins and Van Langevelde 2008; Owen-Smith 2014). In the savannas of Africa, for example, where browse is generally patchily distributed among vast grasslands, grazers dominate the large herbivore assemblage (Mcnaughton and Georgiadis 1986). In contrast, in tropical Asia where grasslands are patchily distributed within forested landscapes, mixed feeders should dominate the large herbivore assemblage. As expected, the majority of large herbivore species in SSEA

Table 4.1 Feeding preferences of groups of large herbivore species in Asia listed by order, family, and genus

Order, family	Species (common name)	Feeding type	Comments	References
Artiodactyla, Bovidae	*Antilope cervicapra* (blackbuck)	Grazer	The blackbuck is primarily a grazer throughout the year, even during periods when grasses have low nutrition. They appear to have evolved to occupy feeding niches in predominantly dry grassland landscapes	Jhala (1997), Isvaran (2007)
	Gazella bennettii (chinkara) *Gazella subgutturosa* (goitered gazelle)	Mixed feeders	These species are also found primarily in arid systems though the chinkara inhabits areas that are lower in elevation, latitude, but higher in temperature than the goitered gazelle. While it is unclear what the proportion of browse and grasses is in the diets of these species, and whether these proportions change seasonally, we know that the small body mass of this species has low quantity, but high quality nutritional requirements. These species, therefore, need to select for high-quality forage, but their arid habitats constrain available forage choices. Studies of gazelles in Africa have shown that the Grant's gazelle is primarily a browser, while the Thomson's gazelle is primarily a grazer (Cerling et al. 2003)	*chinkara*: Bagchi et al. (2003a, b) *goitered gazelle*: Xu et al. (2008, 2012)

(continued)

Table 4.1 (continued)

Order, family	Species (common name)	Feeding type	Comments	References
	Bos gaurus (gaur) *Bos javanicus* (banteng) *Bos mutus* (yak) *Bos sauveli* (kouprey)	Grazers	Among all the ruminants that inhabit Asia, the four species that belong to the large body mass genus *Bos* best exemplify roughage feeders. The *Bos* species are endemic to Asia and the size of their rumens allows maximizing both fiber digestibility and the extraction of energy from a predominantly graminoid diet, especially when graminoids are phenologically old and dry (roughage). These species, however, do browse (for e.g., the gaur has been recorded feeding on numerous browse species including bark of trees), primarily during the dry season	*gaur*: Schaller (1967), Chetri (2003), Gad and Shyama (2011), Ahrestani et al. (2012) *banteng*: Matsubayashi et al. (2007) *yak & kouprey*: Data deficient
	Bubalus arnee (wild buffalo) *Bubalus depressicornis/quarlesi* (anoa) *Bubalus mindorensis* (tamaraw)	Grazers	The lack of field data on the wild buffalo means that the best information available for this species is based on the domestic buffalo, which is primarily a grazer. The wild buffalo has a body mass similar to species of the genus *Bos* and based on observations of domestic buffalo can be classified as a roughage feeder. The range of body mass among the three *Bubalus* species, however, is large (tamaraw: 200 kg–buffalo: 600 kg), and therefore the diets of the species would vary with respect to the proportion of browse in the diet, most probably increasing with decreasing body mass	Data deficient

(continued)

Table 4.1 (continued)

Order, family	Species (common name)	Feeding type	Comments	References
	Boselaphus tragocamelus (nilgai)	Mixed feeder	The nilgai, a species monotypic within its genus, is the world's oldest extant tropical bovid species and the largest antelope species in Asia. It has been reported to be a mixed feeder consistent with what its premaxillary shape suggests. Its diet varies seasonally and includes graminoids, woody vegetation, forbs, sedges, and fruits (for e.g., 34 plant species have been known to germinate from nilgai feces in Keoladeo National Park, India)	Bagchi et al. (2003a, b), Solanki and Naik (1998) Khan (1994), Leslie (2008)
	Tetracerus quadricronis (four-horned antelope)	Mixed feeder	Like the Nilgai, the four-horned antelope is another species monotypic within its genus. It is believed to be a concentrate feeder with a wide latitudinal range and occupies open habitats with low tree density. Its habitat is more forested than that occupied by Asia's gazelles and blackbuck, which hints at its mixed feeder status	Krishna et al. (2009), Leslie and Sharma (2009), Baskaran et al. (2011)
	Capricornis spp. (mainland, red, southern and Himalayan)	Mixed feeders	Although there is data on the diet of the Japanese serow, which appears to be primarily a browser, no data exist for any of serow species in SSEA. Based on their habitat preferences, which includes both forests and grasslands, it is likely that the SSEA serow species are mixed foragers that vary their feeding seasonally, switching from graminoids in the wet season to browse in the dry season	Data deficient. There exists, however, data from other parts of Asia: Ochiai (1999), Song et al. (2005)

(continued)

Table 4.1 (continued)

Order, family	Species (common name)	Feeding type	Comments	References
	Capra spp. (markhor, ibex) *Hemitragus spp.* (tahr spp.) *Naemorhedus spp.* (goral spp.) *Ovis ammon* (argali) *Pseudois nayaur* (bharal) *Budorcas taxicolor* (takin)	Mixed feeders	In general sheep and goat species are considered opportunistic feeders that feed a lot on graminoids, but also on forbs and woody plants. Among all the large herbivore species found in SSEA, these species inhabit the highest altitude regions, which are relatively arid. The extreme seasonal variation in climate and forage availability that these species experience is probably the main reason why they are generalists rather than specialist feeders. Although the takin has been found to be a generalist, it prefers to browse	*bharal*: Mishra et al. (2004), Suryawanshi et al. (2010) *argali*: Namgail et al. (2004), Singh et al. (2010), Wingard et al. (2011) *goral*: Fakhar-I-Abbas et al. (2008) *takin*: Schaller et al. (1986)
Artiodactyla, Cervidae	*Axis axis* (chital) *Axis porcinus* (hog deer) *Axis calamianensis* (Calamian deer) *Axis kuhlii* (Bawean deer)	Mixed feeders	These species primarily feed on grasses during the wet season and have been found to switch to browse during the dry season, presumably when levels of protein in graminoids are too low to support their nutritional needs. These species generally occupy open forested habitats, and are not found in wet forests with little to no grasses in the understory. Although these species graze a lot, they are more selective of which plant parts they consume when compared to the sympatric larger (Cervid) *Rusa* species	*chital*: Moe and Wegge (1997), Ahrestani et al. (2012) *hog deer*: Dhungel and Ogara (1991), Odden et al. (2005), Wegge et al. (2006)

(continued)

Table 4.1 (continued)

Order, family	Species (common name)	Feeding type	Comments	References
	Muntiacus spp. (muntjac spp.)	Mixed feeders	Although these species forage on both grasses and browse, they do so selectively. They are concentrate feeders that select primarily leafy components, often fruit, and ingest very little fibrous forage	Data deficient
	Recervus duvaucelii (swamp deer) *Recervus eldii* (Eld's deer)	Mixed feeders	Both these species are associated with swampy habitat where they feed primarily on grasses, aquatic shrubs, forbs, and sedges	Schaller (1967)
	Rusa unicolor (sambar) *Rusa timorensis* (rusa) *Rusa alfredi|marianna* (Philippine sambar spp.)	Mixed feeders	Occupying a diverse set of habitats, the sambar is the most widely distributed large herbivore species in Asia. Its wide range hints at a generalist feeding strategy, which recent studies seem to suggest is the case. The *Rusa* species are often found in dense forest, a habitat where grasses are normally not abundant and are patchily distributed. In such habitats, *Rusa* species have been found to be primarily browsers	*sambar*: Schaller (1967), Johnsingh and Sankar (1991), Padmalal et al. (2003), Bagchi et al. (2003a, b), Ahrestani et al. (2012)
Artiodactyla, Tragulidae	*Tragulus/Moschiola spp.* (chevrotain spp.)	Browsers	The concentrate feeder label would probably best describe the feeding habits of these species, as the small size of the rumens of these species demand that fibrous forage levels be kept to a bare minimum While not much is known about Asian species, the African chevrotain is understood to eat mainly fallen fruit	Matsubayashi et al. (2003)

(continued)

Table 4.1 (continued)

Order, family	Species (common name)	Feeding type	Comments	References
Artiodactyla, Moschidae	*Moschus spp.* (musk deer)	Mixed feeders	In general, little detailed information exists of these species. However, given that they occupy high altitude habitats that experience extreme seasonality, observations seem to suggest that musk deer are mixed feeders that forage primarily on grasses in alpine meadows in summer, and then more on shrubs, forbs, and lichen in winter	Data deficient
Perrisodactyla, Equidae	*Equus hemionus* (Indian wild ass) *Equus kiang* (kiang)	Grazers	These hindgut fermenters have digestive physiologies with low fermentation rates and digestive efficiencies that lead to short digestion passage times, which allow these species to have high consumption rates and deal with forage with high levels of roughage. Despite the lack of field data on these Asian species, data from their close relatives—the zebras in Africa—suggest that they should be grazers, which field observations seem to corroborate. Both the Asian species are found in semidesert or desert habitats, albeit at contrasting latitudes and altitudes. At present, it remains unknown if they satisfy their nutritional requirements by eating only grasses	Data deficient

(continued)

Table 4.1 (continued)

Order, family	Species (common name)	Feeding type	Comments	References
Perrisodactyla, Rhinocerotidae	*Rhinoceros unicornis* (Indian)	Grazer	Reflective of their feeding habits, the Javan and Sumatran species inhabit dense forest habitats, while the Indian rhino inhabits terrain that has extensive grasslands. Given that their ranges overlapped historically, it is possible that their forage preferences have resulted from resource partitioning as a means to avoid competition in the past	Dinerstein (1991), Pradhan et al. (2008)
	Rhinoceros sondaicus (Javan)	Browser		
	Dicerorhinus sumatrensis (Sumatran)	Browser		
Perrisodactyla, Tapiridae	*Tapirus indicus* (Asian tapir)	Mixed feeder	One of only four global tapir species, and the only one found in Asia, the Asian tapir is confined to dense tropical forests in SE Asia. Given its relatively high quantity forage requirements as a result of its large body mass, it is likely that the tapir is a generalist and a mixed feeder	Data deficient
Proboscidea, Elephantidae	*Elephas maximus* (Asian elephant)	Mixed feeder	Roughage feeders that forage on a wide variety of plant species, which maybe necessary to satisfy their high quantity requirements. Their hindgut fermentation trait syndroms provide short digestion passage rates, which enables them to tolerate high levels of roughage	Sukumar (1990, 1992), Matsubayashi et al. (2006), Pradhan et al. (2008), Ahrestani et al. (2012)

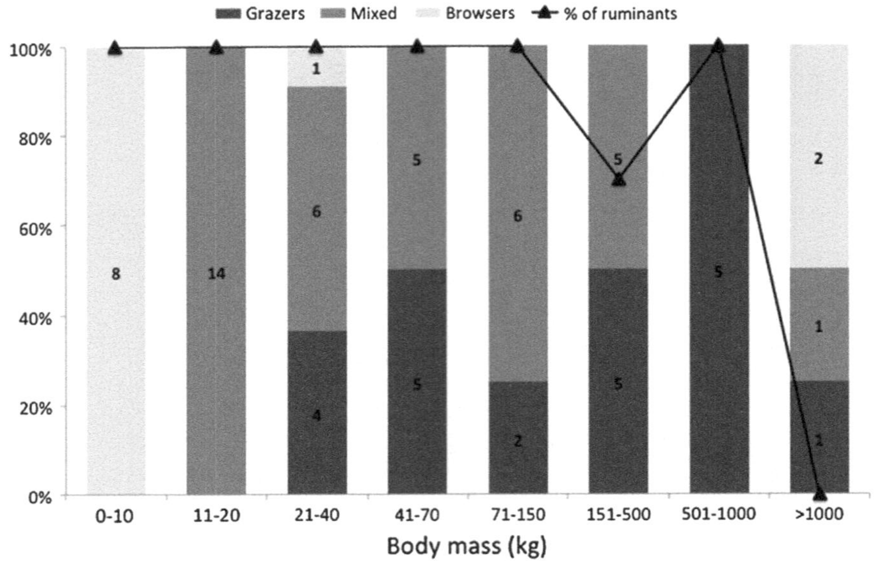

Fig. 4.1 Proportions of 70 large herbivore species found in South and Southeast Asia that are grazers (often bulk foragers), mixed feeders (that utilize both grasses and browse species), and browsers (often concentrate feeders) plotted in relation to body mass. The *line plots* the percentage of species that are ruminants in each body mass class. Numbers within *bars* denote the number of species of grazers, mixed feeders, or browsers in each body mass class. Although the 13 pig species found in the region are in the overall list of large herbivores for the region (see Chapter 1), they are indeed omnivores and were therefore excluded from this analysis

are mixed feeders (Fig. 4.1). However, concentrate feeders (browsers) dominate the smallest size categories, while bulk feeders, predominantly grazers, dominate the largest size categories (>500 kg, Fig. 4.1).

Studies of large herbivore assemblages in other regions have reported that browsers are found across the entire body size range of herbivores (Sponheimer et al. 2003). This is true for the assemblage in SSEA as well—ruminant browsers dominate the lowest body mass classes (Fig. 4.1), while the browsers in the largest body mass class are nonruminants (see Sect. 4.3 for more details). In SSEA, the smallest species are almost exclusively concentrate feeders and primarily frugivores (Fig. 4.1, Table 4.1) and typically occupy forested habitats. For example, the small muntjac and mouse deer species are selective with everything they feed on, selecting leaves and not stems when grazing, and selecting leaves, buds, and flowers when browsing (Barrette 1977). Mouse deer, classified as browsers, are probably the most selective of large herbivore species in SSEA, and represent a distinct frugi-folivore category (Bodmer 1990; Kingdon et al. 2013). They are similar in body mass to the smallest ruminant species found in Africa, the dik-diks, which are known to be concentrate feeders (Manser and Brotherton 1995; Dehority and Odenyo 2003; Kingdon et al. 2013). The tendency for smaller ruminants to be browsers has been previously noted (Case 1979; Gordon and Illius 1994; Clauss et al. 2003), and the increased reliance on high-quality fruits in the smallest bodied species is potentially a consequence of their relatively greater metabolic demands (Bodmer 1990).

The mesoherbivore body mass class (50–500 kg) was comprised primarily of mixed feeders (Fig. 4.1), including both Bovid and Cervid species. Although Fig. 4.1 depicts a low percentage of browsers in the intermediate body mass range, this can be deceiving as many of the mixed feeders in the intermediate body mass categories browse extensively, just not exclusively, like for example, the ubiquitous sambar *Rusa unicolor* (Padmalal et al. 2003; Ahrestani et al. 2012). However, other Cervids such as the chital *Axis axis* appear to adopt a mixed-feeding strategy. For example, chital *Axis axis* in South India were found to graze during the wet season, but then switched to browse during the middle-late dry season (Ahrestani et al. 2012)—the bottleneck period when rainfall is practically nonexistent, graminoid growth is dormant, and grasses have their highest levels of roughage and lowest levels of nutrients (Scoones 1995; Drescher et al. 2006). At present, however, it is unclear which factors regulate the extent to which different mesoherbivore species in SSEA rely on browse relative to graze, and how this changes with body mass, season, and across environmental and vegetation gradients.

While concentrate feeding is confined to the smaller animals, bulk feeding, primarily of grasses, is confined to larger bovid species that are ruminants and equid species that are hindgut fermenters (Fig. 4.1). In general, grazing is more common among species with larger body mass, and in Africa's large herbivore assemblage, grazing ruminants are larger than browsing ruminants (Bell 1971; Case 1979; Bodmer 1990; Van Wieren 1996; Perez-Barberia et al. 2001). It has been argued that the tendency for larger ruminants to be grazers rather than browsers is not a consequence of physiological or digestive limitations (Gordon and Illius 1994), but rather a result of limitations imposed by forage availability and abundance (Van Soest 1994; Clauss et al. 2003). Given their large metabolic requirements, it is easier for large-bodied species to meet their daily forage requirements by consuming grasses that are often more abundantly available than browse (Clauss et al. 2003). However, when browse is available in abundance, browsing ruminants can achieve large body sizes as in the case of moose *Alces alces* and giraffe *Giraffa camelopardalis* (Clauss et al. 2003).

The majority of large-bodied species in SSEA are wild Bovini species, i.e., species belonging to the genus *Bos,* which are similar to domestic cattle and are primarily grazers. Because of their low metabolic requirement to gut capacity ratio, these large Bovini species are able to tolerate high levels of roughage. However, wild Bovini species that inhabit forests, like the gaur and banteng, have been recorded browsing on multiple plant species (Schaller 1967). Gaur have also been observed eating the bark of trees during the dry season (Pasha et al. 2002), which could be related to the lack of nutrition in tropical deciduous forests during the dry season (Ahrestani et al. 2011). The many observations of Bovini species migrating to low-lying regions near rivers and drainage lines during the dry season seem to suggest that these wetter habitats—that typically retain a herbaceous layer in the dry season—act as key foraging habitats for species like the gaur in the dry season (Conry 1989; Ahrestani and Karanth 2014; Prins and Van Oeveren 2014).

4.3 Digestive Physiology Trait Syndromes: Hindgut versus Foregut

The distribution of foregut and hindgut fermenters across the body mass gradient of the large herbivore assemblage in SSEA (2–3070 kg) is illustrated in Fig. 4.1. There is a complete absence of hindgut fermenters in the lower body mass classes and a complete absence of ruminants in the megaherbivore class (>1000 kg). In general, the global megaherbivore guild is dominated by hindgut fermenters except for the giraffe *Giraffa camelopardalis* and hippopotami *Hippopotamus sp.* and it has been argued that the foregut digestive trait syndrome is not the most efficient for species with body mass >1000 kg (Demment and Van Soest 1985; Clauss et al. 2003).

Compared to ruminants, hindgut fermenters can tolerate forage of lower quality primarily because of lower retention times and faster passage rates in their digestive trait syndromes (Van Soest 1994). Their higher faster rates also allow hindgut fermenters to achieve larger body sizes than ruminants. All four megaherbivores in Asia (the Asian elephant and the Sumatran, Javan and Indian rhinoceros) are hindgut fermenters. The Asian elephant is a mixed feeder that consumes both browse and grass throughout the year, or follows a foraging regime that switches between dry season browsing and wet season grazing (Sukumar 1990, 1992). Despite being a mixed and bulk forager—a foraging strategy that is probably necessary to satisfy the metabolic requirements of over three tonnes of body mass—it is possible that the Asian elephant is selective in what it seasonally eats in some areas, similar to what the African elephant does in areas where nutrient levels vary significantly among forage species (Woolley et al. 2009; Owen-Smith and Chafota 2012; Pretorius et al. 2012).

Amongst the rhinocerotids, the Indian rhinoceros *Rhinoceros unicornis* (the Indian rhino) "is a semi-hypsodont, and is able to graze", while the closely related *R. sondaicus* (the Javan rhino) and the more distantly related *Dicerorhinus sumatrensis* (the Sumatran rhino) are brachydont browsers (Prothero et al. 1989). The two lineages (*Rhinoceros* and *Dicerorhinus*) have been present in SSEA since the Pliocene, and *R. sondaicus* appears to be a more recent species than *R. unicornis* (Prothero et al. 1989; De Iongh et al. 2005, also see Chap. 2). The contrasting feeding strategies of the Asian rhinoceros species bear similarities to their African counterparts. The two rhino species in Africa, though possessing similar hindgut digestive trait syndromes, have contrasting diets; the white rhino

Ceratotherium simum is more or less an exclusive grazer and the black rhino *Diceros bicornis* is more or less an exclusive browser (Owen-Smith 1992; Luske et al. 2009; Van Lieverloo et al. 2009). A similar scenario was also present in the North American Micocene ungulate assemblage (Prothero et al. 1989), which had one grazing rhino species sympatric with one browsing rhino species.

Only three hindgut fermenter species below 500 kg inhabit SSEA: two *Equus spp*. (Asiatic and the Tibetan wild ass) that are both grazers, and the Asian tapir that is considered a mixed feeder. Both *Equus* species inhabit dry habitats characterised by sparse and patchily distributed forage of poor quality, conditions that are better handled by hindgut rather than foregut fermentation trait syndromes (Clauss et al. 2003). In contrast to the dry habitats of the *Equus*, the Asian tapir, inhabits mainly moist forests and often rainforests (Clements et al. 2012; Table 4.1). In general, except for its frugivorous habits (Campos-Arceiz et al. 2012), little else is known about the diet of Asian tapir. The other three tapir species in the world are all found in Central and South America, where they often inhabit forests (Fragoso 1997; Galetti et al. 2001). In general, forests do not support high levels of graminoid production and the morphology of the tapir's snout suggests an evolutionary adaptation to frugivory and browsing (Janis 1984; Milewski and Dierenfeld 2013). Much of the research on the tapir species in Central and South America has focused on frugivory, which has shown that the diets of these species in forests, lowland river basins, and marshlands are extremely diverse and include more than 100 plant species (Salas and Fuller 1996; Lizcano and Cavelier 2000; Downer 2001; Tobler 2002; Chalukian et al. 2013).

4.4 Body Size and Habitat Associations

Of the ruminants, members of the Tragulidae and Moschidae are restricted to the smallest size classes, in contrast to the Cervids and Bovids that have larger body mass ranges (Fig. 4.2). Cervids in SSEA do not appear to get as large as Bovids (Fig. 4.2) and in both Cervids and Bovids, the greatest number of species are found in the smaller body mass classes (Fig. 4.2). The largest Cervid in Asia—the Kashmir stag, *Cervus elaphus hanglu*, a species that shares its genus and species lineage with the only deer species found in Africa, the barbary stag *Cervus elaphus barbarus*—has a body mass comparable to the body mass of the *Equus spp.* in Asia; all other Cervid species in Asia are smaller. The nonruminants of SSEA are much larger (>150 kg) with 4 of the 7 nonruminants falling within the megaherbivore category.

Although large herbivores in SSEA occupy a range of different habitats from arid grasslands to rainforests, there are discernible habitat associations within different groups. Tragulids and Moschids, which are predominantly browsers and frugivores, are restricted to forested habitats. Cervids and Bovids occur in a wide

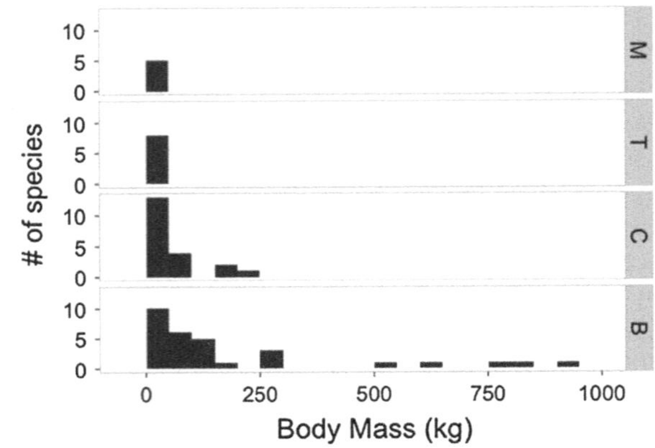

Fig. 4.2 The species of four families of the order Artiodactyla (*M* Moschidae (5 species), *T* Tragulidae (8 species), *C* Cervidae (20 species), *B* Bovidae (30 species)) found in South and Southeast Asia distributed by body mass

array of different habitat types, although the habitat ranges of Cervids tend to be more restricted than those of Bovides. In general, Cervids are nearly never found in extreme dry habitats, presumably because of their greater dependence on moisture when compared to Bovids. Among nonruminants, the Equids are restricted to arid habitats, the single Tapirid to forests and river basins, the Rhinocerotids to forests and grasslands, while the Asian elephant occurs in a range of habitats from tall grasslands to savannas to forests.

The only ruminant found in the dry and hot landscape that the Indian wild ass *Equus hemionus* inhabits, is the chinkara *Gazella benetti*. The chinkara, <20 kg, is smaller by an order of magnitude when compared to the wild ass, ~200 kg, which is a non-ruminant. Constraints imposed by the sparse and patchily distributed forage in these hot and dry habitats seem to allow only the smallest ruminants to persist in these inhospitable habitats. In contrast to the low number of species inhabiting extremely hot and dry habitats, the kiang *Equus kiang*, which inhabits dry and cold high altitude terrain that supports only sparse and patchily distributed forage in conditions of extreme seasonality, is sympatric with multiple ruminant species: mountain goat and sheep species that are generally mixed feeders, and the larger yak, a Bovini species that is considered a grazer. Cold and dry habitats in SSEA, therefore, support a more diverse ruminant assemblage when compared to hot and dry habitats.

4.5 Conclusions and Future Directions

The foraging preferences exhibited by large herbivore species in SSEA, especially at the extreme ends of the body mass gradient, suggest that body mass plays a role in shaping the forage preference of these species (Fig. 4.1). Additionally, the absence of ruminants from the megaherbivore class in this assemblage supports the proposition that the foregut trait syndrome does not commonly support body mass

>1000 kg (Van Soest 1994; Clauss et al. 2003). Existing data suggest that Cervid species in Asia are predominantly mixed feeders, some throughout the year and others seasonally, and that the level of concentrate feeding by deer species increases with decreasing body mass (Table 4.1). The majority of Bovid species <200 kg are mixed feeders too, and the largest Bovids (wild cattle >500 kg) are roughage feeders of primarily grasses (Table 4.1).

From Table 4.1, we see that there is an acute lack of data on the foraging ecology of the majority of large herbivore species in SSEA. The framework—based on interacting causal factors of body mass, digestive trait syndromes, and habitat heterogeneity—that helped understand patterns observed in the foraging by large herbivore assemblages in Africa was established nearly half a century ago (Bell 1970, 1971; Jarman 1974). That framework, however, has yet to be fully tested on a large herbivore assemblage in SSEA. In general, classifying large herbivores either as grazers, browsers, or mixed feeders helps us understand their foraging ecology. However, the seasonal variation in the levels of available nutrients, roughage, and quantity in plant material ultimately determines what a large herbivore consumes over space and time, and forage selection is directed by a complex set of drivers that does not allow for a distinct optimal prediction (Hanley 1982). In the context of prevailing global change, understanding these mechanisms has got a lot more challenging as bush encroachment further modifies grasslands (Mitchard and Flintrop 2013) and increased levels of CO_2 can potentially change the distribution of C_3 and C_4 grasses and woody vegetation worldwide (Chamaille-Jammes and Bond 2010). Finally, though natural history descriptions began in SSEA more than a century ago, it is imperative that ecological research increases in the region as SSEA seriously lags behind in understanding the fundamentals of its natural heritage.

References

Ahrestani FS, Karanth KU (2014) Gaur (Bos gaurus Smith CH, 1827). In: Melletti M, Burton J (eds) Ecology, evolution and behaviour of wild cattle. Cambridge University Press, Cambridge, pp 174–193

Ahrestani FS, Heitkonig IMA, Prins HHT (2011) Herbaceous production in South India-limiting factors and implications for large herbivores. Plant Soil 349:319–330

Ahrestani FS, Heitkonig IMA, Prins HHT (2012) Diet and habitat-niche relationships within an assemblage of large herbivores in a seasonal tropical forest. J Trop Ecol 28:385–394

Bagchi S, Goyal SP, Sankar K (2003a) Habitat separation among ungulates in dry tropical forests of Ranthambhore National Park Rajasthan. Trop Ecol 44:175–181

Bagchi S, Goyal SP, Sankar K (2003b) Niche relationships of an ungulate assemblage in a dry tropical forest. J Mammal 84:981–988

Barrette C (1977) Some aspects of the behaviour of muntjacs in Wilpattu National Park. Mammalia 41:1–34

Baskaran N, Kannan V, Thiyagesan K et al (2011) Behavioural ecology of four-horned antelope (Tetracerus quadricornis de Blainville, 1816) in the tropical forests of southern India. Mamm Biol 76:741–747

Beekman JH, Prins HHT (1989) Feeding strategies of sedentary large herbivores in East Africa, whith emphasis on the African buffalo, Syncerus caffer. Afr J Ecol 27:129–147
Beguin J, Pothier D, Cote SD (2011) Deer browsing and soil disturbance induce cascading effects on plant communities: a multilevel path analysis. Ecol Appl 21:439–451
Bell RHV (1970) The use of the herb layer by grazing ungulates in the Serengeti. In: Watson A (ed) Animal populations in relation to their food resources, tenth symposium of tile British ecological society. Blackwell Scientific Publications, Oxford, pp 111–123
Bell RHV (1971) Grazing ecosystem in serengeti. Sci Am 225:86
Bodmer RE (1990) Ungulate frugivores and the browser-grazer continuum. Oikos 319–325
Bodmer R, Ward D (2006) Frugivory in large mammalian herbivores. In: Danell K, Duncan P, Bergström R, Pastor J (eds) Large herbivore ecosystem dynamics and conservation. Cambridge University Press, Cambridge, pp 232–260
Bouchenak-Khelladi Y, Verboom GA, Hodkinson TR et al (2009) The origins and diversification of C-4 grasses and savanna-adapted ungulates. Glob Change Biol 15:2397–2417
Campos-Arceiz A, Traeholt C, Jaffar R et al (2012) Asian tapirs are no elephants when it comes to seed dispersal. Biotropica 44:220–227
Case TJ (1979) Optimal body size and an animal's diet. Acta Biotheor 28:54–69
Cerling TE, Wang Y, Quade J (1993) Expansion of C4 ecosystems as an indicator of global ecological change in the late Miocene. Nature 361:344–345
Cerling TE, Harris JM, Macfadden BJ et al (1997) Global vegetation change through the Miocene/Pliocene boundary. Nature 389:153–158
Cerling TE, Harris JM, Passey BH (2003) Diets of east african bovidae based on stable isotope analysis. J Mammal 84:456–470
Chalukian SC, Soledad De Bustos M, Leonidas Lizarraga R (2013) Diet of lowland tapir (Tapirus terrestris) in El Rey National Park, Salta, Argentina. Integr Zool 8:48–56
Chamaille-Jammes S, Bond WJ (2010) Will global change improve grazing quality of grasslands? A call for a deeper understanding of the effects of shifts from C-4 to C-3 grasses for large herbivores. Oikos 119:1857–1861
Chetri M (2003) Food habits of gaur (*Bos gaurus*) and livestock (cows and buffaloes) in Parsa Wildlife Sanctuary, central Nepal. Himalayan J Sci 1:31–36
Clauss M, Frey R, Kiefer B et al (2003) The maximum attainable body size of herbivorous mammals: morphophysiological constraints on foregut, and adaptations of hindgut fermenters. Oecologia 136:14–27
Clauss M, Hofmann RR, Streich WJ et al (2008) Higher masseter muscle mass in grazing than in browsing ruminants. Oecologia 157:377–385
Clauss M, Hofmann RR, Streich WJ et al (2010) Convergence in the macroscopic anatomy of the reticulum in wild ruminant species of different feeding types and a new resulting hypothesis on reticular function. J Zool 281:26–38
Clements GR, Rayan DM, Aziz SA et al (2012) Predicting the distribution of the Asian tapir in Peninsular Malaysia using maximum entropy modeling. Integr Zool 7:400–406
Conry PJ (1989) Gaur (*Bos gaurus*) and development in Malaysia. Biol Conserv 49:47–65
Damuth J, Janis CM (2011) On the relationship between hypsodonty and feeding ecology in ungulate mammals, and its utility in palaeoecology. Biol Rev 86:733–758
De Iongh HH, Prins HHT, Van Strien N et al (2005) Some observations on the presence of one-horned rhinos in the bas reliefs of the Ankor Wat temple complex. Pachyderm 96–100
Dehority BA, Odenyo AA (2003) Influence of diet on the rumen protozoal fauna of indigenous African wild ruminants. J Eukaryot Microbiol 50:220–223
Demment MW, Van Soest PJ (1985) A nutritional explanation for body-size patterns of ruminants and nonruminant herbivores. Am Nat 125:641–672
Dhungel SK, Ogara BW (1991) Ecology of the hog deer in Royal Chitwan National Park, Nepal. Wildlife Monographs 1–40
Dinerstein E (1991) Seed dispersal by greater one-horned rhinoceros (Rhinoceros unicornis) and the flora of rhinoceros latrines. Mammalia 55:355–362

Downner CC (2001) Observations on the diet and habitat of the mountain tapir (Tapirus pinchaque). J Zool 254:279–291

Drescher M, Heitkonig IMA, Van Den Brink PJ et al (2006) Effects of sward structure on herbivore foraging behaviour in a South African savanna: an investigation of the forage maturation hypothesis. Aust Ecol 31:76–87

Duncan AJ, Poppi DP (2008) Nutritional ecology of grazing and browsing ruminants. In: Gordon IJ, Prins HHT (eds) Ecology of browsing and grazing. Springer, Berlin, pp 89–116

Fakhar-I-Abbas F, Akhtar T, Mian A (2008) Food and feeding preferences of Himalayan gray goral (Naemorhedus goral bedfordi) in Pakistan and Azad Jammu and Kashmir. Zoo Biol 27:371–380

Fragoso JMV (1997) Tapir-generated seed shadows: scale-dependent patchiness in the Amazon rain forest. J Ecol 85:519–529

Gad SD, Shyama SK (2011) Diet composition and quality in Indian bison (*Bos gaurus*) based on fecal analysis. Zoolog Sci 28:264–267

Galetti M, Keuroghlian A, Hanada L et al (2001) Frugivory and seed dispersal by the lowland tapir (Tapirus terrestris) in southeast Brazil. Biotropica 33:723–726

Gordon IJ (2003) Browsing and grazing ruminants: are they different beasts? For Ecol Manage 181:13–21

Gordon IJ, Illius AW (1988) Incisor arcade structure and diet selection in ruminants. Funct Ecol 2:15–22

Gordon IJ, Illius AW (1994) The functional significance of the browser-grazer dichotomy in African ruminants. Oecologia 98:167–175

Gordon IJ, Illius AW (1996) The nutritional ecology of African ruminants: a reinterpretation. J Anim Ecol 65:18–28

Hanley TA (1982) The nutritional basis for food selection by ungulates. J Range Manag 35:146–151

Hewison AJM, Morellet N, Verheyden H et al (2009) Landscape fragmentation influences winter body mass of roe deer. Ecography 32:1062–1070

Hobbs NT, Baker DL, Bear GD et al (1996) Ungulate grazing in sagebrush grassland: mechanisms of resource competition. Ecol Appl 6:200–217

Hofmann RR, Stewart DRM (1972) Grazer and browser: a classification based on the stomach-structure and feeding habits of East African ruminants. Mammalia 36:226–240

Hofmann RR (1973) The ruminant stomach (stomach structure and feeding habits of East African game ruminants). East Afr Monogr Biol 2:1–364

Hofmann RR (1989) Evolutionary steps of ecophysiological adaptation and diversification of ruminants: a comparative view of their digestive system. Oecologia 78:443–457

Hofmann RR, Streich WJ, Fickel J et al (2008) Convergent evolution in feeding types: salivary gland mass differences in wild ruminant species. J Morphol 269:240–257

Isvaran K (2007) Intraspecific variation in group size in the blackbuck antelope: the roles of habitat structure and forage at different spatial scales. Oecologia 154:435–444

Janis C (1984) Tapirs as living fossils. In: Eldredge N, Stanley SM (eds) Living fossils. Springer, New York, pp 80–86

Janis C (2008) An evolutionary history of browsing and grazing ungulates. In: Gordon IJ, Prins HHT (eds) Ecology of browsing and grazing. Springer, Berlin, pp 21–45

Janis CM, Constable EC, Houpt KA et al (2010) Comparative ingestive mastication in domestic horses and cattle: a pilot investigation. J Anim Physiol Anim Nutr 94:E402–E409

Jarman PJ (1974) Social organization of antelope in relation to their ecology. Behaviour 48:215

Jhala YV (1997) Seasonal effects on the nutritional ecology of blackbuck Antelope cervicapra. J Appl Ecol 34:1348–1358

Johnsingh AJT, Sankar K (1991) Food plants of chital, sambar and cattle on Mundunthurai plateau, Tamil Nadu, South India. Mammalia 55:57–66

Kaiser TM, Brasch J, Castell JC et al (2009) Tooth wear in captive wild ruminant species differs from that of free-ranging conspecifics. Mamm Biol 74:425–437

Khan JA (1994) Food habits of ungulates in dry tropical forests of Gir Lion Sanctuary, Gujarat, India. Acta Theriol 39:185–193

Kingdon J, Happold D, Butynski T et al (2013) Mammals of Africa (Volume VI) A&C Black

Krishna YC, Clyne PJ, Krishnaswamy J et al (2009) Distributional and ecological review of the four horned antelope, Tetracerus quadricornis. Mammalia 73:1–6

Langvatn R, Hanley TA (1993) Feeding-patch choice by red deer in relation to foraging efficiency—an experiment. Oecologia 95:164–170

Leslie DM (2008) Boselaphus tragocamelus (Artiodactyla: Bovidae). Mamm Species 813:1–16

Leslie DM, Sharma K (2009) Tetracerus quadricornis (Artiodactyla: Bovidae). Mamm Species 843:1–11

Lizcano DJ, Cavelier J (2000) Daily and seasonal activity of the mountain tapir (Tapirus pinchaque) in the Central Andes of Colombia. J Zool 252:429–435

Luske BL, Mertens T, Lent PC et al (2009) Impact of the black rhinoceros (Diceros bicornis minor) on a local population of Euphorbia bothae in the Great Fish River Reserve, South Africa. Afr J Ecol 47:509–517

Manser MB, Brotherton PNM (1995) Environmental constraints of the foraging behavior of a dwarf antelope (Madoqua kirkii). Oecologia 102:404–412

Matsubayashi H, Bosi E, Kohshima S (2003) Activity and habitat use of lesser mouse-deer (Tragulus javanicus). J Mammal 84:234–242

Matsubayashi H, Lagan P, Sukor JRA (2006) Utilization of Macaranga trees by Asian elephants (Elephas maximus) in Borneo. Mammal Study 115–118

Matsubayashi H, Lagan P, Sukor JRA (2007) Herbal seed dispersal by the banteng (Bos javanicus) in a Bornean tropical rain forest. Malayan Nat J 59:297–303

Mcnaughton SJ, Georgiadis NJ (1986) Ecology of African grazing and browsing mammals. Annu Rev Ecol Syst 17:39–65

Milewski AV, Dierenfeld ES (2013) Structural and functional comparison of the proboscis between tapirs and other extant and extinct vertebrates. Integr Zool 8:84–94

Mishra C, Van Wieren SE, Ketner P et al (2004) Competition between domestic livestock and wild bharal Pseudois nayaur in the Indian Trans-Himalaya. J Appl Ecol 41:344–354

Mitchard ETA, Flintrop CM (2013) Woody encroachment and forest degradation in sub-Saharan Africa's woodlands and savannas 1982–2006. Philos Trans R Soc B-Biol Sci 368

Moe SR, Wegge P (1997) The effects of cutting and burning on grass quality and axis deer (Axis axis) use of grassland in lowland Nepal. J Trop Ecol 13:279–292

Mysterud A (2000) Diet overlap among ruminants in Fennoscandia. Oecologia 124:130–137

Namgail T, Fox JL, Bhatnagar YV (2004) Habitat segregation between sympatric Tibetan argali Ovis ammon hodgsoni and blue sheep Pseudois nayaur in the Indian Trans-Himalaya. J Zool 262:57–63

Ochiai K (1999) Diet of the Japanese serow (Capricornis crispus) on the Shimokita Peninsula, northern Japan, in reference to variations with a 16-years interval. Mammal Study 24:91–102

Odden M, Wegge P, Storaas T (2005) Hog deer Axis porcinus need threatened tallgrass floodplains: a study of habitat selection in lowland Nepal. Anim Conserv 8:99–104

Owen-Smith N (1992) Megaherbivores. Cambridge University Press, Cambridge

Owen-Smith N (2014) Spatial ecology of large herbivore populations. Ecography 37:416–430

Owen-Smith N, Chafota J (2012) Selective feeding by a megaherbivore, the African elephant (Loxodonta africana). J Mammal 93:698–705

Padmalal U, Takatsuki S, Jayasekara P (2003) Food habits of sambar Cervus unicolor at the Horton Plains National Park, Sri Lanka. Ecol Res 18:775–782

Pasha MKS, Areendran G, Sankar K et al (2002) Debarking of teak Tectona grandis Linn. f. by gaur Bos gaurus H. Smith during summer in a tropical dry deciduous habitat of central India. J Bombay Nat Hist Soc 99:238–244

Perez-Barberia FJ, Gordon IJ, Nores C (2001) Evolutionary transitions among feeding styles and habitats in ungulates. Evol Ecol Res 3:221–230

Pradhan NMB, Wegge P, Moe SR et al (2008) Feeding ecology of two endangered sympatric megaherbivores: Asian elephant Elephas maximus and greater one-horned rhinoceros Rhinoceros unicornis in lowland Nepal. Wildl Biol 14:147–154

Pretorius Y, Stigter JD, De Boer WF et al (2012) Diet selection of African elephant over time shows changing optimization currency. Oikos 121:2110–2120

Prins HHT, Gordon IJ (2008) Grazers and browsers in a changing world. In: Gordon IJ, Prins HHT (eds) Ecology of browsing and grazing. Springer, Berlin, pp 1–20

Prins HHT, Olff H (1998) Species-richness of African grazer assemblages: towards a functional explanation. In: Newbery DM, Prins HHT, Brown ND (eds) Dynamics of tropical communities. British ecological society symposium, vol 37. Blackwell Science, Oxford, pp 449–490

Prins HHT, Van Langevelde F (2008) Assembling a diet from different places. In: Prins HHT, Van Langevelde F (eds) Resource ecology: spatial and temporal dynamics of foraging. Springer, Dordrecht, pp 129–158

Prins HHT, Van Oeveren H (2014) Bovini as keystone species and landscape architects. In: Melletti M, Burton J (eds) Ecology, evolution and behaviour of wild cattle. Cambridge University Press, Cambridge, pp 21–29

Prothero DR, Guérin C, Manning E (1989) The history of the Rhinocerotoidea. In: Prothero DR, Schoch RM (eds) The evolution of perissodactyls. Oxford University Press, Oxford

Redjadj C, Darmon G, Maillard D et al (2014) Intra- and interspecific differences in diet quality and composition in a large herbivore community. Plos One 9

Rerat A (1978) Digestion and absorption of carbohydrates and nitrogenous matters in the hindgut of the omnivorous nonruminant animal. J Anim Sci 46:1808–1837

Salas LA, Fuller TK (1996) Diet of the lowland tapir (Tapirus terrestris L) in the Tabaro River valley, southern Venezuela. Can J Zool-Revue Canadienne De Zoologie 74:1444–1451

Schaller GB (1967) The Deer and the Tiger: a study of wildlife in India. University of Chicago Press, Chicago

Schaller GB, Qitao T, Wenshi P et al (1986) Feeding behabior of Sichuan takin. Mammalia 50:311–322

Scoones I (1995) Exploiting heterogeneity—Habitat use in cattle in dyrland Zimbabwe. J Arid Environ 29:221–237

Seagle SW (2003) Can ungulates foraging in a multiple-use landscape alter forest nitrogen budgets? Oikos 103:230–234

Singh NJ, Yoccoz NG, Lecomte N et al (2010) Scale and selection of habitat and resources: Tibetan argali (Ovis ammon hodgsoni) in high-altitude rangelands. Can J Zool-Revue Canadienne De Zoologie 88:436–447

Solanki GS, Naik RM (1998) Grazing interactions between wild and domestic herbivores. Small Ruminant Res 27:231–235

Song Y-L, Gong H-S, Zeng Z-G et al (2005) Food habits of Serow. Chin J Zool 40:50–56

Sponheimer M, Lee-Thorp JA, DeRuiter DJ et al (2003) Diets of southern African Bovidae: Stable isotope evidence. J Mammal 84:471–479

Sukumar R (1990) Ecology of the Asian elephant in southern India. 2. Feeding habits and crop raiding patterns. J Trop Ecol 6:33–53

Sukumar R (1992) The Asian elephant: ecology and management. Cambridge University Press, Cambridge

Suryawanshi KR, Bhatnagar YV, Mishra C (2010) Why should a grazer browse? Livestock impact on winter resource use by bharal Pseudois nayaur. Oecologia 162:453–462

Tobler MW (2002) Habitat use and diet of Baird's tapirs (Tapirus bairdii) in a montane cloud forest of the Cordillera de Talamanca, Costa Rica. Biotropica 34:468–474

Van Lieverloo RJ, Schuiling BF, De Boer WF et al (2009) A comparison of faecal analysis with backtracking to determine the diet composition and species preference of the black rhinoceros (Diceros bicornis minor). Eur J Wildl Res 55:505–515

Van Soest PJ (1994) Nutritional ecology of the ruminant. Cornell University Press, Cornell

Van Wieren SE (1996) Browsers and grazers: foraging strategies in ruminants. Digestive strategies in ruminants and nonruminants 119–146

Wegge P, Shrestha AK, Moe SR (2006) Dry season diets of sympatric ungulates in lowland Nepal: competition and facilitation in alluvial tall grasslands. Ecol Res 21:698–706

Wingard GJ, Harris RB, Pletscher DH et al (2011) Argali food habits and dietary overlap with domestic livestock in Ikh Nart Nature Reserve, Mongolia. J Arid Environ 75:138–145

Woolley LA, Millspaugh JJ, Woods RJ et al (2009) Intraspecific strategic responses of African elephants to temporal variation in forage quality. J Wildl Manag 73:827–835

Xu W, Qiao J, Liu W et al (2008) Food habits of goitered gazelles (Gazella subgutturosa sairensis) in northern Xinjiang. Acta Theriologica Sinica 28:280–286

Xu W, Xia C, Lin J et al (2012) Diet of Gazella subgutturosa (Guldenstaedt, 1780) and food overlap with domestic sheep in Xinjiang, China. Folia Zool 61:54–60

Chapter 5
Frugivory and Seed Dispersal by Large Herbivores of Asia

Sachin Sridhara, Kim McConkey, Soumya Prasad and Richard T. Corlett

Abstract Seed dispersal is a key ecological process with ~50–80 % of all tropical plants depending on animals to provide this service. Wide-ranging and large-bodied species are believed to play a disproportionately important role in the seed dispersal process. Although mounting evidence demonstrates a strong role for large herbivores in seed dispersal, our broad knowledge of this interaction remains surprisingly rudimentary. In this chapter, we partially bridge this knowledge gap for South and Southeast Asia (SSEA). We start by synthesizing known information on how the sensory ecology of these animals aid in locating fruit. Next, we review the fruits consumed and dispersed by distinct large herbivore groups, and the influence of their digestive physiology on seed dispersal patterns. We collated more than 300 records of frugivory and/or seed dispersal for 27 species of large herbivores. The data suggests that smaller species of large herbivores may be constrained in the size of fruits they can consume while larger species are not. An assessment of overlap in the traits of fruits consumed indicates much more dissimilarity between groups of large herbivores than previously thought. Finally, we examine the implications of

S. Sridhara (✉)
James Cook University, 14-88, McGregor Road, Smithfield 4878,
Cairns, QLD, Australia
e-mail: sachin.sridhara@gmail.com

K. McConkey
School of Natural Sciences and Engineering, National Institute of Advanced Studies,
Indian Institute of Science Campus, Bangalore 560012, India
e-mail: kimmcconkey@gmail.com

S. Prasad
School of Life Sciences, Jawaharlal Nehru University, JNU New Campus,
New Delhi 110067, India
e-mail: prasadsoumya@gmail.com

S. Prasad
Nature Science Initiative, 36, Curzon Road, Dehradun 248001, India

R.T. Corlett
Xishuangbanna Tropical Botanical Garden, Chinese Academy of Sciences,
Menglun, Mengla, Yunnan 666303, China
e-mail: corlett@xtbg.org.cn

© Springer Science+Business Media Dordrecht 2016

F.S. Ahrestani and M. Sankaran (eds.), *The Ecology of Large Herbivores in South and Southeast Asia*, Ecological Studies 225,
DOI 10.1007/978-94-017-7570-0_5

this interaction both from a historical perspective and for the future. Despite the broad-scale synthesis achieved, we conclude that information on seed dispersal services provided by large herbivores in SSEA remains largely incomplete.

Keywords Large herbivores · Ungulates · Seed dispersal · Frugivory · Plant animal interactions

5.1 Introduction

Seed dispersal is a key ecological process influencing the spatial patterns and population dynamics of plants (Wang and Smith 2002). Low rates of seedling recruitment under the canopies of parent trees make seed dispersal essential for many plant species (Nathan and Muller-Landau 2000; Matthesius et al. 2011). Disruptions to the seed dispersal process can have profound consequences for ecosystem function and resilience (Terborgh 2013). In tropical ecosystems, 50–80 % of plants are dispersed by vertebrate fauna (Jordano 2000). However, vertebrate frugivores are highly variable in their effectiveness as seed dispersers. Within the community of seed dispersers, certain frugivore species tend to have disproportionately important functional roles (Jordano et al. 2007; Donatti et al. 2011). Large and highly mobile frugivores often have non-redundant seed dispersal roles in the habitats they occupy (Lundberg and Moberg 2003). Due to constraints on the size of seeds that can be processed by smaller frugivores, large-bodied frugivores can consume fruits that are accessible only to a limited number of frugivores, such as large or well protected fruits (Janzen and Martin 1982; Kitamura et al. 2002). Large frugivores also disperse very large quantities of seeds and over much longer distances compared to smaller frugivores within communities (Janzen and Martin 1982; Spiegel and Nathan 2007; Lundberg and Moberg 2003).

Large herbivores constitute some of the most wide-ranging fruit-eating animals in the world, with the incidence of frugivory being greatest among tropical species (Bodmer 1990). Tropical Asia supports over 80 species of large herbivores (>2 kg) belonging to nine families (Groves and Grubb 2011, see Table 1.1) and occupying a wide range of habitats, from arid zones to rainforests (Wilson and Mittermeier 2011). Asia is a hot spot for deer radiation (Geist 1998, also see Chap. 2), has the highest diversity of extant pig species (Meijaard et al. 2011), and the only forest-dependent rhinoceros species (van Strien et al. 2008). This high diversity and abundance of large herbivores may have been a key causative factor for the differences in the flora and fauna between tropical Asia and other tropical regions (Corlett 2007). Fruits are consumed by species from all herbivore families to varying degrees (Bodmer 1990; Corlett 1998), yet relationships between large herbivores and the seeds they disperse are poorly understood.

The relative role of large herbivores in seed dispersal can be understood when compared with the rest of the disperser community. However, comprehensive assessments of seed dispersal at community scales, especially for diverse tropical regions, are very limited. In an undisturbed tropical dry forest with an intact faunal assemblage in India, large herbivores had the greatest contribution to seed dispersal of the plant community; dispersing 31 % of the 67 plants available (Prasad 2011). In a disturbed tropical wet forest in Malaysia, large herbivores interacted with 26 % of 49 plant species (Yasuda et al. 2005), while in a relatively undisturbed tropical wet forest site in Thailand, large herbivores dispersed only 6 % of the plant species, although they interacted with 18 % of the plant community (Kitamura et al. 2002). One report from an African wet forest indicates that large herbivores interact with over 50 % of the plant community (Gautier-Hion et al. 1980, 1985). However, except for Prasad (2011), few studies have used approaches that allowed robust sampling of frugivory by large herbivores which are mostly terrestrial frugivores feeding on fallen fruit. Consequently, these studies may have under-estimated the contribution of large herbivores to seed dispersal processes. Additionally, most sites in tropical Asia have already suffered a decline of large fauna due to hunting and habitat degradation, and thus large herbivores are likely to be under represented in these communities. While existing community-wide datasets from tropical Asia, though incomplete, suggest variability in the relative role of large herbivores as dispersers, they nevertheless highlight a community-wide influence. Given the enormous potential of large herbivores in SSEA to play a functionally unique role in seed dispersal, a better understanding of their interactions with associated plant communities is essential.

Insights into the relationships between large herbivores and fruits in Asia are also vital due to the widespread population declines of many species caused by hunting, habitat destruction and competition from domestic livestock (Corlett 2007, see Chap. 11). Hunting in many parts of Asia has depressed large herbivore populations to critically low levels or extirpated them locally (Robinson and Bennett 2000; Corlett 2007). Additionally, the habitats of Asian large herbivores are among the most fragmented in the world (Riitters et al. 2000; Karanth et al. 2010). The consequences of such large-scale disturbances for seed dispersal can be dramatic, from preventing plant migration in the face of climate change to a total collapse of dispersal function (McConkey et al. 2012; Corlett and Westcott 2013). Mitigating such effects requires a deeper understanding of the role of large herbivores as seed dispersers. In this chapter, we examine the nature of these interactions by reviewing known information on fruits consumed and seeds dispersed by large herbivores. First, we summarize studies examining the role of sensory cues in large herbivores to locate fruit in the forest. Next, we review fruit selection, the nature and outcome of fruit processing by large herbivores, and examine the relationship between fruit size and herbivore body size. Finally, we conclude by discussing the effects historical and recent declines in large herbivores have had at the community level, in light of the interactions we describe above.

5.2 Data Compilation

We compiled a database of fruits known to be consumed by large herbivores across tropical Asia and noted whether the seeds within them were either dispersed or destroyed. Data were collated from published articles, books, reports, theses, unpublished results and personal observations of researchers. Information on fruit traits such as length, width and weight of fruits and seeds, and fruit type were obtained from both primary and secondary sources. Secondary sources included digital floras, information from herbariums and databases maintained by individual researchers or institutions. Fruit and seed measurements were restricted to two dimensions—maximum length and width. When multiple measures existed or were provided as a range, the mean or mid-point was recorded, respectively. The diameter of globular fruits and seeds was taken as both length and width. Plants that were not identified to a species level, but had associated fruit trait data, were retained for the analysis. Herbivore action on seeds was recorded as dispersed or destroyed when available. If dispersal or predation records were ambiguous, fruits were recorded as being "consumed" only.

Herbivore traits were obtained from a combination of primary and secondary sources. A single mean value was computed when ranges or multiple values were available. Traits recorded included body weight (kg), home range size (ha), daily movement range (m), digestive physiology, and IUCN status. All binomial names follow the IUCN listing.

5.3 How Do Large Herbivores Locate and Select Fruit?

It is unlikely that large herbivores rely solely (or even largely) on sensory cues when locating fruits. Fruit crops last days or weeks, so returning to previously rewarding locations is a strategy that is probably used by all frugivores (Corlett 2011). A combination of spatial and temporal memory would allow long-lived animals to return annually to preferred fruiting plants (Campos-Arceiz and Blake 2011). However, sensory information is needed for precise location of individual fruits and for the detection of newly ripened fruit crops. As far as is known, all large herbivore species have the "typical" forms of all the major mammalian senses: vision, touch, hearing, smell and taste.

Vision enables the herbivore to locate and identify food resources, including fruits. All large herbivore species that have been tested appear to be typical dichromats, with two spectrally distinct cone types. They are most sensitive in the blue to yellow-green part of the human visual spectrum and lack discrimination ability at the orange-red end (Ahnelt and Kolb 2000; Peichl 2005; Schiviz et al. 2008; Corlett 2011; Jacobs 2012). Fruits that are red or orange and eaten by large herbivores are likely to be eaten most by trichromatic primates or tetrachromatic birds (Corlett 2011). In low light the cones are inactive and rods take over, making vision monochromatic, with peak sensitivity in the blue–green. As in most mammals, rods dominate the retinas of all herbivores that have been studied. As well as colour, herbivores can potentially use brightness (lightness) to distinguish fruits from their background, under both high and low light conditions, but the relative importance of this is unknown.

Visual acuity may be more important than colour or brightness discrimination in locating fruit, but lack of standardization in the methods by which this is measured makes comparisons among taxa difficult. In general, it appears that acuity is low in large herbivores (Corlett 2011). Visual acuity is relatively higher in open-country species, such as horses, where it is presumably essential for predator avoidance, and lower in forest species, such as pigs and mouse deer (Schiviz et al. 2008; Sugnaseelan et al. 2013). Even at close range, domestic pigs failed to discriminate visual cues <20 mm across (Zonderland et al. 2008). In contrast, Asian elephants can apparently discriminate cues 5 mm across at a distance of 2 m (i.e. at the tip of the trunk) (Shyan-Norwalt et al. 2010).

If an animal cannot see a fruit, it may be able to detect the fruit by either touch, the sound the fruit makes when falling, or by the smell of the fruit. The vibrissae (tactile hairs or whiskers) around the mouths of all herbivores are tactile sense organs (Ahl 1986) that may help detect fruits on the ground, although this has not been studied. A pig's entire snout seems to have the same function, and fallen fruits hidden in litter are presumably detected snout-first. Hearing is potentially useful for not only detecting the sound of falling fruits, but also the noise from feeding arboreal frugivores that many terrestrial frugivores follow (Prasad and Sukumar 2010). The ability to accurately localize sounds seems to vary considerably across large herbivores, being high in pigs and elephants, but low in domestic cattle (Heffner and Heffner 1992).

Olfaction is often assumed to have a dominant role in the mammalian sensory system, with the exception of primates (Corlett 2011), but comparisons between senses are difficult to make. Asian elephants are better at odour discrimination than humans, mice and macaques (Rizvanovic et al. 2013) and indirect evidence suggests that this may also be true for pigs and deer (Graves 1984; Rizvanovic et al. 2013). The sense of taste offers the last opportunity for rejecting a fruit before swallowing (Corlett 2011). Large herbivores appear to have the same five basic tastes as humans (sweet, salty, sour, umami and bitter) (Ginane et al. 2011; Jiang et al. 2012). However, the details of what is detected as sweet or, in particular, bitter, are known to vary considerably across mammalian taxa and it is possible that this influences fruit choice in herbivores.

Finally, a diet based on sensory cues alone would inevitably be nutritionally unbalanced, since these cues are imperfect proxies for nutritional value (Corlett 2011; Cazetta et al. 2012). Preferences for colours, odours and tastes need to be fine-tuned by the post-ingestive physiological feedbacks that enable an animal to eat more of what it needs and less of what it does not need. This "diet learning" ability has been widely demonstrated in domestic herbivores and can be assumed to be present in their wild relatives too (Provenza 1995; Yearsley et al. 2006).

5.4 Fruit (or Seed) Selection, Processing and Deposition

In addition to variation in sensory ecology outlined above, differences in body size, digestive physiology, nutritional requirements, and habitat specificity may influence fruit choice by large herbivores. The nature of seed processing, seed retention times and consequently the seed shadows generated by large herbivores for endozoochorously dispersed seeds are primarily influenced by the digestive anatomy and physiology of herbivores, which shows tremendous variation among herbivore families. The outcome of this plant–herbivore interaction determines whether the herbivore is a disperser or a seed predator, the distances to which seeds are transported and the deposition of seeds to particular habitats (Brodie et al. 2009a, b; Velho et al. 2012a, b; Jadeja et al. 2013). In this section, we summarize relationships between fruit traits and the outcome of seed processing by large herbivores in relation to the digestive anatomy and physiology of large herbivores.

Based on their digestive anatomy and physiology, Asian large herbivores can be broadly classified into three groups which reflect similarities in seed processing: (1) foregut fermenters or ruminants (mouse–deer, muntjac, deer, antelope and wild cattle), (2) simple-stomach frugivores (pigs, babyrousa) and (3) simple-stomach folivores (tapirs, wild ass, rhinoceros and elephant). Although processing varies within these groups, the overlap in characteristics merits a synthesis among the distinct groups.

5.4.1 Foregut Fermenters (Tragulidae, Cervidae and Bovidae)

In SSEA, extant ruminants belonging to three families—Tragulidae (eight species), Cervidae (20 species), and Bovidae (30 species)—constitute the majority of large herbivores. We were unable to find information on frugivory and seed dispersal by Moschidae (musk deer, 5 species) and so do not consider them any further here. Ruminants range in size from 2 to 1000 kg; chevrotains (Tragulidae) and muntjacs (Cervidae) occupy the lower end of this spectrum (2–20 kg), while wild cattle (Bovidae; 240–850 kg) are comparable in size to rhinoceroses (Table 5.1). It has been suggested that the digestive physiology of ruminants may have evolved for a frugivorous diet, later diversifying for folivory (Bodmer 1989; Gagnon and Chew 2000; Clauss et al. 2003).

Theories based on energetics, digestive physiology and allometry predict that smaller ruminants should consume more fruits than larger species (Jarman 1974; Hofmann 1989; Wenninger and Shipley 2000; Clauss et al. 2003). Specifically, energy requirements per unit mass increase with declining body size, necessitating small ruminants to ingest food items of high quality, such as fruits. Additionally, their smaller digestive tracts have faster passage rates associated with lower digestion efficiency (Hofmann and Stewart 1972; Demment and van Soest 1985;

Table 5.1 Traits of large herbivores recorded to consume fruits, listed by increasing weight; IS–IUCN status: DD–data deficient, LC–least concern, NT–near threatened, VU–vulnerable, EN–endangered, CE–critically endangered; SC–number of species whose fruits were consumed; WT–body weight in kg; HR–home range in hectares; DMR–daily movement rate in m; HAB–habitat type; DA–digestive anatomy: R-3C–ruminant with three chambered stomach, R-4C–ruminant with four chambered stomach, S-FR–simple-stomach frugivore, S-FO–simple-stomach folivore

Species	Family	IS	SC	WT	HR	DMR	HAB	DA	Ref
Tragulus javanicus	Tragulidae	DD	22	2	4.85	546	Forest	R-3C	1, 2
Moschiola indica	Tragulidae	LC	6	3	–	–	Forest	R-3C	3
Tragulus napu	Tragulidae	LC	10	7	2.5	–	Forest	R-3C	2
Muntiacus atherodes	Cervidae	LC	18	16	6.2	–	Forest	R-4C	1
Muntiacus muntjak	Cervidae	LC	78	20	6	–	Forest	R-4C	1, 3, 4
Tetracerus quadricornis	Bovidae	VU	2	20	–	–	Open forest	R-4C	5
Gazella bennettii	Bovidae	LC	10	23	–	–	Grassland	R-4C	6
Antilope cervicapra	Bovidae	NT	1	30	766	–	Grassland	R-4C	7
Axis porcinus	Cervidae	EN	11	37	51.25	252	Grassland	R-4C	8
Sus cebifrons	Suidae	CE	2	45	–	–	Forest	S-FR	9
Sus celebensis	Suidae	NT	1	55	–	–	Generalist	S-FR	9
Sus barbatus	Suidae	VU	31	70	–	500	Forest	S-FR	9
Babyrousa babyrussa	Suidae	VU	8	70	–	–	Forest	S-FR	10
Axis axis	Cervidae	LC	40	72	169	–	Open forest	R-4C	3, 11
Sus verrucosus	Suidae	EN	2	90	–	–	Open forest	S-FR	9
Rucervus eldii	Cervidae	EN	8	100	814.5	–	Open forest	R-4C	12, 13
Babyrousa togeanensis	Suidae	EN	6	100	–	–	Forest	S-FR	9
Sus scrofa	Suidae	LC	62	135	1800	8400	Generalist	S-FR	9, 14
Rusa unicolor	Cervidae	VU	73	200	900	–	Forest	R-4C	3, 11, 15
Boselaphus tragocamelus	Bovidae	LC	15	240	8800	2150	Open forest	R-4C	16
Equus hemionus	Equidae	EN	15	250	2800	7500	Desert	S-FO	17
Tapirus indicus	Tapiridae	EN	16	350	1275	2160	Forest	S-FO	18, 19
Dicerorhinus sumatrensis	Rhinocerotidae	CE	18	750	2500	7500	Forest	S-FO	20

(continued)

Table 5.1 (continued)

Species	Family	IS	SC	WT	HR	DMR	HAB	DA	Ref
Bos gaurus	Bovidae	VU	23	850	11200	4000	Forest	R-4C	11, 21
Rhinoceros sondaicus	Rhinocerotidae	CE	5	1350	14500	–	Forest	S-FO	22
Rhinoceros unicornis	Rhinocerotidae	VU	14	2350	5000	–	Grassland	S-FO	23, 24
Elephas maximus	Elephantidae	EN	84	4000	41700	4000	Generalist	S-FO	25, 26, 27

1-Heydon (1994), 2-Yasuda et al. (2005), 3-Prasad (2011), 4-Kitamura et al. (2002), 5-Baskaran et al. (2011), 6-Dookia and Jakher (2007), 7-Jadeja et al. (2013), 8-Dhungel and O'Gara (1991), 9-Meijaard et al. (2011), 10-IUCN species account, 11-Schaller (1967), 12-Aung et al. (2001), McShea et al. (2001), 14-Mitchell et al. (2009), 15-Sankar (1994), 16-Jamal (1994), 17-Ghasemi et al. (2012), 18-Campos-Arceiz et al. (2012), 19-O'Farrill et al. (2013), 20-Flynn (1978), 21-Krishnan (1972), 22-Corbet and Hill (1992), 23-Dinerstein and Wemmer (1988), Dinerstein (1991), 25-Samansiri and Weerakoon (2008), 26-Kitamura et al. (2007), 27-Campos-Arceiz and Blake (2011)

Clauss et al. 2003). Conversely, fruits may not be dominant in the diet of larger ruminants, which are more efficient at processing more abundant lower quality food items such as grass and leaves (Clauss et al. 2013). Nevertheless, frugivory is reported from several species of ruminants (Table 5.2) and medium-sized ruminants can be seasonally very frugivorous (John Singh 1981). There are fewer reports of frugivory by larger ruminants such as wild cattle (genera *Bos* and *Bubalis*), nilgai (*Boselaphus tragocamelus*) or sambar (*Rusa unicolor*) compared to small and medium-sized ruminants such as muntjac (*Muntiacus muntjac*) and chital (*Axis axis*). From available records, it appears that the extent of frugivory may be higher in small to medium-sized ruminants (e.g. tragulids, muntjacs and *Axis* deer).

All ruminants house symbiotic bacteria in their fore-stomachs. While cervids and bovids have a four-chambered stomach, the third stomach chamber in tragulids is poorly developed or completely absent (Agungpriyono et al. 1992). Ruminants have strong molars and seeds can be destroyed during repeated chewing (Bodmer 1989). While some ruminants have been documented to swallow fruits whole (Chen et al. 2001; Prasad et al. 2006; Jadeja et al. 2013), others may chew fruits too large to swallow (Feer 1995). The orifice between the first and second stomach chambers constrains the size of food particles that can pass through the fore-stomach, resulting in two distinct seed dispersal mechanisms. First, seeds that are small enough to pass through the fore-stomach (typically <1 mm) are defecated intact (Demment and van Soest 1985; Mouissie et al. 2005; Jadeja et al. 2013). Second, larger seeds, typically drupes (fleshy fruit which have seeds encased in a stony wall, providing strong protection) that can survive repeated mastication during rumination are spat out after being stored in the rumen (first stomach chamber) for several hours (Chen et al. 2001; Prasad et al. 2006). These two distinct mechanisms of dispersal are perhaps associated with significantly different seed retention times, survival probabilities, and germination success, all of which eventually determine the efficacy of ruminants as dispersers. Based on the limited number of both plant and ruminant species that have been examined for damage or germination,

Table 5.2 Traits of fruit and seed consumed or dispersed by large herbivores of Asia. Numbers indicate the total in each category

Group	Life form of plants	Fruit type	Fruit colour	Largest fruit consumed (mm)	Largest seed dispersed (mm)
Tragulidae	Tree 32	Drupe 15; Berry 5; Capsule 3; Pod 2	Green 10; Yellow 3; Brown 3	80 × 80	23 × 17
Muntjacs	Tree 78; Liana 6; Shrub 3; Palm 1	Drupe 30; Berry 17; Capsule 10; Fig. 6; Pod 1	Yellow 30; Red 12; Black 12; Purple 7; Orange 6	87 × 84	80 × 80
Large cervids	Tree 53: Shrub 6; Vine 4; Liana 3	Drupe 30; Berry 17; Pod 8; Fig. 2	Yellow 11; Green 8; Red 6; Black 5	548 × 19	35 × 17
Bovidae	Tree 27; Shrub 5; Vine 3; Herb 2	Drupe 12; Pod 8; Berry 7	~3 each of Black, Green, Yellow, Red	548 × 19	9 × 7
Babirusa	Tree 11; Vine 1	Drupe 6; Complex fruit 3; Berry 2; Pod 2	Yellow, Green, Black, Brown ~ 2 each	350 × 125	48 × 28
Pigs	Tree 47; Shrub 8; 4 of Palm & Grass; 3 of Vine & Herb	Drupe: 23; Dry fruit 15; Berry 12; Complex fruit 6; Pod 5	Red 18; Yellow 11; Brown 8; Green 5; Black 4; Orange 4; Purple 4; White 3; Blue 2; Pink 1	350 × 125	48 × 28
Elephants	Tree 45; Herb 19; Vine 10; Grass 8; Shrub 6; Palm 2	Pod 22; Berry 20; Drupe 16; Complex fruit 9; Grass 7; Capsule 4; Dry fruit 1	Yellow 13; Brown 10; Red 9; Green 5; Black 3; Purple 3; Orange 1	375 × 175	39 × 26
Rhinoceros	Tree 19; Herb 6; Vine 4; Shrub 3	Berry 9; Drupe 6; Capsule 5; Complex fruit 4; Pod 4	Brown 7; Red 4; Yellow 4; Green 3; Purple 2; Orange 1; White 1	300 × 200	30 × 23
Tapirs	Tree 15	Complex fruit 6; Capsule 4; Drupe 4; Berry 2	Green 4; Yellow 3; Brown 2; 1 each of Red, Purple, Black and Orange	132 × 100	39 × 26
Wild ass	Herb; Shrub; Tree	NA	NA	NA	NA

ruminants appear to play a key role in the dispersal of certain tropical plants, typically with drupes (Chen et al. 2001; Prasad et al. 2006; Brodie et al. 2009a, b). However, the fate of most seeds consumed by ruminants (i.e. whether they are dispersed or destroyed) remains unknown. There is a need to examine the exact nature of the interaction between ruminants and fruits, particularly for newly described ruminant species from tropical Asia (Van Dung et al. 1993; Rabinowitz et al. 1999; Groves and Grubb 2011).

5.4.1.1 Tragulidae

The small body size, crepuscular habits, elusive behaviour, and tendency to prefer dense undergrowth (Sridhara et al. 2013) has precluded reliable observations of frugivory by tragulids until the more recent use of camera-traps (Prasad et al. 2010). From the literature we were able to tabulate fruits from 32 tree species in the diets of tragulids. Accounts of frugivory were restricted to three tragulid species, although eight occur in Asia. Records of fruits consumed by tragulids were restricted to trees, mostly drupaceous and coloured green. Dispersal was confirmed for 6 species (all drupes), and seed predation recorded for one dry-fruited species (Yasuda et al. 2005). The largest fruits consumed and seeds dispersed by tragulids were smaller than for other ruminant groups (Table 5.2).

5.4.1.2 Cervidae

We found frugivory records for only two species of muntjac, although nine species are recognized in SSEA. For these, we were able to record instances of fruit consumption of 88 plant species. Fruits consumed were largely from trees, predominantly drupaceous, and were mostly yellow. A total of 15 species were recorded as being dispersed (11 drupes, 3 berries), and the largest fruits consumed and dispersed were similar in size (Table 5.2). Seeds that were consistently destroyed (7 species—4 drupes, 2 berries and 1 capsule) ranged in size from 10 × 8 to 25 × 15 mm.

We documented a total of 67 plant species whose fruit were consumed by larger cervids belonging to three genera: *Rucervus*, *Rusa* and *Axis*. Accounts of frugivory were restricted to four species, although 10 large cervid species occur in tropical Asia. Fruits consumed were mostly from trees, largely drupaceous, and predominantly yellow or green (Table 5.2). Dispersal records were confined to 24 plant species (15 drupes, 4 berries, 2 figs and 1 capsule). Seeds consistently destroyed (2 drupes, 2 berries and 1 capsule) varied from 15 × 9 to 25 × 15 mm.

5.4.1.3 Bovidae

Fruits from 38 plant species were documented to be consumed by bovids. Records of frugivory were restricted to one species each from the genera *Gazella*, *Naemorhedus*,

Tetracerus, *Antilope*, *Capricornis*, *Boselaphus* and *Bos*, although several other species of bovids are found in tropical Asia. Fruits consumed were mostly from trees, drupaceous and equally likely to be black, green, yellow, or red. Twelve species were recorded as being dispersed (including 5 pods, 2 berries, 1 each of drupe and capsule), while no information was available on destruction of seeds.

5.4.1.4 Seed Deposition

In general, there is surprisingly little information on the seed deposition patterns by ruminants. Brodie et al. (2009a, b) found that muntjacs (*Muntiacus muntjac*) alone deposit seeds of *Choerospondias axillaris* in open microhabitats where their germination success is the highest, although sambar (*Cervus unicolor*) and gibbons (*Hylobates lar*) also disperse *C. axillaris*. Gut passage rates or seed retention times, which affect deposition patterns, are unknown for all but one species of ruminant, chital (*Axis axis*). Chital retain the seeds of *Phyllanthus emblica* for 7–27 h before regurgitating them (Prasad et al. 2006). Studies on wild and domesticated ruminants from Europe and America indicate that gut passage rates range from 20 h to 5 days for medium to large ruminant species (Janzen et al. 1985; Schmidt et al. 2004). The uneven retention time of seeds combined with the large home ranges of medium–large ruminants (<114 km^2—Table 5.1) is expected to result in scattered seed deposition patterns and long seed dispersal distances (Vellend et al. 2003; Myers et al. 2004).

5.4.2 Simple-Stomach Frugivores (Suidae)

In Asia, the family *Suidae* comprises 13 species in three genera (IUCN 2014). Pigs and their relatives are primarily frugivorous, but are unique in comparison with other ungulates in supplementing their diet with a wide range of other food types including animal matter.

Both pigs and babirusa are of medium-size, with weights ranging from 45–135 kg (pigs) and 70–100 kg (babirusa) (Table 5.1). The strong jaws and teeth, and irregular molar surfaces of pigs and babirusa make them well suited for crushing seeds (Meijaard et al. 2005). Domestic pigs have a well-developed caecum and spiral colon that are major sites for microbial digestion, with efficient cellulose digestion (Leus and Macdonald 1997). However, wild suid digestive physiology has rarely been investigated. Reports indicate that they are more folivorous than domesticated pigs (Leus and Macdonald 1997). The diet of *Sus scrofa* varies greatly among the habitats and regions in which it occurs (Ickes et al. 2001); both this species and *S. barbatus* appear to prefer fruit and consume it in abundance when available. Similarly, the digestive physiology of babirusa suggests that it is primarily frugivorous, or a frugivore-folivore (Leus 1994). Indeed, babirusa do show a preference for fruit (Leus 1994; Clayton 1996) and are more specialized in digesting fruit components and less efficient in digesting grass fibre than pigs (Leus 1994). Babirusa have enlarged stomachs compared to pigs as more fermentation occurs here than in the large intestine.

Published accounts also indicate significant variability in whether *Sus* are considered dispersers (Hamann and Curio 1999; Donatti et al. 2011) or predators of seeds (Campos and Ojeda 1997; Curran and Leighton 2000; Lynes and Campbell 2000; Meijaard et al. 2005), even for congeneric plant species (Campos and Ojeda 1997; Lynes and Campbell 2000). While the seeds of some fruit types (e.g. dry) are consistently destroyed (Curran and Leighton 2000), seed treatment is probably type- and species-specific and there may be significant differences in seed handling. Seeds from berries (10 of 12 species dispersed), pods (4 of 5 species) and complex fruit (6 of 6 species) were more likely to be dispersed than destroyed, while seeds from dry fruits (15 species) were always destroyed. Plant species with drupe fruit were equally likely to have their seeds dispersed (11 species) as destroyed (12 species). Seed predation by *Sus* can have profound effects on seed recruitment (Curran and Leighton 2000), while its potential role as a seed disperser in most Asian habitats may be less significant, albeit understudied (Hamann and Curio 1999). In temperate regions, fruit forms 32–42 % of the diet of *S. scrofa* (Diong 1982; Thomson and Challies 1988), while in *S. barbatus* stomachs investigated in Asia, crushed seeds were often dominant (Caldecott 1991; Meijaard et al. 2005; Wulffraat 2006). There have been an increasing number of studies indicating an important dispersal role for introduced pigs outside of their native range, particularly where alternative native dispersers are scarce (Diong 1982; Fedriani and Delibes 2009; Matías et al. 2010; Donatti et al. 2011; O'Connor and Kelly 2012).

5.4.2.1 Babirusa

Information about frugivory by babirusa is largely anecdotal and the only evidence of seed dispersal comes from stomach contents (Clayton 1996). We found 21 records of frugivory for *B. babyrussa* and *B. togeanensis*, while no information exists for *B. celebensis*. Most records of frugivory were of drupaceous fruits from trees, equally likely to be coloured yellow, green, black or brown. Even if babirusa destroys the seeds of some species, they are one of the largest fruit-eating animals in their range, exceeded in size only by the anoa (*Bubalus depressicornis*) (Whitten et al. 1987).

5.4.2.2 Pigs

Available information on frugivory in *Sus* comes largely from the widespread *Sus scrofa* and *S. barbatus,* with limited records for *S. celebensis, S. cebifrons,* and *S. verrucosus.* In total, fruits of 97 species were recorded as being consumed, of which 30 species were reported as destroyed and 32 dispersed. Fruits were mostly from trees and were predominantly drupaceous and red in colour. Generally, the seeds dispersed by *Sus* are small to medium sized (n = 12 species, with fruit length <10 mm), but seeds up to 48 × 28 mm have been recorded as dispersed.

5.4.2.3 Seed Deposition

With gut passage rates of 2–4 days (O'Connor and Kelly 2012) and home range sizes of 6–34 km^2 (Saunders and Kay 1991; Caley 1997; Mitchell et al. 2009), *Sus scrofa* (and perhaps other pig species) may be effective long distance seed dispersers for some plant species (O'Connor and Kelly 2012). Further, defecation of seeds can be spaced over multiple days, producing scattered seed deposition patterns (O'Connor and Kelly 2012), which can be beneficial for seedling recruitment. *Sus barbatus* has unique ranging behaviour, whereby periodic mass migrations of individuals occur every few years, probably to coincide with predictable fruiting in *Dryobalanops* forests in the Malayan peninsula (Kawanishi et al. 2008) or seasonally fruiting *Dinochloa* bamboo in Borneo (Davies and Payne 1982). At other times the populations are mostly small and exhibit local movements only, although movements may increase as populations expand to exploit food sources (Kawanishi et al. 2008). During annual migrations, *S. barbatus* may cover distances of 8–22 km per month, and 250–650 km in all (Caldecott 1991) and thus have the potential to disperse seeds over long distances.

5.4.3 Simple-Stomach Folivores (Elephantidae, Rhinocerotidae, Equidae and Tapiridae)

Unlike ruminants (4.1) and pigs (4.2), simple-stomach folivores belong to four different families with different phylogenetic histories: Elephantidae (1 Asian species), Rhinocerotidae (3 Asian species), Equidae (1 species in tropical Asia) and Tapiridae (1 Asian species) (see Table 1.1; also refer to Chap. 2 for further details). However, these species display similarities in their digestive systems and consequently show some similarities in their seed dispersal behaviour. Although collectively called "hindgut fermenters", fermentation in these animals actually occurs in the mid-gut, which results in distinctly different seed processing compared to foregut fermenters or simple-stomach frugivores. These large herbivores have simple stomachs, with symbiotic bacteria housed in the caecum and colon (Clemens and Maloiy 1982). The breakdown of cellulose is less efficient compared to foregut fermenters and faecal particle size is significantly larger, with faeces containing poorly digested forage (Fritz et al. 2009). Thus, fruits and seeds are poorly digested and often pass undamaged through their guts. Since fermentation occurs in the caecum and colon, these animals can process low-nutrient food rapidly. This rapid food processing gives them an advantage at large body sizes, as they are able to take in significantly larger quantities of food (Clauss et al. 2003). Perissodactyls (rhinoceroses, equids, and tapirs) have a compartmentalized colon (Stevens and

Hume 1995), which makes their digestive systems more complex than elephants. Although fruit is typically not a predominant food item in the diet of elephants, rhinoceroses, wild asses and tapirs, the regular appearance of seeds in their dung indicate that they do consume a wide range of fruit, sometimes in significant quantities (Table 5.2).

5.4.3.1 Elephants

Asian elephants (4000 kg) are the largest fruit-eating animals in Asia (Table 5.1). The extent of frugivory exhibited by elephants is highly variable (Campos-Arceiz and Blake 2011). Their mild masticatory action results in little or no seed damage in the mouth, although digestive seed predation in the gut has been reported (Campos-Arceiz et al. 2012). However, they are probably primarily seed dispersers.

Elephants consume fruits from a diverse range of life forms, but mostly from trees and herbs (Table 5.2), reflecting their diverse habitat use. Elephants were recorded to consume fruits from 84 species in our review. Types of fruit eaten were diverse, but most were pods, berries or drupes, and the seeds were usually dispersed. Fruits were mostly yellow, green or brown (Table 5.2). Elephants consume the largest fruits among large herbivores in Asia and can disperse seeds as large as 39 × 28 mm.

5.4.3.2 Perissodactyls (Rhinoceroses, Tapirs and Equids)

Rhinoceroses

Rhinoceroses (750–2350 kg), second in size only to elephants, are predominantly folivorous. However, seeds and fruits are regularly found in their dung and latrines (Flynn 1978; Dinerstein 1991). The rarity of most rhinoceros species has prevented detailed investigation of the proportion of fruit in their diet. Frugivory has nevertheless been reported, and fruit of 37 plant species are recorded as consumed by Asian rhinoceroses. These fruits are mostly from trees, usually dull coloured (green, brown or yellow), and are either berries or drupes. The relatively less diverse range of fruits consumed by rhinoceroses perhaps reflects their current rarity and habitat specialization.

Tapirs

Tapirs (350 kg) have been documented to disperse and destroy seeds. The dentition of tapirs is well suited for seed crushing (Campos-Arceiz et al. 2012) and seed fragments have been reported from dung of Neotropical tapir species (Janzen 1981). Like pigs, their function as dispersers or seed predators is variable. Captive tapirs have been documented to spit or swallow seeds whole and also defecate both viable and destroyed seeds (Campos-Arceiz et al. 2012). However, the extent of frugivory in tapirs in the wild is unclear, making any comparison with other simple-stomached herbivores difficult.

In our review, we found records of tapirs feeding on fruits from 21 species of plants. These fruits were all from trees, yellow, green or brown, and were complex fruits, berries or drupes. Tapirs are extremely variable in their seed treatment and often both disperse and destroy seeds from the same species (Campos-Arceiz et al. 2012). Of the 21 species we recorded as consumed by tapirs, five usually had the seeds destroyed, four had seeds destroyed as well as dispersed, two were dispersed and the remaining 10 were observations of frugivory only. Further, tapirs have difficulty handling large fruits and frequently spit seeds from large fruits (Campos-Arceiz et al. 2012). *Durio zibethinus* seeds (39 × 26 mm) are the largest reported to be dispersed by Asian tapirs (Holden et al. 2003; Campos-Arceiz et al. 2012).

Equids

The Asiatic wild ass (250 kg) is the smallest simple-stomached herbivore. Equids have small stomachs relative to their body size and tend to process fruit slowly, limiting the quantity of food that can be consumed (Giffin and Gore 1989). The Asiatic wild ass has been studied in Israel, Iran and India, and seeds are frequently reported in dung and seedlings have been found near dung piles (Shah 1993; Peled 2010; Ghasemi et al. 2012), but detailed reports on extent of frugivory are unavailable. Similarly, there is no information on seed predation, although digestive seed predation has been reported for domestic horses (Janzen et al. 1985). In general, asses in the tropics are found only in desert regions, limiting their fruit consumption to berries, capsules and pods of herbs, shrubs and sometimes trees.

5.4.3.3 Megafaunal Fruit

It has been proposed that the largest fruit within tropical plant communities evolved in response to frugivory by "megafauna"—the largest terrestrial vertebrate frugivores (>1000 kg; Guimaraes et al. 2008). Analyses of elephant fruit and plant communities in South America suggest that megafaunal fruit fall into two categories: fruits 4–10 cm in diameter with up to five large seeds, and fruits >10 cm diameter with numerous small seeds (Guimarães et al. 2008). In tropical Asia, elephants and rhinoceroses are able to swallow and disperse the largest seeds in the habitats they occupy (Cochrane 2003; Corlett 2011). In our review, fruit that fit the "megafaunal fruit" syndrome featured prominently in the diets of elephants (19 of 50 plant species for which fruit width was available), rhinoceroses (10 of 24 plant species) and tapirs (9 of 13 plant species). The role of tapirs, however, in the dispersal of large seeds is variable; from studies on captive individuals, seeds of large fruits were less likely to be swallowed and more likely to be damaged, and only small-seeded species were regularly defecated in a viable condition (Campos-Arceiz et al. 2012). Fruits that fit the "megafaunal fruit" syndrome are known to be dispersed by smaller frugivores as well as other dispersal mechanisms such as wind or water (Cochrane 2003), and in a staggered successive manner with back-up dispersal mechanisms being invoked when megafauna fail to disperse the fruit (Sekar and Sukumar 2013). It remains to be tested if these fruit evolved in response to features of the sensory and digestive systems of large simple-stomached herbivores. No data on dispersal of megafaunal fruits exist for Asian equids, but in the Neotropics introduced horses appear to have replaced some of the seed dispersal services that were lost with the extinction of the megafauna around 10,000 years ago (Janzen and Martin 1982).

5.4.3.4 Seed Deposition

Large simple-stomached herbivores process vast quantities of forage within a day, consequently exhibiting the longest known daily displacements rates and home ranges for terrestrial mammals. They also have long gut passage times (rhinoceroses 61–122 h, elephants 40–122 h, tapirs 63–235 h, ass 34–39 h), although this may not scale with body size, since the retention times of smaller foregut herbivores overlap (2–96 h, S. Prasad unpublished data). These large herbivores also undertake long, directed seasonal movements to access water, grasses in different stages of growth, or salt licks (Campos-Arceiz et al. 2008). Estimated seed dispersal distances for tapirs (Campos-Arceiz and Blake 2011) and elephants (Campos-Arceiz et al. 2008), indicate that these animals are effective long-distance seed dispersers. Asses and rhinoceroses are likely to be similarly capable of long-distance seed dispersal, although no data exist; Asiatic wild ass have daily movements of 4–11 km, and home ranges of 8–19 km^2, while rhinoceroses have home ranges of 5–50 km^2 and may travel up to 10 km within a day (van Strien et al. 2008).

Both elephants and rhinoceroses deposit seeds within large dung piles, which can provide suitable conditions for seedling recruitment although this advantage may be partially offset by high seedling density and associated competition (Campos-Arceiz and Blake 2011). Rhinoceroses and tapirs frequently deposit seeds in latrines (Dinerstein 1991; O'Farrill et al. 2013), which can have both positive and negative consequences for seed recruitment (González-Zamora et al. 2012; O'Farrill et al. 2013). While these are untested in rhinoceroses, latrines of *R. unicornis* were associated with distinct floras (Dinerstein 1991).

5.4.4 Relationship Between Body Size of Herbivores and Size of Fruit Consumed

Seed size and weight are key determinants of recruitment success and seedling survival, consequently influencing the reproductive success of plants (Moles and Westoby 2006). Because frugivores can exert selective pressures on fruit and seed size (Forget et al. 2007), the constraint of frugivore body size on fruit traits needs closer examination. For instance, fruits with very large seeds are accessed only by a few frugivores that are often non-redundant in dispersal function, whereas fruits with smaller seeds usually have a wider suite of consumers and dispersers (Gautier-Hion et al. 1985; Kitamura et al. 2002). Indeed, recent evidence suggests that the functional extinction of large frugivores can cause rapid evolution in trees towards smaller seed sizes (Galetti et al. 2013) or shifts in plant communities in ecological time (Velho et al. 2012a, b).

Here, we examine the nature of the relationship between body size of large herbivores and the largest fruit they can consume. Fruit size is highly correlated to seed size and is therefore a good measure of the limits imposed by gape width of large herbivores on the size of seeds potentially dispersed (Forget et al. 2007). We analyzed fruit and seed width (seed width = second largest axis of the fruit or seed), which often determines if a fruit or seed is swallowed (Corlett 1998). Each species of plant and herbivore was considered to be an independent data point. We used generalized linear models and quantile regressions to examine whether the average and maximum width of fruit and seed consumed were constrained by the body size of large herbivores. To linearize the allometric relationship we log-transformed the predictor (i.e. body weight of herbivores) and used the log link function for Gaussian error distributions. We checked for over-dispersion and examined residuals graphically to check for normality and heteroscedasticity. All analyses were performed in R 3.0.1 (R-Core-Team 2013).

Results indicate that body size may limit the size of fruits that can be consumed by large herbivores (Fig. 5.1). Although maximum fruit length, seed length, and width of seed consumed increased with body weight of large herbivores, the only significant relationship was with fruit width, suggesting that body size (correlated with gape width) of large herbivores can limit the sizes of fruits consumed (Fig. 5.1). Results from the quantile regressions however, suggest that maximum fruit length alone shows a positive relationship with body weight (Fig. 5.2a). Fruit width, seed length and seed width however, do not show any pattern with body size (Fig. 5.2b–d) for the upper quantile (0.95). Because these three traits show a positive relationship for the other quantiles used (0.25, 0.5, 0.75) we could perhaps infer that minimum fruit sizes consumed by herbivores increases with body size, and the fruit sizes consumed by smaller herbivores are on an average smaller than their larger bodied counterparts. Although larger bodied herbivores appear to

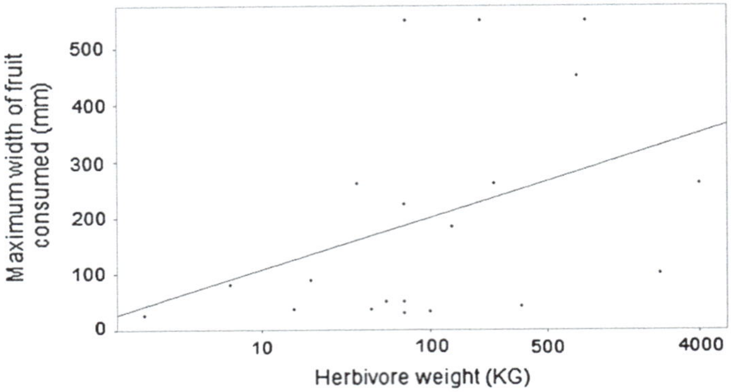

Fig. 5.1 Relationship between body weight of large herbivores in tropical Asia, and the maximum width of fruits consumed

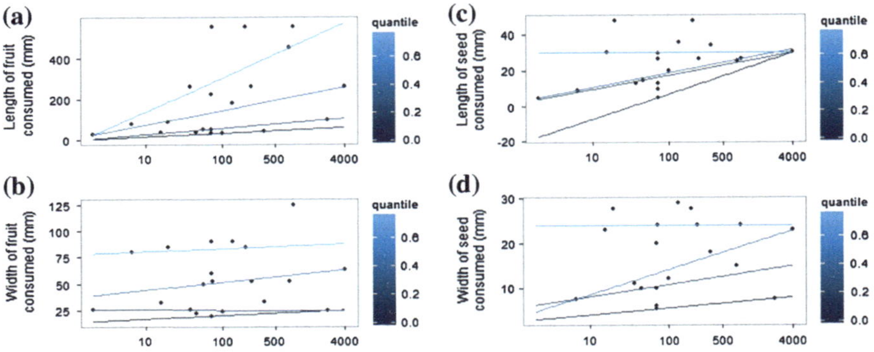

Fig. 5.2 Quantile regressions of fruit traits **a** length of fruit consumed, **b** width of fruit consumed, **c** length of seed consumed and **d** width of seed consumed as a function of herbivore body weight. The quantile regressions were carried out using all available data, but for ease of interpretation we only plot maximum fruit and seed dimensions as points. The four different lines in each graph represent the 25th, 50th, 75th and 95th quantiles

consume fruits that span a wider range of fruit lengths than smaller bodied herbivores (Fig. 5.2a), overall our results do not suggest a strong link between disperser body mass and fruit or seed size likely to consumed or dispersed. More data are needed to better elucidate these links, and we advise caution against dismissing the importance of larger bodied herbivore species until such data has been assembled. Nevertheless, plant species dependent on large herbivores that are especially threatened (Table 5.1) may still be vulnerable in the absence of redundant dispersers (Corlett and Westcott 2013).

5.4.5 *Overlap in Fruit Choice Among Asian Herbivores*

We also assessed the similarity in the traits of fruits consumed by the different families of large herbivores found in Asia. We compiled data on the digestive system, maximum and minimum fruit length and width, maximum and minimum seed length and width, maximum and minimum fruit and seed weight, fruit type and fruit colour associated with each of the large herbivore species. This data was collated for seven of the nine families of large herbivores; Equidae and Moschidae were excluded due to the lack of information on fruit traits. We used the package "FD" in software R to compute the dissimilarity in traits. Gower's dissimilarity index was used since it allows for categorical traits (fruit type and colour, in this case). Numbers close to 0 indicate greater similarity in fruit choice while numbers close to 1 indicate dissimilarity.

In general, our results suggest a degree of dissimilarity in the fruit choice by the different families of large herbivores, providing the first family-wise comparisons at a continental scale. Results indicate that fruit choice is most similar between cervids and elephants, followed by cervids and bovids (Table 5.3). The similarity between cervids and bovids may reflect the overlap in their digestive physiology and body size. The similarity between elephants and cervids, however, is surprising given the differences in their digestive physiology and sensory capabilities. Fruit choice may overlap based on the limited fruit characteristics we found and used in the analysis, perhaps due to sympatry of cervids and elephants throughout SSEA until recent times. Incorporating currently unknown information such as nutritional characteristics of fruits and the relative abundances of these fruits in the diet may alter the dissimilarity metrics when reassessed in the future. This similarity may also be due to the incomplete nature of the dataset. Nevertheless, our results, although preliminary, call into question the assumed non-redundancy of large herbivores as seed dispersers, specifically small–medium species in the families Cervidae and Bovidae. Tapirs are consistently dissimilar to most families, but it is unclear whether this reflects scarcity of information on frugivory or the uniqueness in their fruit choice. As other animals including bears, civets, macaques, and orangutans, also compete with large herbivores for fruits on the ground, fruit characteristics are

Table 5.3 Dissimilarity in the traits of fruits consumed by the different families of large herbivores of Asia

	Tragulidae	Cervidae	Bovidae	Suidae	Tapiridae	Rhinocerotidae	Elephantidae
Tragulidae	0.00	0.63	0.63	0.60	0.49	0.67	0.69
Cervidae	0.63	0.00	0.28	0.44	0.72	0.58	0.23
Bovidae	0.63	0.28	0.00	0.47	0.75	0.48	0.42
Suidae	0.60	0.44	0.47	0.00	0.58	0.64	0.53
Tapiridae	0.49	0.72	0.75	0.58	0.00	0.75	0.61
Rhinocerotidae	0.67	0.58	0.48	0.64	0.75	0.00	0.55
Elephantidae	0.69	0.23	0.42	0.53	0.61	0.55	0.00

Values towards 0 indicate similarity in traits, while values towards 1 indicate dissimilarity as likely to have been shaped by these frugivores too. Finally, these results must be interpreted cautiously due to the incomplete nature of the dataset. Studies from more landscapes covering the entire suite of frugivores can help verify these patterns across regional, continental and global scales.

5.5 Other Modes of Seed Dispersal by Large Herbivores

Apart from locating fruits actively, the consumption of fruits may be a passive activity when other plant parts (e.g. foliage, twigs) are being eaten. Janzen (1984) proposed that small fruits interspersed in the foliage may be consumed inadvertently, even though the animals are not actively seeking them. Because small fruits may not provide sufficient rewards to the consumers, foliage serves the function of the fruit by "packaging" seeds, which potentially leads to endozoochorous dispersal (Janzen 1984). Known as the "foliage-as-fruit" hypothesis, this mechanism of dispersal is relevant to large herbivores, whose diet often consists predominantly of foliage. Several grasses, herbs and a few trees seemingly fit the requirements for this mechanism of seed dispersal in Asia (R. Corlett pers. obs.), although the hypothesis remains largely untested. Preliminary evidence supporting this hypothesis was found for elephants in Sri Lanka (Samansiri and Weerakoon 2008), three ruminants, and wild pigs in India (Middleton and Mason 1992), but was rejected for rhinoceroses in Nepal (Dinerstein 1991). More studies that test the hypothesis are necessary to assess its extent and significance for plant species.

Seeds of herbs and grasses can also be transported by adhering to the pelage or fur of animals. This mode of dispersal, referred to as epizoochory, has been reported for large herbivores because the body hair on these mammals enables the transport of certain kinds of seeds. European populations of *Sus scrofa* and certain temperate deer species effectively disperse seeds that attach to hair (Couvreur et al. 2004; Schmidt et al. 2004). However, there are no reports of epizoochory for any of the large herbivores from tropical Asia, although fruits and seeds with adaptations for external attachment are widespread in open habitats, where they often attach to human clothing (R. Corlett pers. obs). Reports suggest that pigs transport seeds of *Rafflesia* in their hooves (MacKinnon et al. 1996).

5.6 Large Herbivore Distributions and Consequences for Seed Dispersal: Past, Present and Future

The Asian tropics still harbour a number of large herbivores, but the abundance and range of most extant species have contracted over historical and recent times (Groves and Grubb 2011, Chap. 11), and several additional species have become extinct since the Late Pleistocene (also see Chap. 2). In this section, we summarize past and present distributions of herbivores, and assess the range of dispersal services that might have been lost in the Asian tropics.

All tropical terrestrial ecosystems, including several islands, have lost at least one megafaunal species (>44 kg) since the Pleistocene (Corlett 2011, also see Chap. 2). Tropical Asia has experienced significant megafaunal losses at rates that are higher than some other continents (Corlett 2011). The drivers of this region-wide extinction of megafauna remain unclear, with both climate change and the arrival of *Homo* species being plausible explanations (Louys 2012). Herbivorous mammals, including elephants, stegodons, rhinoceroses, horses, bovids, and tapirs, constitute the largest group of extinct megafauna. Since the Late Pleistocene, at least two species each of elephants, stegodons, rhinoceroses and bovids, one species of giraffe, and at least one species of horse have gone extinct from south Asia (Chauhan 2008). The number of Late Pleistocene rhinoceros species is uncertain (Antoine 2012), but the distribution of the family has certainly declined (Corlett 2011). Similarly, at least one species each of elephant, stegodon, rhinoceros, horse and bovid, in addition to a giant tapir, have been lost from Southeast Asia (Mishra et al. 2010; Corlett 2011). All of these extinct species are likely to have consumed at least some fruit, like their extant relatives (Corlett 2010). Since their diversity, distributions and abundances were greater than today (Corlett 2010), they may have played a significant role in the evolution of modern fruits.

At present, ruminants are widely distributed, from the arid zones of western India to the rainforests of Indonesia and Philippines, but are predominantly found in woody habitats, ranging from savannah to rainforest (Eisenberg and Seidensticker 1976), and attain their highest densities in tropical dry forests and woody savannas (Karanth and Sunquist 1992). Most ruminants occur in woody environments and avoid disturbed environments, possibly restricting their functional roles to these habitats. A few ruminants, however, may occur at high densities in human dominated landscapes (e.g. Nilgai *Boselaphus tragocamelus*). *Sus* (pigs) is the most diverse and widely distributed genus among suids. *Sus* species are scattered across tropical Southeast Asia, primarily in forested habitats, with some species being able to persist in grasslands and cultivated areas. Most pig species partly forage in disturbed regions, and can occupy both forest and non-forest habitats (Table 5.1). The disturbance tolerance of suids varies considerably and only *S. scrofa* is a true habitat generalist and potentially a disperser of seeds across modified habitats. The three *Babyrousa* (babirusa) species are confined to Sulawesi or nearby islands (Indonesia), where they persist in forests and occasionally cultivated regions. *Porcula* comprises a single species found in tall grasslands in India and no frugivory information exists for this species. Elephants are found in a range of habitats, from alluvial grasslands to rainforest and are distributed widely from India to Indonesia. With the exception of the Indian Rhino, *Rhinoceros unicornis*, all Asian rhinoceroses have highly restricted ranges within the rain forests of southeast Asia (IUCN 2014). The Indian rhinoceros is found in grasslands and riverine forests in northern India and Nepal, a tiny fraction of its historical and Pleistocene ranges. The Asiatic wild ass is restricted to deserts and very dry regions of western India, while tapirs are found only in the rainforests of Southeast Asia (IUCN 2014).

In summary, with few exceptions, most large herbivores are currently restricted to specific habitats within the continent, perhaps limiting the spatial extent of their seed dispersal function. However, elephants, often within a population, continue to utilize a range of habitats and probably remain responsible for long-range seed dispersal in a large mosaic of heterogeneous habitats for a wide range of plant species.

Apart from native large herbivores, free-ranging or feral buffalo, cows, horses and goats occur in many parts of tropical Asia (Dinerstein and Wemmer 1988; Middleton and Mason 1992). Although these feral and introduced animals have been documented to have detrimental impacts on plant populations (Scowcroft and Hobdy 1987; Moriarty 2004), they also function as seed dispersers (Middleton and Mason 1992; Giordani 2008), including dispersal of large fruits (Dinerstein and Wemmer 1988). Several Asian herbivores, including spotted deer (*Axis axis*), sambar (*Rusa unicolor*), blackbuck (*Antilope cervicapra*), and wild boar (*Sus scrofa)* have been introduced to other continents and become invasive in their introduced ranges (Moriarty 2004). These populations of introduced large herbivores may assist in the spread of both invasive and native plants (e.g. Moriarty 2004).

Most Asian large herbivores species are threatened by some combination of hunting, habitat loss, fragmentation, logging, competition from livestock, and hybridization with domestic species (IUCN 2013, Chaps. 1 and 11). Populations of many species are declining and their distributions are contracting, with larger species consistently more threatened than smaller ones (Table 5.1). Some smaller herbivore species, especially those that are habitat-specialists with very restricted ranges, are also at risk (Groves and Grubb 2011). It is not surprising therefore, that studies predicting future distributions of large herbivores in Asia present a grim prospect of further declines in their ranges and densities (Karanth et al. 2010; Corlett 2011). However, the impacts of these declines on seed dispersal processes are poorly understood. It has been suggested the loss of large mammals, including herbivores, can lead to declines in dispersal of large-seeded species, leading to a slow shift in forest community composition towards small-seeded, bird- and wind-dispersed species (Wright et al. 2007). Indeed, excessive harvest of muntjac (*Muntiacus muntjac*) in Thailand resulted in the decline of both seedling densities and seed dispersal of *Cheorospondias axillaris* (Brodie et al. 2009a, b).

Large, wide-ranging, habitat generalists such as elephants and pigs are especially important. They are likely to be regular dispersers of seeds across heterogeneous habitats and there is probably very low redundancy in their dispersal role in some habitats (Campos-Arceiz and Blake 2011). In the short term, plants may decline simultaneously with their key seed dispersers as has been noted in an elephant (Cochrane 2003) and deer (Brodie et al. 2009a, b) dispersed species. Long-term consequences of herbivore declines on plants with no redundant seed dispersers may include range contraction and local extinctions (Corlett and Westcott 2013). However, plants that are seemingly dependant on megafaunal dispersers may have evolved mechanisms to be dispersed by other frugivores. For instance, rodents and introduced livestock disperse seeds of plants thought to be entirely dependant on megafauna extinct 10,000 years ago (Janzen and Martin 1982; Guimarães et al. 2008). Recent evidence from India also demonstrates that small and medium-sized herbivores, specifically native ruminants such as deer can provide "back-up" dispersal services to megafauna-dispersed trees, albeit to a much lesser extent (Sekar and Sukumar 2013). Additionally, these small to medium-sized herbivores, although not capable of accessing the wide range of fruits available to much larger megafauna, have been documented to disperse more plant species than their larger counterparts in both wet and dry tropical forests of Asia (Prasad 2011). We therefore suggest a rethinking of the importance placed on small and medium-sized herbivores as seed dispersers. While conservation efforts have often focussed on charismatic megafauna and should continue to do so, the unique and diverse dispersal services provided by small- and medium-sized herbivores should also be integrated into conserving planning.

5.7 Conclusion

We set out to produce a quantitative summary of frugivory and seed dispersal by large herbivores in tropical Asia. However, we found very little quantitative information on plant–herbivore interactions in the region. Our understanding of the extent of frugivory by herbivores, especially for smaller species, is incomplete. We also understand very little about fruit choice, seed processing and gut retention times of seeds dispersed by herbivores. These gaps in our understanding severely restrict our ability to assess dispersal services lost due to extinctions of large herbivores, and the potential future impacts if herbivore species currently under threat continue to decline. Most importantly, with the current limitations in our understanding of seed dispersal by large herbivores in tropical Asia, we are unable to predict shifts in herbivore–plant interactions for the future under changing climates and land-use. A robust understanding of plant–herbivore interactions in heterogeneous landscapes is critical to plan management strategies that address the anticipated impacts of climate change and associated drivers on biodiversity (McConkey et al. 2012; Corlett and Westcott 2013). Large herbivores can effectively disperse seeds very long distances away from parent trees (Vellend et al. 2003; Myers et al. 2004), potentially facilitating migration of plants at rates suitable to track climatic changes as predicted for the future (Corlett and Westcott 2013).

It has been suggested that lost ecological services may be reinstated by reintroducing species into their former ranges, or by introducing ecologically equivalent species (taxon substitutes) into areas that have experienced extinctions of fauna involved in key ecological processes (Ehrlich and Mooney 1983). Faunal reintroductions to address lost seed dispersal services have been undertaken at a few sites (e.g. Aldabra tortoises in Mauritius—Hansen et al. 2008). However, before attempting this in tropical Asia we need a better understanding of the seed dispersal services provided by large herbivores and of the community-level consequences of species reintroductions, especially novel introductions. Additionally, given the high human pressures and the continued use of forests in tropical Asia, it is imperative that we mitigate these threats (e.g. fragmentation or hunting) that have led to the decline of large herbivores, before attempting reintroductions.

We hope this review will initiate further enquiry into seed dispersal services provided by large herbivores in tropical Asia, and extend the emphasis of research beyond documenting their distributions and abundances to studying their ecological function.

Acknowledgements We thank the following people for providing data on fruits consumed by herbivores and other information: Asmita Sengupta, Chanpen Saralamba, Freddy Pattiselanno, Firmann Aldy, Matt Heydon, Meghna Krishnadas, Robert Steinmetz, Stephan Wulffraat, Naveed Khan. Farshid Ahrestani, Mahesh Sankaran, David J. Chivers, and David A. Westcott made useful comments on the manuscript.

References

Agungpriyono S, Yamamoto Y, Kitamura N et al (1992) Morphological study on the stomach of the lesser mouse deer (Tragulus javanicus) with special reference to the internal surface. J Vet Med Sci/Japan Soc Vet Sci 54:1063–1069

Ahl A (1986) The role of vibrissae in behavior: a status review. Vet Res Commun 10:245–268

Ahnelt PK, Kolb H (2000) The mammalian photoreceptor mosaic-adaptive design. Prog Retin Eye Res 19:711–777

Antoine P-O (2012) Pleistocene and Holocene rhinocerotids (Mammalia, Perissodactyla) from the Indochinese Peninsula. CR Palevol 11:159–168

Aung M, Mcshea WJ, Htung S et al (2001) Ecology and social organization of a tropical deer (Cervus eldi thamin). J Mammal 82:836–847

Baskaran N, Kannan V, Thiyagesan K et al (2011) Behavioural ecology of four-horned antelope (*Tetracerus quadricornis* 'de Blainville, 1816) in the tropical forests of southern India. Mammalian Biology-Zeitschrift für Säugetierkunde 76:741–747

Bodmer RE (1989) Frugivory in Amazon ungulates. Dissertation, University of Cambridge

Bodmer RE (1990) Ungulate frugivores and the browser-grazer continuum. Oikos 57:319–325

Brodie JF, Helmy OE, Brockelman WY et al (2009a) Bushmeat poaching reduces the seed dispersal and population growth rate of a mammal-dispersed tree. Ecol Appl 19:854–863

Brodie JF, Helmy OE, Brockelman WY et al (2009b) Functional differences within a guild of tropical mammalian frugivores. Ecology 90:688–698

Caldecott J (1991) Eruptions and migrations of bearded pig populations. Bongo 18:233–243

Caley P (1997) Movements, activity patterns and habitat use of feral pigs (*Sus scrofa*) in a tropical habitat. Wildlife Res 24:77–87

Campos-Arceiz A, Blake S (2011) Megagardeners of the forest–the role of elephants in seed dispersal. Acta Oecol 37:542–553

Campos-Arceiz A, Larrinaga AR, Weerasinghe UR et al (2008) Behavior rather than diet mediates seasonal differences in seed dispersal by Asian elephants. Ecology 89:2684–2691

Campos-Arceiz A, Traeholt C, Jaffar R et al (2012) Asian tapirs are no elephants when it comes to seed dispersal. Biotropica 44:220–227

Campos CM, Ojeda RA (1997) Dispersal and germination of *Prosopis flexuosa* (Fabaceae) seeds by desert mammals in Argentina. J Arid Environ 35:707–714

Cazetta E, Galetti M, Rezende EL et al (2012) On the reliability of visual communication in vertebrate-dispersed fruits. J Ecol 100:277–286

Chauhan PR (2008) Large mammal fossil occurrences and associated archaeological evidence in Pleistocene contexts of peninsular India and Sri Lanka. Quatern Int 192:20–42

Chen J, Deng XB, Bai ZL et al (2001) Fruit characteristics and *Muntiacus muntijak vaginalis* (Muntjac) visits to individual plants of *Choerospondias axillaries*. Biotropica 33:718–722

Clauss M, Frey R, Kiefer B et al (2003) The maximum attainable body size of herbivorous mammals: morphophysiological constraints on foregut, and adaptations of hindgut fermenters. Oecologia 136:14–27

Clauss M, Steuer P, Müller DWH et al (2013) Herbivory and body size: allometries of diet quality and gastrointestinal physiology, and implications for herbivore ecology and dinosaur gigantism. PLoS ONE 8:e68714

Clayton LM (1996) Conservation biology of the babirusa, *Babyrousa babyrussa* in Sulawesi, Indonesia. Dissertation, University of Oxford

Clemens E, Maloiy G (1982) The digestive physiology of three East African herbivores: the elephant, rhinoceros and hippopotamus. J Zool 198:141–156

Cochrane EP (2003) The need to be eaten: balanites wilsoniana with and without elephant seed-dispersal. J Trop Ecol 19:579–589

Corbet GB, Hill JE (1992) The mammals of the Indomalayan region: a systematic review. Oxford University Press, Oxford

Corlett RT (1998) Frugivory and seed dispersal by vertebrates in the oriental (Indomalayan) region. Biol Rev Camb Philos Soc 73:413–448

Corlett RT (2007) The impact of hunting on the mammalian fauna of tropical Asian forests. Biotropica 39:292–303

Corlett RT (2010) Megafaunal extinctions and their consequences in the tropical Indo-Pacific. Terra Australis 32:117–131

Corlett RT (2011) How to be a frugivore (in a changing world). Acta Oecol 37:674–681

Corlett RT, Westcott DA (2013) Will plant movements keep up with climate change? Trends Ecol Evol 28:482–488

Couvreur M, Christiaen B, Verheyen K et al (2004) Large herbivores as mobile links between isolated nature reserves through adhesive seed dispersal. Appl Veg Sci 7:229–236

Curran L, Leighton M (2000) Vertebrate responses to spatiotemporal variation in seed production of mast-fruiting Dipterocarpaceae. Ecol Monogr 70:101–128

Davies G, Payne J (1982) A faunal survey of Sabah World Wildlife Fund Malaysia

Demment MW, Van Soest PJ (1985) A nutritional explanation for body-size patterns of ruminant and nonruminant herbivores. Am Nat 125:641–672

Dhungel SK, O'gara BW (1991) Ecology of the hog deer in Royal Chitwan National Park, Nepal. Wildlife Monographs 119:3–40

Dinerstein E (1991) Seed dispersal by greater one-horned rhinoceros (*Rhinoceros unicornis*) and the flora of Rhinoceros latrines. Mammalia 55:355–362

Dinerstein E, Wemmer CM (1988) Fruits Rhinoceros eat: dispersal of *Trewia nudiflora* (Euphorbiaceae) in lowland Nepal. Ecology 69:1768–1774

Diong CH (1982) Population biology and management of the feral pig (*Sus scrofa* L.) in Kipahulu Valley, Maui. Dissertation, University of Hawaii

Donatti CI, Guimaraes PR, Galetti M et al (2011) Analysis of a hyper-diverse seed dispersal network: modularity and underlying mechanisms. Ecol Lett 14:773–781

Dookia S, Jakher G (2007) Food and feeding habit of Indian Gazelle (Gazella bennettii), in the Thar Desert of Rajasthan. Indian Forester 133:1327–1340

Ehrlich PR, Mooney HA (1983) Extinction, substitution, and ecosystem services. Bioscience 33:248–254

Eisenberg JF, Seidensticker J (1976) Ungulates in southern Asia: a consideration of biomass estimates for selected habitats. Biol Conserv 10:293–308

Fedriani JM, Delibes M (2009) Seed dispersal in the Iberian pear, *Pyrus bourgaeana*: a role for infrequent mutualists. Ecoscience 16:311–321

Feer F (1995) Seed dispersal in African forest ruminants. J Trop Ecol 11:683–689

Flynn R (1978) The Sumatran rhinoceros in the Endau-Rompin National Park of Peninsular Malaysia. Malayan Naturalist 4:5–12

Forget P-M, Dennis A, Mazer S et al (2007) Seed allometry and disperser assemblages in tropical rainforests: a comparison of four floras on different continents. In: Dennis AJ, Schupp EW, Green R, Westcott DA (ed) Seed dispersal: theory and its application in a changing world. CAB International Publishing, Wallingford, UK, pp 5–36

Fritz J, Hummel J, Kienzle E et al (2009) Comparative chewing efficiency in mammalian herbivores. Oikos 118:1623–1632

Gagnon M, Chew AE (2000) Dietary preferences in extant African Bovidae. J Mammal 81: 490–511

Galetti M, Guevara R, Côrtes MC et al (2013) Functional extinction of birds drives rapid evolutionary changes in seed size. Science 340:1086–1090

Gautier-Hion A, Duplantier JM, Quris R et al (1985) Fruit characters as a basis of fruit choice and seed dispersal in a tropical forest vertebrate community. Oecologia 65:324–337

Gautier-Hion A, Emmons L, Dubost G (1980) A comparison of the diets of three major groups of primary consumers of Gabon (primates, squirrels and ruminants). Oecologia 45:182–189

Geist V (1998) Deer of the world: their evolution, behaviour, and ecology Stackpole Books. Mechanicsburg, USA

Giffin JM, Gore T (1989) Horse owner's veterinary handbook Howell Book House. New York, USA

Ginane C, Baumont R, Favreau-Peigné A (2011) Perception and hedonic value of basic tastes in domestic ruminants. Physiol Behav 104:666–674

Giordani L (2008) The role of goats in germination and dispersal of *Mimosa luisana* Brandegee (Leguminosae-Mimosoideae) seeds in Tehuacán-Cuicatlán valley, Puebla State, Mexico. Dissertation, Norwegian University of Life Sciences

Ghasemi A, Hemami M, Iravani M Senn J (2012) Seed dispersal by Persian wild ass (Eqqus hemionus onager) in Qatruiyeh National Park, south central Iran. Conference Paper

González-Zamora A, Arroyo-Rodríguez V, Oyama K et al (2012) Sleeping sites and latrines of spider monkeys in continuous and fragmented rainforests: implications for seed dispersal and forest regeneration. PLoS ONE 7:e46852

Graves H (1984) Behavior and ecology of wild and feral swine (Sus scrofa). J Anim Sci 58: 482–492

Groves C, Grubb P (2011) Ungulate taxonomy John Hopkins University Press. Baltimore, USA

Guimarães PR Jr, Galetti M, Jordano P (2008) Seed dispersal anachronisms: rethinking the fruits extinct megafauna ate. PLoS ONE 3:e1745

Hamann A, Curio E (1999) Interactions among frugivores and fleshy fruit trees in a Philippine submontane rainforest. Conserv Biol 13:766–773

Hansen DM, Kaiser CN, Müller CB (2008) Seed dispersal and establishment of endangered plants on oceanic islands: the Janzen-Connell model, and the use of ecological analogues. PLoS ONE 3:e2111

Heffner RS, Heffner HE (1992) Visual factors in sound localization in mammals. J Comp Neurol 317:219–232

Heydon MJ (1994) The Ecology and management of Rainforest Ungulates in Sabah, Malaysia. Technical Report. Institute of Tropical Ecology, Aberdeen University

Hofmann R (1989) Evolutionary steps of ecophysiological adaptation and diversification of ruminants: a comparative view of their digestive system. Oecologia 78:443–457

Hofmann R, Stewart D (1972) Grazer or browser: a classification based on the stomach-structure and feeding habits of East African ruminants. Mammalia 36:226–240

Holden J, Yanuar A, Martyr DJ (2003) The Asian tapir in Kerinci Seblat National Park, Sumatra: evidence collected through photo-trapping. Oryx 37:34–40

Ickes K, Dewalt SJ, Appanah S (2001) Effects of native pigs (*Sus scrofa*) on woody understorey vegetation in a Malaysian lowland rain forest. J Trop Ecol 17:191–206

IUCN (2013) The IUCN Red List of Threatened Species. Version 2013-09. http://www.iucnredlist.org. Downloaded on 10 October 2013

IUCN (2014) The IUCN Red List of Threatened Species. Version 2014-04. http://www.iucnredlist.org. Downloaded on 24 April 2014

Jacobs GH (2012) The evolution of vertebrate color vision. Adv Exp Med Biol 739:156–172

Jadeja S, Prasad S, Quader S et al (2013) Antelope mating strategies facilitate invasion of grasslands by a woody weed. Oikos 122:1441–1452

Jamal A (1994) Food habits of ungulates in dry tropical forests of Gir Lion Sanctuary, Gujarat, India. Acta Theriol 39:185–193

Janzen D, Demment M, Robertson J (1985) How fast and why do germinating guanacaste seeds (*Enterolobium cyclocarpum*) die inside cows and horses? Biotropica 17:322–325

Janzen DH (1981) Digestive seed predation by a Costa Rican Baird's tapir. Biotropica 13:59–63

Janzen DH (1984) Dispersal of small seeds by big herbivores: foliage is the fruit. Am Nat 123:338–353

Janzen DH, Martin PS (1982) Neotropical anachronisms: the fruits the gomphotheres ate. Science 215:19–27

Jarman P (1974) The social organisation of antelope in relation to their ecology. Behaviour 215–267

Jiang P, Josue J, Li X et al (2012) Major taste loss in carnivorous mammals. Proc Natl Acad Sci 109:4956–4961

John Singh A (1981) Importance of fruit in the diet of chital in dry season. J Bombay Nat Hist Soc 78:594

Jordano P (2000) Fruits and frugivory. In: Fenner M (ed) Seeds: the ecology of regeneration in plant communities. CAB International, Wallingford, pp 125–166

Jordano P, Garcia C, Godoy JA et al (2007) Differential contribution of frugivores to complex seed dispersal patterns. Proc Natl Acad Sci 104:3278–3282

Karanth KK, Nichols JD, Karanth KU et al (2010) The shrinking ark: patterns of large mammal extinctions in India. Proc Roy Soc B: Biol Sci 277:1971–1979

Karanth KU, Sunquist ME (1992) Population structure, density and biomass of large herbivores in the tropical forests of Nagarahole, India. J Trop Ecol 8:21–35

Kawanishi K, Sunquist ME (2008) Food habits and activity patterns of the Asiatic golden cat (Catopuma temminckii) and dhole (Cuon alpinus) in a primary rainforest of Peninsular Malaysia. Mammal Study 33:173–177

Kitamura S, Yumoto T, Poonswad P et al (2002) Interactions between fleshy fruits and frugivores in a tropical seasonal forest in Thailand. Oecologia 133:559–572

Kitamura S, Yumoto T, Poonswad P et al (2007) Frugivory and seed dispersal by Asian elephants, Elephas maximus, in a moist evergreen forest of Thailand. J Trop Ecol 23:373–376

Krishnan M (1972) An ecological survey of mammals in India: the Gaur. J Bombay Nat Hist Soc 69:322–349

Leus K (1994) Foraging behaviour, food selection and diet digestion of *Babyrousa babyrussa* (Suidae, Mammalia). Dissertation, University of Edinburgh

Leus K, Macdonald AA (1997) From babirusa (Babyrousa babyrussa) to domestic pig: the nutrition of swine. Proc Nutr Soc 56:1001–1012

Louys J (2012) The future of mammals in Southeast Asia: Conservation insights from the fossil record. In: Louys J (ed) Paleontology in ecology and conservation. Springer, Hiedelberg, pp 227–238

Lundberg J, Moberg F (2003) Mobile link organisms and ecosystem functioning: implications for ecosystem resilience and management. Ecosystems 6:0087–0098

Lynes B, Campbell S (2000) Germination and viability of mesquite (*Prosopis pallida*) seed following ingestion and excretion by feral pigs (*Sus scrofa*). Trop Grasslands 34:125–128

Mackinnon K, Hatta G, Halim H et al (1996) The ecology of Kalimantan Oxford University Press, Oxford

Matías L, Zamora R, Mendoza I et al (2010) Seed dispersal patterns by large frugivorous mammals in a degraded mosaic landscape. Restor Ecol 18:619–627

Matthesius A, Chapman H, Kelly D (2011) Testing for Janzen-Connell effects in a West African montane forest. Biotropica 43:77–83

Mcconkey KR, Prasad S, Corlett RT et al (2012) Seed dispersal in changing landscapes. Biol Conserv 146:1–13
Mcshea WJ, Aung M, Poszig D et al (2001) Forage, habitat use, and sexual segregation by a tropical deer (Cervus eldi thamin) in a dipterocarp forest. J Mammal 82:848–857
Meijaard E, D'huart J, Oliver W (2011) Family Suidae (Pigs). Handbook of the mammals of the world, vol 2
Meijaard E, Sheil D, Nasi R et al (2005) Life after logging: reconciling wildlife conservation and production forestry in Indonesian Borneo CIFOR, Bogor, Indonesia
Middleton BA, Mason DH (1992) Seed herbivory by nilgai, feral cattle, and wild boar in the Keoladeo National Park, India. Biotropica 538–543
Mishra S, Gaillard C, Hertler C et al (2010) India and Java: contrasting records, intimate connections. Quatern Int 223:265–270
Mitchell J, Dorney W, Mayer R et al (2009) Migration of feral pigs (Sus scrofa) in rainforests of north Queensland: fact or fiction? Wildl Res 36:110–116
Moles AT, Westoby M (2006) Seed size and plant strategy across the whole life cycle. Oikos 113:91–105
Moriarty A (2004) The liberation, distribution, abundance and management of wild deer in Australia. Wildl Res 31:291–299
Mouissie A, Van Der Veen CE, Veen GC et al (2005) Ecological correlates of seed survival after ingestion by fallow deer. Funct Ecol 19:284–290
Myers JA, Vellend M, Gardescu S et al (2004) Seed dispersal by white-tailed deer: implications for long-distance dispersal, invasion, and migration of plants in eastern North America. Oecologia 139:35–44
Nathan R, Muller-Landau HC (2000) Spatial patterns of seed dispersal, their determinants and consequences for recruitment. Trends Ecol Evol 15:278–285
O'connor S-J, Kelly D (2012) Seed dispersal of matai (*Prumnopitys taxifolia*) by feral pigs (*Sus scrofa*). N Z J Ecol 36:228
O'farrill G, Galetti M, Campos-Arceiz A (2013) Frugivory and seed dispersal by tapirs: an insight on their ecological role. Integrative zoology 8:4–17
Peichl L (2005) Diversity of mammalian photoreceptor properties: adaptations to habitat and lifestyle? Anat Rec Part A: Discover Mol Cell Evol Biol 287:1001–1012
Peled T (2010) Reintroductions as an ecosystem restoration tool: a case study of reintroduced ungulates as vectors for seed-dispersal. Dissertation, Ben-Gurion University of the Negev
Prasad S (2011) Seed dispersal in the tropical dry forests of Mudumalai, Southern India. Dissertation, Indian Institute of Science
Prasad S, Krishnaswamy J, Chellam R et al (2006) Ruminant-mediated seed dispersal of an economically valuable tree in indian dry forests. Biotropica 38:679–682
Prasad S, Pittet A, Sukumar R (2010) Who really ate the fruit? a novel approach to camera trapping for quantifying frugivory by ruminants. Ecol Res 25:225–231
Prasad S, Sukumar R (2010) Context-dependency of a complex fruit–frugivore mutualism: temporal variation in crop size and neighborhood effects. Oikos 119:514–523
Provenza FD (1995) Postingestive feedback as an elementary determinant of food preference and intake in ruminants. J Range Manage Arch 48:2–17
R-Core-Team (2013) R: A Language and Environment for Statistical Computing. R Development Core Team, Vienna, Austria
Rabinowitz A, Myint T, Khaing ST et al (1999) Description of the leaf deer (*Muntiacus putaoensis*), a new species of muntjac from northern Myanmar. J Zool 249:427–435
Robinson JG, Bennett EL (2000) Hunting for sustainability in tropical forests. Columbia University Press
Riitters K, Wickham J, O'neill R et al (2000) Global-scale patterns of forest fragmentation. Conserv Ecol 4:3
Rizvanovic A, Amundin M, Laska M (2013) Olfactory discrimination ability of Asian elephants (*Elephas maximus*) for structurally related odorants. Chem Senses 38:107–118

Samansiri K, Weerakoon DK (2008) A study on the seed dispersal capability of Asian elephants in the northwestern region of Sri Lanka. Gajah 28:19–24
Sankar K (1994) The ecology of three large sympatric herbivores (chital, sambar and nilgai) with special reference for reserve management in Sariska Tiger Reserve. University of Rajasthan, Jaipur
Saunders G, Kay B (1991) Movements of Feral Pigs (*Sus Scrofa*) at Sunny Corner, New South Wales. Wildlife Research 18:49–61
Schaller GB (1967) The deer and the tiger. University of Chicago Press
Schiviz AN, Ruf T, Kuebber-Heiss A et al (2008) Retinal cone topography of artiodactyl mammals: influence of body height and habitat. J Comp Neurol 507:1336–1350
Schmidt M, Sommer K, Kriebitzsch W-U et al (2004) Dispersal of vascular plants by game in northern Germany. Part I: Roe deer (*Capreolus capreolus*) and wild boar (*Sus scrofa*). Eur J Forest Res 123:167–176
Scowcroft PG, Hobdy R (1987) Recovery of goat-damaged vegetation in an insular tropical montane forest. Biotropica 19:208–215
Sekar N, Sukumar R (2013) Waiting for Gajah: an elephant mutualist's contingency plan for an endangered megafaunal disperser. J Ecol 101:1379–1388
Shah N (1993) Ecology of wild ass (*Equus hemionus khur*) in Little Rann of Kutch. Dissertation, Baroda University
Shyan-Norwalt MR, Peterson J, Milankow King B et al (2010) Initial findings on visual acuity thresholds in an African elephant (Loxodonta africana). Zoo biology 29:30–35
Spiegel O, Nathan R (2007) Incorporating dispersal distance into the disperser effectiveness framework: frugivorous birds provide complementary dispersal to plants in a patchy environment. Ecol Lett 10:718–728
Sridhara S, Edgaonkar A, Kumar A (2013) Understorey structure and refuges from predators influence habitat use by a small ungulate, the Indian chevrotain (Moschiola indica) in Western Ghats, India. Ecol Res 28:427–433
Stevens CE, Hume ID (1995) Comparative physiology of the vertebrate digestive system. Cambridge University Press, Cambridge
Sugnaseelan S, Prescott NB, Broom DM et al (2013) Visual discrimination learning and spatial acuity in sheep. Appl Anim Behav Sci 147:104–111
Terborgh J (2013) Using Janzen-Connell to predict the consequences of defaunation and other disturbances of tropical forests. Biol Conserv 163:7–12
Thomson C, Challies C (1988) Diet of feral pigs in the podocarp-tawa forests of the Urewera Ranges. N Z J Ecol 11:73–78
Van Dung V, Giao PM, Chinh NN et al (1993) A new species of living bovid from Vietnam. Nature 363:443–445
van Strien NJ, Steinmetz R, Manullang B, Sectionov Han KH, Isnan W, Rookmaaker K, Sumardja E, Khan MKM Ellis S (2008) Rhinoceros sondaicus. The IUCN Red List of Threatened Species 2008: e.T19495A8925965. http://dx.doi.org/10.2305/IUCN.UK.2008.RLTS.T19495A8925965.en
Velho N, Isvaran K, Datta A (2012a) Rodent seed predation: effects on seed survival, recruitment, abundance, and dispersion of bird-dispersed tropical trees. Oecologia 169:995–1004
Velho N, Ratnam J, Srinivasan U et al (2012b) Shifts in community structure of tropical trees and avian frugivores in forests recovering from past logging. Biol Conserv 153:32–40
Vellend M, Myers JA, Gardescu S et al (2003) Dispersal of Trillium seeds by deer: implications for long-distance migration of forest herbs. Ecology 84:1067–1072
Wang BC, Smith TB (2002) Closing the seed dispersal loop. Trends Ecol Evol 17:379–386
Wenninger PS, Shipley LA (2000) Harvesting, rumination, digestion, and passage of fruit and leaf diets by a small ruminant, the blue duiker. Oecologia 123:466–474
Whitten T, Mustafa M, Henderson GS (1987) The Ecology of Sulawesi Gadja Mada. University Press, Yogjakarta
Wilson DE Mittemeier RA (2011) Handbook of the Mammals of the World-Volume 2. Hoofed Mammals. Lynx Edicions

Wright SJ, Hernandéz A, Condit R (2007)The bushmeat harvest alters seedling banks by favoring lianas, large seeds, and seeds dispersed by bats, birds, and wind. Biotropica 39:363–371

Wulffraat S (2006) Diet studies of Bearded Pigs in Kayan Mentarang National Park, Borneo, Indonesia. Suiform Soundings??? 6:17–18

Yasuda M, Miura S, Ishii N et al (2005) Fallen fruits and terrestrial vertebrate frugivores: a case study in a lowland tropical rain forest in Peninsular Malaysia. In: Forget P-M, Lambert JE, Vander Hulme P E, Wall SB (eds) Seed fate: predation, dispersal and seedling establishment. CABI Publishing, Wallingford, pp 151–174

Yearsley JM, Villalba JJ, Gordon IJ et al (2006) A theory of associating food types with their postingestive consequences. Am Nat 167:705–716

Zonderland JJ, Cornelissen L, Wolthuis-Fillerup M et al (2008) Visual acuity of pigs at different light intensities. Appl Anim Behav Sci 111:28–37

Chapter 6
Behavioural Ecology of a Grassland Antelope, the Blackbuck *Antilope cervicapra*: Linking Habitat, Ecology and Behaviour

Yadvendradev V. Jhala and Kavita Isvaran

Abstract The blackbuck (*Antelope cervicapra*), a medium-sized antelope endemic to the Indian subcontinent, is found in diverse shortgrass and savanna habitats. It displays extreme sexual dimorphism and highly variable social and mating systems. The behavioural ecology, demography and nutritional physiology of the blackbuck appear to be closely adapted to a stochastic low-resource environment and predation pressure. The chapter, based on extensive coverage of blackbuck habitats across India and intensive long-term data from a typical blackbuck habitat, elucidates these adaptations and explores links between habitat, ecology and behaviour. Blackbuck populations fluctuate mostly in response to catastrophic events, such as floods and droughts. They show high fecundity and a fast numerical response to predation and food availability. Combined predation on calves and adults by jackals, wolves and feral dogs can potentially regulate their population below nutritional carrying capacity. Predation by wolves and sex differences in mating strategies seem to shape the demographic structure of blackbuck. Male blackbuck have shorter life spans than females and males show a mortality peak in midlife corresponding to a peak in mate-competition. Blackbuck are primarily grazers and can survive on seasonally low-quality diets (crude protein <3 %) by catabolizing proteins, and reducing movement and forage intake during summer. Blackbuck respond to an arid environment by producing concentrated alkaline urine and dry feces. However, they need to drink when forage water content is below 30 %. Blackbuck social organisation is highly variable and appears to be strongly influenced by habitat. Group sizes increase with habitat openness and resource availability. A comparison of peninsular Indian ungulates reveals that blackbuck in

Y.V. Jhala
Wildlife Institute of India, Dehradun, Uttarakhand, India
e-mail: jhalay@wii.gov.in

K. Isvaran (✉)
Centre for Ecological Sciences, Indian Institute of Science, Bengaluru, India
e-mail: kavita@ces.iisc.ernet.in

© Springer Science+Business Media Dordrecht 2016

F.S. Ahrestani and M. Sankaran (eds.), *The Ecology of Large Herbivores in South and Southeast Asia*, Ecological Studies 225,
DOI 10.1007/978-94-017-7570-0_6

semi-arid grasslands show, by far, the largest group sizes. Blackbuck mating system similarly shows a close association with habitat. Like the large group sizes, the rare and spectacular lek mating system is associated with high local animal numbers and open grassland habitat. In most populations, local numbers are lower and resource-based mating territoriality is the dominant mating system. Taken together, the behaviour, nutritional ecology, breeding biology and demography of the blackbuck are suggestive of an antelope highly specialised to the shortgrass semi-arid biome.

Keywords Antelope ecology · Behavioural plasticity · *Antilope cervicapra*

6.1 Introduction

Antelope, ungulates belonging to several clades within the family Bovidae, are closely associated with open, semi-arid savanna habitats. Species diversity in several antelope clades is thought to represent a remarkable recent evolutionary radiation closely related to the spread of savanna habitats globally (Hassanin and Douzery 1999; Janis 2008; Bouchenak-Khelladi et al. 2009). Current antelope diversity and densities are particularly high in savannas (McNaughton and Georgiadis 1986; du Toit and Cumming 1999), habitats which are also often associated with spectacular behaviour such as large social group sizes, vast migrations and mating aggregations (Jarman 1974; Fryxell 1995).

Most of our information on the links between habitat and the ecology and behaviour of large herbivores, including antelope, comes from Africa. There are fewer antelope in Asia, but these species are found in a wide range of habitats. One such Asian antelope, the blackbuck *Antilope cervicapra*, is a good model for studying the ecology of an herbivore in dry grasslands. The blackbuck, closely related to gazelles (Hassanin and Douzery 1999), is a medium-sized (30–40 kg), group-living grazer endemic to the Indian subcontinent (Ranjitsinh 1989). Blackbuck are highly sexually dimorphic; whereas females are dully coloured and hornless, males show black and white contrasting coloration and long spiralling horns. This antelope is found in diverse habitats and shows unusually wide variation in behaviour, both among and within populations, thus allowing us to examine links between habitat, ecology and behaviour.

In this chapter, we first describe the population ecology of blackbuck in a typical semi-arid grassland habitat. Next, we describe the foraging ecology of blackbuck and possible physiological and behavioural adaptations to feeding on grass, and related issues such as adaptations to limited water. We then describe the social organisation and mating system of blackbuck and the influence of density and the habitat on social and mating behaviour. Finally, we present a synthesis outlining the links between habitat, ecology and behaviour and future directions.

6.2 Population Ecology of Blackbuck

6.2.1 High Local Abundance and Distinct Patterns in Population Growth

Blackbuck populations, like most ungulates, are capable of irrupting when offered protection from hunting and given access to abundant resources (Caughley 1977). Such a population response makes them locally abundant, which can potentially be a cause of conflict with local farmers (Jhala 1993a). Furthermore, such responses suggest that blackbuck evolved under conditions of highly stochastic environmental and/or high predation pressure. Stochastic environments may favour relatively high fecundity that allows ungulate populations to respond fast to stochastic increases in resource availability. Strong predation pressure can maintain ungulate populations below nutritional carrying capacity and when predation pressure is reduced, ungulate populations may show rapid growth towards the carrying capacity of the environment (Errington 1946; Bowyer et al. 2014). Here, we present long-term data from Velavadar National Park in the Bhal region of Gujarat state in India , a major stronghold of the species, that has helped unravel the mechanism of population regulation in blackbuck.

The Bhal area is a flat alluvial plain of about 2500 km^2 formed by silt deposits from the rivers Narmada, Sabarmati, Mahi and Tapi, which empty into the Gulf of Cambay and the Arabian Sea. The water table in the Bhal is only a few metres below the surface and is saline. Combined with a low rainfall regime (mean $\pm$ SD = 518 $\pm$ 215 mm), the Bhal, till recently, was essentially a semi-arid grassland interspersed with marsh lands and bare saline patches. The Bhal habitat is, thus, probably similar to that in which the blackbuck evolved $\sim$1.9–3.4 million years ago (Hassanin and Douzery 1999), during a dry spell, when forests gave way to shortgrass systems over much of India (Mani 1974). Blackbuck numbers and density have historically been high, and herds of several hundred to over a thousand animals were a common sight till the last century (Ranjitsinh 1989). Over the past 80 years, much of the Bhal region has been converted to agricultural fields, with rainfed seasonal cropping. More recently, it has also become a hub of industrial development. An exotic shrubby tree, the mesquite (*Prosopis juliflora*), was introduced in the Bhal about 100 years ago to provide much-needed fuel wood for local communities. *P. juliflora* has since become invasive and has converted the semi-arid grasslands to thorn forest habitat. The last remaining grassland habitat in the Bhal region now occurs within the only protected area in the region, the Velavadar Blackbuck National Park (21°56′ N 72°10′ E), and covers an area of approximately 34 km^2. Grasslands dominated by *Dicanthium-Sporobolus* communities, saline areas dominated by *Suaeda* communities, *P. juliflora* scrubland and tidal mudflats constitute the National Park. The exotic *P. juliflora* is managed within the National Park; its seedlings are manually removed every year from the grassland habitat, while the scrub habitat is thinned and managed as a savanna.

The open nature of the Bhal landscape and the preference of blackbuck for open grassland habitat permitted the use of the total-count approach for estimating population sizes. Blackbuck have a high-intensity grazing peak during the early morning hours, when all individuals are on their feet and intently grazing with very slow movement (Jhala 1991, 1997). It was during this time period that total counts were conducted using six to eight teams of observers allocated to different demarcated regions of the Park. Counts were done on foot, four-wheel drive vehicles and from horseback, and were repeated on three consecutive mornings in winter (December–February) of each year. Observations on movement of blackbuck herds, their resting sites and grazing areas were obtained a week before the total counts, which enabled planning and demarcating observation blocks and team allocations. The maximum number recorded among the three counts was considered as the annual population estimate, and the difference between the three was 4.2–8.8 %. Due to good a priori knowledge of blackbuck herd locations, counting different areas of the Park simultaneously with trained observers, timing counts to coincide with high-intensity foraging peaks of blackbuck and by completing the count within a couple of hours, we were reasonably certain that our estimates provided a fairly accurate estimate of the total abundance of blackbuck in the Park.

The Bhal region is prone to climatic extremes, with droughts and floods being a regular feature. The blackbuck population fluctuates in response to both droughts and floods, which are both major sources of catastrophic mortality. After the establishment of the National Park, the Velavadar population has undergone three such severe declines (Fig. 6.1a). After each decline, the blackbuck population responded with a growth spurt, with a realised instantaneous growth rate (r) that ranged between 0.2 and 0.27 ($P < 0.03$, $R^2 = 0.91$–0.98). However, after the severe drought of 1986–87, although the blackbuck population recovered, it did not achieve the earlier population size of $\sim$2200. Instead it stabilised at approximately 1200 individuals (Fig. 6.1a). Regression of natural $\log_e$-transformed population size versus time (1989–2001) (Skalski et al. 2005) shows that the *slope* of the line, which estimates instantaneous rate of population growth (r) (Caughley 1977), is not different from zero. This suggests that the population was stationary between 1989 and 2001 and regulated well below the historical carrying capacity of Velavadar National Park. Top–down processes, specifically mortality through predation, rather than bottom–up processes (primarily limited resources increasing mortality and depressing reproductive rates) are likely to be responsible for regulating blackbuck populations below historical carrying capacities. While predators like wolves were earlier persecuted and largely absent from the Velavadar area, due to protection, these predators had, by this time, established healthy populations in the region (Jhala 1993b). In parallel, good management practices in the Velavadar National Park, which included managing unpalatable weeds and promoting palatable grasses, are likely to have actually enhanced the nutritional carrying capacity. Thus, increased predation rather than reduced resource availability is likely to explain the pattern of a relatively stationary population below the historical carrying capacity.

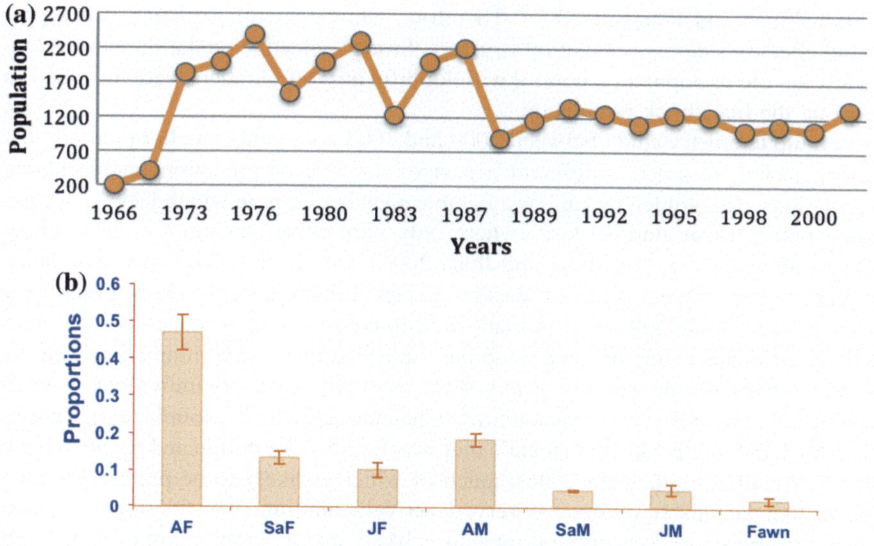

Fig. 6.1 **a** Population size of blackbuck in Velavadar National Park from 1965 to 2001. Between 1988 and 2001 the population was estimated by systematic total count by YVJ. A regression fitted to $\log_e$(population) versus years for the data from 1989 to 2001 suggests that the population was in stationary phase, which is well below the historical carrying capacity of the Park. The *slope* of the regression line, which estimates instantaneous growth rate (r), was −0.0022 (SE = 0.01), with $R^2 = 0.0065$ and $P = 0.52$; suggesting that the population was stationary between 1989–2001. **b** The proportion of the blackbuck population of Velavadar National Park in different age and sex classes between 1998 and 2001 during the stable age distribution. AF—adult female, SaF—subadult female, JF—juvenile female, AM—adult male, SaM—subadult male, JM—juvenile male. Error *bars* are standard errors between years

6.2.2 *The Role of Predation in Controlling the Blackbuck Population*

The Indian peninsular wolf, *Canis lupus pallipes,* is currently the main predator of adult blackbuck in relatively natural semi-arid habitats, and together with the golden jackal (*Canis aureus*), is also the main predator of blackbuck calves (Jethva and Jhala 2004). Feral dogs, if not controlled, are also a major threat to blackbuck, both adults and calves. Wolves in the Bhal area were intensively studied using radio telemetry between 1994 and 2002 (Jhala 1993b; Jethva and Jhala 2004). During this time, a single pack consisting of four to six adults, used Velavadar National Park as part of its territory. The diet of wolves was almost exclusively blackbuck, and wolves were observed to feed on average at 4-day intervals and consume 1.8 kg/animal/day (Jethva and Jhala 2004). Since wolves are territorial, and the small Velavadar National Park (34 km^2 during the period of the study) with an abundant blackbuck population was within the territory of a single wolf pack, the numerical response of wolves to this localised food source was limited (Jhala

1993b; Mech and Peterson 2003). Therefore, although wolf predation had a profound effect in shaping population structure through selective predation on breeding males, its role in population control was doubtful as wolf numbers were too small to regulate the blackbuck population.

A radio telemetry study between 2000 and 2003 on another smaller predator, the golden jackal, revealed a different aspect of the role of predation on blackbuck populations. The golden jackal has a flexible social system, in which jackals occupy home ranges of around 30 km^2, where only core areas (<5 km^2) of their home ranges are exclusive (Aiyadurai and Jhala 2006). During times of super abundance of food, i.e. the calving peaks of blackbuck, jackals congregate in the National Park from as far as 30–40 km away. A mark-recapture population estimate, together with density estimates using distance sampling, indicated that jackal numbers within the National Park during calving peaks were as high as 38–40 individuals with a density of 1.01 (SE = 0.21) jackal/km^2 (Chauhan and Jhala unpublished). During blackbuck calving peaks, the jackal's diet was found to be dominated by blackbuck (33 %, Aiyadurai and Jhala 2006), much of which is likely to be predated hidden calves that jackals have been observed actively hunting. Considering wolf and jackal densities and consumption rates, it is likely that a combination of jackal and wolf predation on calves resulted in regulating blackbuck population below the nutritional carrying capacity. Such regulation did not occur prior to the 1987 crash possibly because, previously, predators were persecuted and even controlled in the region, and therefore, their role in population regulation of blackbuck would have likely been minimal. Unfortunately, no reliable information on predator densities exists prior to our work.

6.2.3 High Female Fecundity and a Distinct Seasonality in Calving

Demographic parameters, including sex ratio, age structure and calf to adult female ratio were also estimated for the Velavadar blackbuck population. Between 1998 and 2001, representative blackbuck herds from across Velavadar National Park comprising over 40 % of the population, were videographed each year in the winter. Since the herds were spatially distinct and were videographed during the grazing peak while conducting the total counts, there was minimal chance of recording the same individuals and/or herds more than once, and therefore, the sample was considered as "without replacement" (Skalski et al. 2005). The video data were analysed in slow motion and each individual blackbuck was aged and sexed based on methods outlined by Mungall (1978) and Jhala (1991) and refined based on observations of captive blackbuck of known ages (Fig. 6.1b).

The video data were also used to estimate fecundity. The ratio of young to reproductive female has been found to be 0.37 in the winter. However, since blackbuck calves have a "lying out" stage, wherein the calf is hidden for most of the

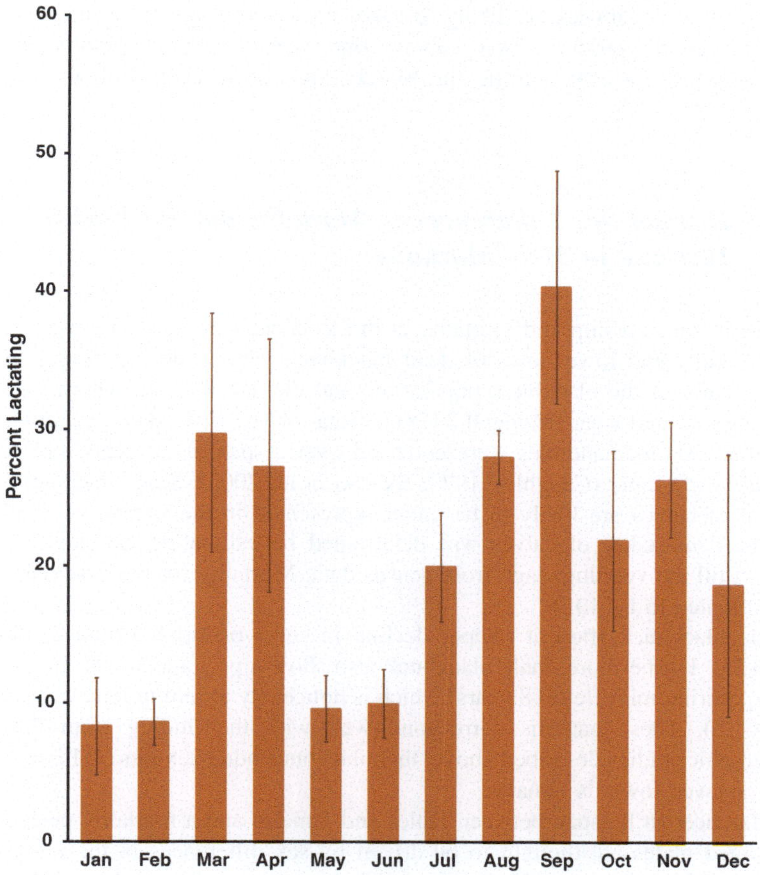

Fig. 6.2 Percent of adult female blackbuck in the population observed to be lactating in each month. The error *bars* are standard errors

day and does not accompany the mother on foraging bouts (Mungall 1978), it is likely that the number of calves is underestimated by observation data. Blackbuck give birth to a single calf and can calve throughout the year. We determined fecundity and seasonality of calving by determining the proportion of lactating females (Priyadarshini 2005) in the population (Fig. 6.2); this parameter is likely to be a better estimator of calving since observation of young calves is difficult.

The blackbuck population at Velavadar shows two clear calving peaks, one just after the monsoon (September), and the other just prior to the nutritionally stressed summer (March–April). Assuming that lactating udders can be visually detected for up to 2 months after parturition, we averaged the proportion of adult lactating females every 2 months across 3 years. The total annual percentage of lactating females is 130 %, i.e. on average, a blackbuck female produces calves 1.3 times per year. Most adult females are likely to calve once a year but those in good body condition could produce two calves, one during the post-monsoon peak (September) and another during the March–April peak, spaced about 6 months apart.

6.2.4 Distinct Sex Differences in Mortality and an Earlier Increase in Male Mortality

Patterns in survivorship and mortality at the Velavadar population were also estimated. Skulls and lower jaws of dead blackbuck were collected during the stationary phase of the blackbuck population, and the age at death determined from tooth eruption and wear (Mungall 1978). A total of 167 male skulls/mandibles and 200 female skulls/mandibles were collected over a span of 10 years and aged to generate a life table (Caughley 1977; Skalski et al. 2005). Since the fragile skull material of calves are likely to be under represented in the sample of skulls and mandibles, mortality of calves was determined by estimating calf numbers that survived till the yearling stage from census data. Mortality for the first year of life was estimated to be 40 %.

Male blackbuck show a steeper decline in survivorship (l_x) than do females (Fig. 6.3a). Furthermore, male blackbuck also show a prominent peak in mortality rate (q_x) during mid-life (6–8 years) which is noticeably absent in female blackbuck (Fig. 6.3b). These patterns correspond well with the finding from the other demographic studies described above, that blackbuck adult sex ratio (M:F = 0.39) is highly skewed towards females.

Differences in lifespan between males and females and a mortality peak among prime-aged males are thought to be driven by sex differences in the intensity of sexual selection (Trivers 1972; Stearns 1992; Clutton-Brock and Isvaran 2007). The Velavadar blackbuck population shows a lek mating system, which is characterised by a strong male mating skew and intense male–male competition for access to females (Isvaran and Jhala 2000). In this mating system, due to the high variance in male mating success and, thus, the potential for a few males to experience extreme mating gains, males are expected to exhibit costly behavioural traits, such as energetically expensive displays, frequent fights, and behaviour that may increase their vulnerability to predation. Lekking male blackbuck appear to engage in costly behaviour (Sect. 6.4 below). They may also be more vulnerable to predation, compared with individuals (females and non-breeding males) in groups. Grouping in large herbivores is most often a defence against predation. Female blackbuck are largely found in herds (see Sect. 6.4) and even males, when not defending mating territories, form groups. Wolves, the dominant predators of blackbuck in Velavadar National Park, are selective predators (Mech 1970) and predate vulnerable prey. Compared with herds, territorial males may face a higher risk of wolf predation by being relatively solitary at predictable locations, for example, standing on dung

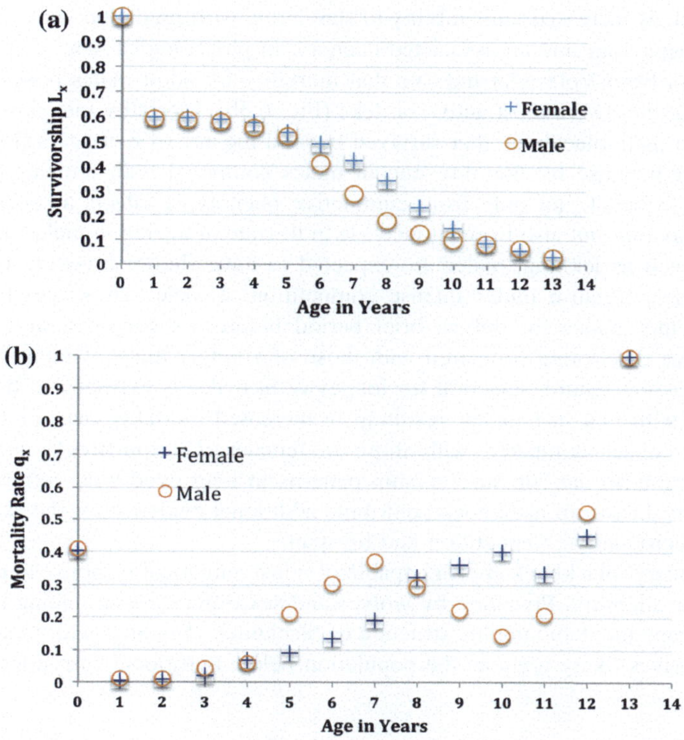

Fig. 6.3 **a** Survivorship, and **b** Mortality rate of blackbuck in Velavadar National Park based on life table analysis of 167 male and 200 female blackbuck samples

piles and attracting female blackbuck and wolves alike. Adult male blackbuck dominate wolf diets in the Velavadar population (Jhala 1993b; Jethva and Jhala 2004) and wolves were often observed killing male blackbuck on the lek (Jhala pers. obs.). So, overall, breeding male blackbuck are expected to have higher mortality rates than females and non-breeding males of the same age.

Furthermore, males are expected to be reproductively successful for a relatively short period of time compared to females. Given the intense competition for lek territories, it seems likely that only males that have substantial body weight and horn length manage to secure prime territories on the lek; such characteristics develop in blackbuck after the age of over 4 years. It is also likely that given the intense competition and costly behaviour, males are likely to maintain lek territories only for a few years before being outcompeted by younger males. Thus, the lek mating system may explain the sex differences both in longevity and in the pattern in mortality rate during their lifespan: during their prime reproductive years when males are likely to occupy territories on the lek their mortality rate increases due to factors such as increased vulnerability to starvation, wolf predation and disease due to costly behaviour and an associated decrease in physiological condition.

The data from Velavadar indicate that mortality for adult males peaked during the age that they were most active on leks (Fig. 6.3b). Mortality rate subsequently declined in male blackbuck that survived beyond the age of 8–9 years (Fig. 6.3b). The earlier increase in mortality rate in males compared with females has been suggested to result, not only from cumulative phenotypic effects associated with costly behaviour, but also from an increase in the rate of ageing in males. In mating systems, such as lekking, males are expected to have shorter breeding life spans than females, because under intense competition, a male can successfully outcompete other males for only a brief period before his condition and physical performance deteriorate compared with those of younger males. Given the reduction in breeding tenure, selection for longevity in males is expected to be relaxed compared with that in females, resulting in increased mortality rates and reduced lifespan in males compared with those in females (Clutton-Brock and Isvaran 2007). Long-term data on survivorship patterns in long-lived wild vertebrates are rare and the data from blackbuck contribute additional evidence for the association between reproductive competition and lifespan.

In summary, blackbuck appear capable of a fast numerical response to predation and food availability. Predation by wolves and sex differences in mating strategies seem to shape the demographic structure of blackbuck. Together with predation by jackals, wolves likely regulate the population below nutritional carrying capacity.

6.3 Foraging Ecology of Blackbuck

6.3.1 A Grazer Adapted to Steep Seasonal Fluctuations in Nutritional Quality of Forage

Blackbuck prefer short grasslands (<50 cm), and avoid wooded habitats and tall grasslands (Jhala 1991). Blackbuck diet primarily consists of grasses and therefore, there is profound seasonality in its nutritional ecology. Blackbuck have access to high quantity and quality forage during the monsoon (July–September) and early winter (October–December) coinciding with periods of grass growth. However, after seeding, above ground grass parts lose nutritive quality rapidly during late winter (January–February) and in summer (March–June), seasons when blackbuck experience nutritional bottlenecks as their diet becomes less digestible and with very low protein content. The protein content of the blackbuck diet drops significantly (<4 %) in summer, well below the maintenance requirement for ruminants which is between 5.5–9 % (Robbins 1983) (Fig. 6.4). During this period, protein digestibility is negative, i.e. blackbuck lose more protein via feces than they can obtain from the forage (Fig. 6.4). The digestibility of dry matter declines from a

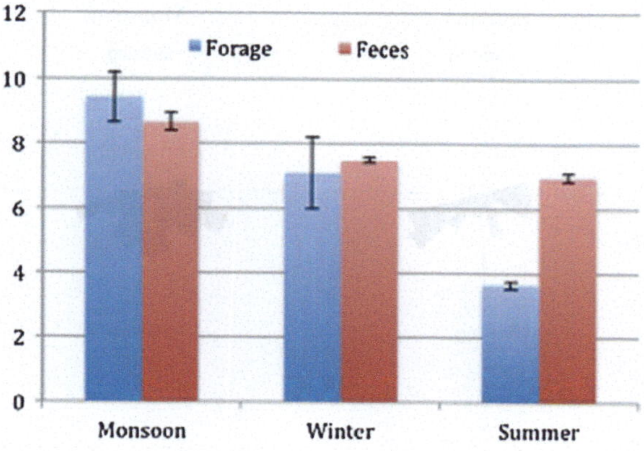

Fig. 6.4 Seasonal percent crude protein content of blackbuck forage and feces in Velavadar National Park (Jhala 1997). Error *bars* are standard errors

high of 76.5 % during the monsoon to a low of 32 % during summers. Blackbuck adapt to this low-quality diet by reducing intake from over 130 $g/kg^{0.75}$ during the monsoon to less than 20 $g/kg^{0.75}$ during summers (Jhala 1997). Blackbuck foraging behaviour also reflects this reduced intake. Foraging time is highest in the monsoon season when both forage quality and moisture content are high. Time invested in foraging is similar between winter and summer seasons (Figs. 6.5 and 6.6a).

Blackbuck movement patterns are also in consonance with their nutritional ecology. The daily distance covered by blackbuck ranges between a maximum of 5.7 and a minimum of 1.5 km. During the winter, blackbuck cover larger distances than in the monsoon or summer (Fig. 6.6b). During the monsoon, the forage is lush with high nutrient content, and therefore blackbuck forage for longer periods of time, but they do not need to move much. In contrast, in the summer, forage quality and quantity are poor and blackbuck adopt a strategy of reducing intake as they are in a negative protein balance. In such a situation, it would be best to conserve body reserves and move little. During the winter, the overall forage quality is low, but by selectively foraging blackbuck can potentially maintain a positive protein digestibility and build up body reserves for the summer. To selectively forage, blackbuck would need to move greater distances during this season when compared with other times of the year.

Like most tropical ungulates, the body condition of blackbuck cycles from good (during monsoon and winters) to poor (during late summers) due to the utilisation of body fat and muscle proteins. However, blackbuck face a more prolonged period of low nutrition in comparison with other tropical ungulates studied in Africa as the dry spell in India lasts for over 9 months in most blackbuck habitats when compared with about 4–6 months in Africa. Also, blackbuck are unusual in that they are relatively small (with correspondingly higher energy requirements) to have specialised on a low-quality forage source i.e. grasses (Hofmann 1989).

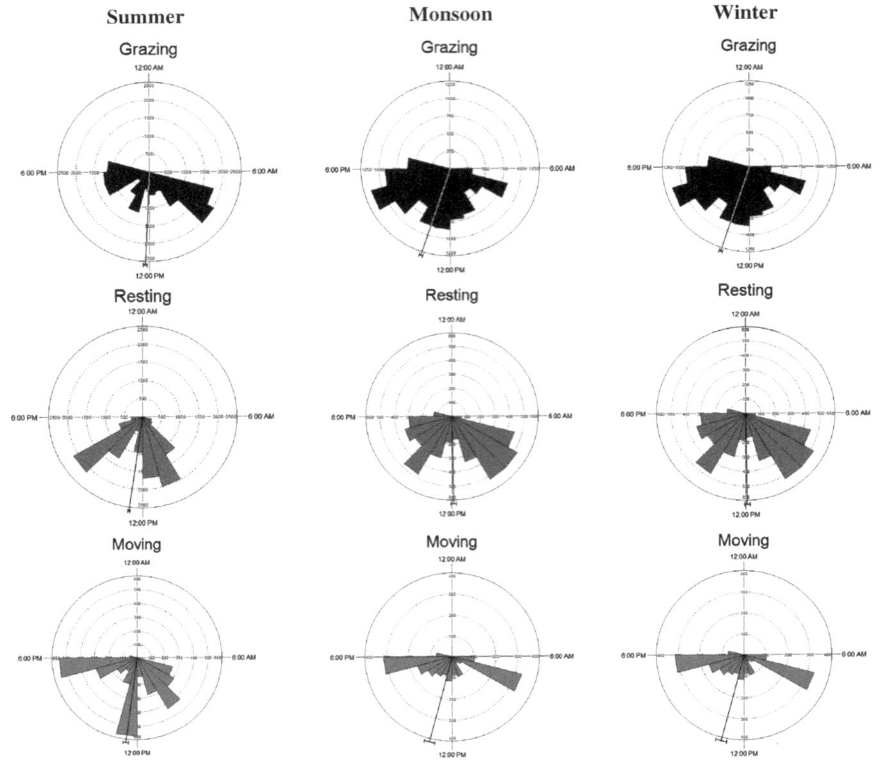

Fig. 6.5 Variation in daily activity pattern across seasons of blackbuck in the Bhal determined by 6 days of continuous scan sampling from dawn to darkness in each season of six different groups ranging from 30 to 128 individuals

6.3.2 Adaptations to Limited Water Availability

Blackbuck are adapted to survive on low-quality grassland systems but are limited in their distribution by the availability of perennial surface water. Thus, unlike the chinkara, *Gazella bennetti*, blackbuck are not found in true deserts but attain their highest densities in semi-arid grassland-scrub systems. Blackbuck have evolved mechanisms for water conservation and when deprived of water, they can increase the concentration of urea in their urine and reabsorb water from their feces (Jhala et al. 1992). The concentration of urea and uric acid in the bloodstream of water-deprived blackbuck increased by 70 % and 42 %, respectively, while their packed cell volume remained the same. While on water deprivation, blackbuck were observed to produce a highly concentrated alkaline urine with a specific gravity of 1.058 and a pH of 9.5, comparable to that of water-deprived Dorcas gazelle (Jhala et al. 1992). These adaptations of the blackbuck are similar to those observed in true desert animals like the camel and gazelle. However, like most

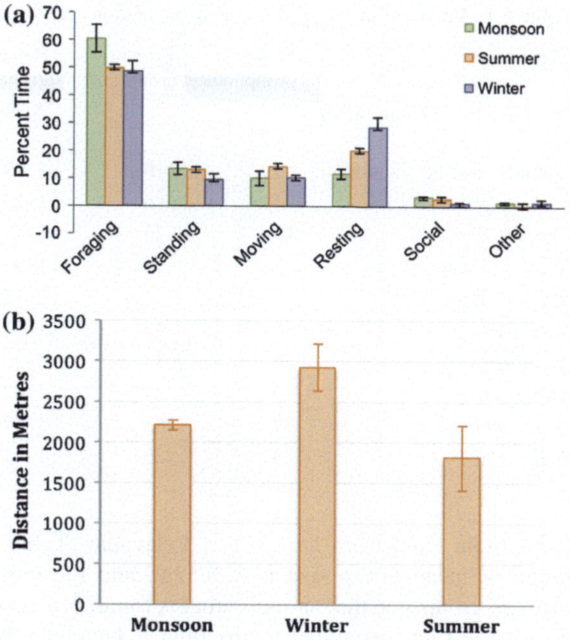

Fig. 6.6 **a** Seasonal time activity budget and **b** Daily distance moved by blackbuck in different seasons in the Bhal (error bars are standard errors). Six herds (30–128 individuals) were followed from dawn to darkness for 6 days in each season. Scan sampling was used to measure behaviour and the movement path of herds was estimated by a hand held global positioning system unit

desert animals, the blackbuck too need to drink when preformed water in the diet is very low (less than 30 %). Such conditions are often encountered in late winter and summer by blackbuck in most habitats, and therefore, blackbuck need to have occasional access to drinking water during these seasons.

6.4 Social Organisation and Behavioural Flexibility

6.4.1 Social Groups: Wide Variation in Group Size Linked to Habitat Structure and Resources

Blackbuck are typically found in groups. The two main exceptions are: territorial males, not all but those that defend dispersed, solitary territories; and females with very young calves, who appear to leave groups to nurse their calves and/or remain in the proximity of their calves. Calves typically have very small daily ranges in their first few weeks of life (Mungall 1978). The main kinds of grouping associations seen are all-male groups, female groups which contain females of all ages and juvenile males and mixed-sex groups with females and males of all ages (Mungall 1978; Ranjitsinh 1989).

Blackbuck group sizes vary tremendously both within and among populations. For example, group sizes are reported to range from 1 to 13 in Guindy National

Table 6.1 Variation in group size across populations of blackbuck in India

Population	Mean group size	Typical group size	Range in group size	Reference
Mahavir Harina Vanasthali			2–30	Srinivasulu and Srinivasulu (1999)
Mahavir Harina Vanasthali	3	16	1–69	Isvaran (2007)
Guindy National Park	4	6	1–23	Isvaran (2007)
Nannaj	19	76	1–183	Isvaran (2007)
Point Calimere	8	26	1–83	Isvaran (2007)
Point Calimere			2–129	Nair (1976)
Rehekuri	6	15	1–31	Isvaran (2007)
Savainagar	6	13	1–40	Isvaran (2007)
Rollapadu	10	38	1–82	Isvaran (2007)
Tal Chappar	37	162	1–325	Isvaran (2007)
Velavadar	53	223	1–419	Isvaran (2007)

Park, India , and from 1 to 423 in Velavadar (Table 6.1). Across populations, the range of mean group size is 3.3–52.9, and the range of typical group size (the average group size that an individual is found in) is 5.6–223.4 (Isvaran 2007). Why do we see such variation? If grouping is beneficial why do we not see uniformly large group sizes across all populations?

The variation among and within populations is most likely maintained by heterogeneity in the trade-offs associated with grouping, arising from variation in ecological conditions (Isvaran 2007). Grouping behaviour is thought to be favoured in open plains species, such as blackbuck, primarily because of its anti-predation benefits (Jarman 1974). Grouping may reduce the risk of predation through several ways. Larger groups may detect a predator earlier, predators find it more difficult to focus on and attack an individual in larger groups (confusion-effect), and the probability of attack reduces in larger groups because of the shared risk (dilution-effect) (Krause and Ruxton 2002). Grouping may also reduce the cost of being vigilant, allowing individuals to take advantage of the vigilance shown by other group members and thereby reducing their individual vigilance. The time saved by reducing individual vigilance could be allocated to other important activities, such as feeding. The main cost associated with grouping appears to be competition for food, and individuals in groups may have to range farther to find sufficient food (Terborgh and Janson 1986).

Evidence from observing individuals in groups of different sizes within a blackbuck population supports the idea that grouping is favoured by anti-predation benefits but constrained by competition for food. In the Velavadar population, the probability of detecting a threat increased with group size (benefit), the time spent vigilant was negatively associated with group size (benefit), and the time spent feeding was positively associated with group size (benefit). However, the distance moved by a group increased with group size (cost) (Fig. 6.7, Isvaran 2007).

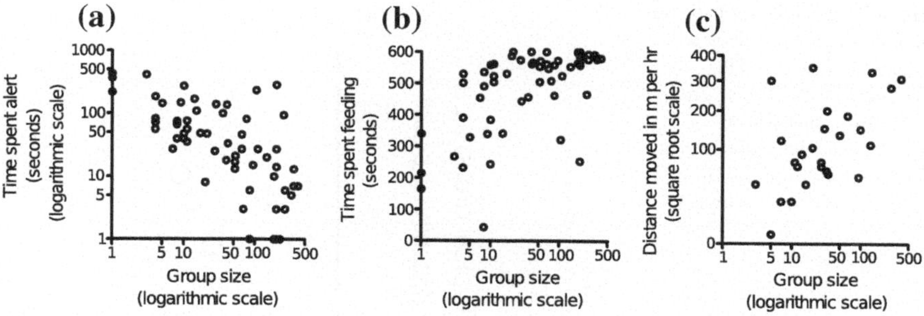

Fig. 6.7 Trade-offs involved in grouping: associations between group size and **a** Time spent alert, **b** Time spent feeding and **c** Distance moved while foraging (reproduced from Isvaran 2007 with kind permission from Springer)

Ecological conditions that influence these costs and benefits can change optimal group sizes and therefore, result in variation in group sizes. A study examining variation in group size among populations found that habitat structure and forage abundance are two ecological conditions that influence trade-offs associated with grouping and thereby, favour variation in group sizes (Isvaran 2007). Group sizes appear to increase with habitat openness, with the largest groups found in areas with large, open habitat patches (Fig. 6.8). Habitat structure is thought to influence group size by affecting the degree to which grouping can provide anti-predation benefits (Brashares and Arcese 2002). Specifically, the anti-predation benefits to grouping are thought to increase with increasing habitat openness. This is because grouping in large open habitats may reduce predation risk by increasing the ability to detect a predator and providing confusion and dilution effects. However, in more closed wooded habitats grouping may not provide as strong anti-predation benefits because individuals may not be able to monitor other individuals in the group very well and may also hinder each other's ability to flee. In support of the idea that anti-predation benefits of grouping are not as strong in more closed habitats, behavioural observations show that after controlling for group size, individuals in closed habitats are more vigilant and detect potential predators earlier. These findings suggest that they do not rely as much on group-related anti-predation benefits as do individuals in open habitats.

Surprisingly predator presence, which might be expected to influence anti-predation benefits, was not found to be associated with group size variation. Typical group size was not larger in populations that live with wolves, predators of adult blackbuck. However, the role of other sources of predation risk, including village dogs and humans, has not yet been assessed. Forage abundance has also been shown to be important (Fig. 6.9) and it appears to influence group size by modulating the cost of competing for forage. Forage abundance was positively related to group size suggesting that where forage is low, the costs of competing for forage outweigh any anti-predation benefits, favouring smaller group sizes (Isvaran 2007).

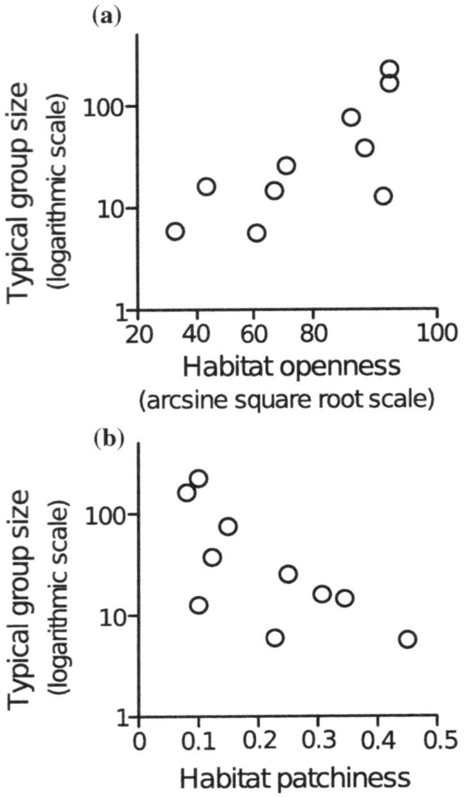

Fig. 6.8 Association between habitat structure **a** habitat openness and **b** habitat patchiness and typical group size across blackbuck populations (reproduced from Isvaran 2007 with kind permission from Springer)

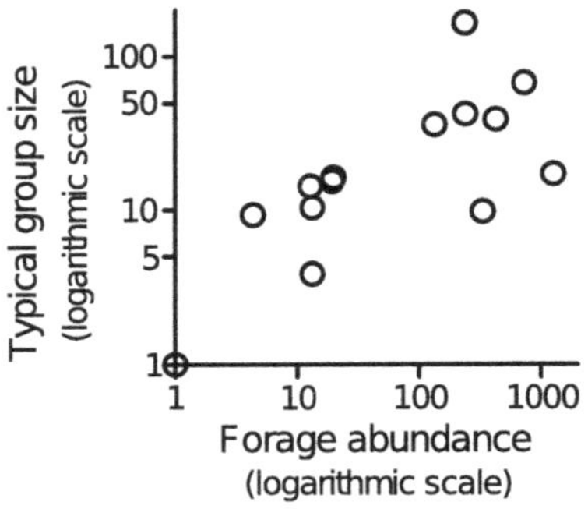

Fig. 6.9 Association between forage abundance and typical group size across blackbuck populations (reproduced from Isvaran 2007 with kind permission from Springer)

To summarise, group sizes in blackbuck appear to reflect a trade-off between anti-predation benefits and the costs of competing for forage. Habitat structure and resource abundance appear to modulate this trade-off. The role of other costs and benefits involved in the trade-off, such as the cost of transmitting diseases in larger groups, and the influence of other ecological variables, such as the risk posed by

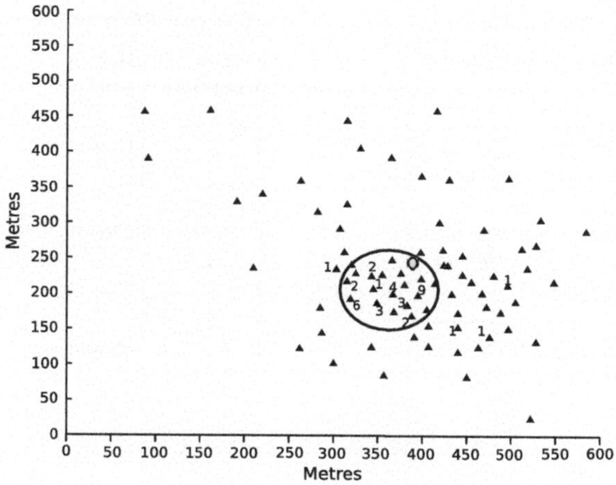

Fig. 6.10 Mating success is greatest in the centre of a blackbuck lek. Each solid *triangle* is the approximate centre of a lek territory, the *circle* indicates the lek centre, and the numbers indicate the number of matings seen at that territory (reproduced from Isvaran and Jhala 2000 with kind permission from Koninklijke Brill NV)

village dogs and humans, is still not known. Further, the importance of interactions between individuals in influencing observed group sizes in natural populations is not fully understood. Observed group sizes appear to be larger than those predicted by the shape of the relationships between costs/benefits and group size (Fig. 6.10); understanding the factors that influence the difference between optimal and stable group sizes in natural populations is an exciting research area.

6.4.2 Extra-Ordinary Group Sizes and a Role for Semi-arid Savannas?

The largest blackbuck groups (numbering to several hundreds) are seen in open semi-arid grasslands (Isvaran 2007). We hypothesise that this is an outcome of (a) habitat structure: open habitat which promotes grouping as an anti-predator strategy; (b) resource abundance: high productivity such that the costs of competition for food are relatively low; and (c) diet: primarily a grazer that can exploit the high productivity of these habitats. A similar association between group size, diet and habitat is seen in African antelope where too spectacular animal aggregations are seen primarily in grasslands and savannas (Jarman 1974; Brashares et al. 2000). Here, we synthesise data on group sizes of peninsular Indian ungulates to examine whether similar associations are seen between group size, diet and habitat across different species (Table 6.2). The data are limited and so only some preliminary inferences can be made.

The most striking result is that blackbuck in open semi-arid grasslands show by far the largest group sizes. This may be, in part, because most of the other studies were conducted in relatively closed habitats—dry deciduous and dry thorn forests—where group sizes are expected to be limited and individuals are expected to rely on

Table 6.2 Group size, diet and habitat association in ungulate species found in peninsular India

Family	Species	Mean group size	Number of populations	Habitat structure	Diet	Reference
Bovidae	*Tetracerus quadricornis*	1.1	2	Closed	Browser selective	Khan et al. (1995); Biswas and Sankar (2002)
Cervidae	*Muntiacus muntjak*	1	1	Closed	Browser selective	Barrette (2004)
Cervidae	*Axis porcinus*	1.8	1	Closed	Grazer selective	Dhungel and O'Gara (1991)
Cervidae	*Cervus unicolor*	2.8	4	Closed	Mixed selective	Varman and Sukumar (1995); Khan et al. (1995); Biswas and Sankar (2002); Bagchi et al. (2008)
Bovidae	*Bos gaurus*	7.2	2	Closed	Mixed unselective	Varman and Sukumar (1995); Biswas and Sankar (2002)
Bovidae	*Gazella bennetti*	2.4	2	Intermediate	Browser selective	Khan et al. (1995); Bagchi et al. (2008)
Bovidae	*Gazella bennetti*	5.7	4	Open	Browser selective	Dookia and Jakher (2013)
Suidae	*Sus scrofa*	3.2	2	Intermediate	Browser selective	Karanth and Sunquist (1992); Biswas and Sankar (2002)
Bovidae	*Boselaphus tragocamelus*	2.4	2	Intermediate	Mixed selective	Biswas and Sankar (2002); Bagchi et al. (2008)
Cervidae	*Axis axis*	5.2	4	Intermediate	Grazer selective	Varman and Sukumar (1995); Biswas and Sankar (2002); Bagchi et al. (2008); Dave and Jhala (2011)
Bovidae	*Antilope cervicapra*	5.25	4	Intermediate	Grazer selective	Isvaran (2007)
Bovidae	*Antilope cervicapra*	25	5	Open	Grazer selective	Isvaran (2007)

Mean group size—if two or more studies report mean group sizes, then the overall mean of individual study means was calculated. If studies for a species included populations in habitats with qualitatively different structure, the mean group size estimates were segregated by habitat before calculating the overall mean group size. Mixed—includes both grass and browse in diet. Selective—selects particular plant parts, e.g., leaves camouflage rather than early detection and flight, to escape predators. Of all the species (other than blackbuck) reported in Table 6.2, only the chinkara *Gazella bennetti* and the nilgai *Boselaphus tragocamelus* are found in more open habitats. However, data on group sizes for these species from more open habitats are relatively scarce. Apart from habitat structure, diet also appears to be very important. The chinkara, the closest relative of blackbuck, even when found in open habitat forms much smaller groups (Rahmani 1990; Dookia and Jakher 2013) than do blackbuck, as is expected for a selective browser. Similarly, examining the species seen in relatively closed habitats, selective browsers, such as barking deer *Muntiacus muntjak* and four horned antelope *Tetraceros quadricornis*, show smaller group sizes than mixed feeders (whose diet includes both grass and browse), such as sambhar *Cervus unicolor*, selective grazers, such as chital *Axis axis*, or unselective grazers, such as gaur *Bos gaurus* (which feed on fairly coarse forage, also see Chap. 4).

Examining closely related species further highlights potential influences of body size, diet and habitat structure on group size. The hog deer *Axis porcinus* is predominantly a grazer like the closely related chital, but is smaller and its habitat, dense tall grasslands, is much more closed than the open dry deciduous forests with grassy clearings that the chital is found in. Interestingly, hog deer are reported to form larger groups when they congregate in short grasslands, where visibility is relatively high. As another example, the four horned antelope, a more selective browser and found in more wooded habitats than its close relative, the nilgai, also forms smaller groups than the nilgai. Detailed studies on the social behaviour and ecology of these species and variation in their group sizes in relation to ecological conditions are needed to determine the ecological processes driving these associations.

6.4.3 Mating System Based on Territoriality

Blackbuck show two broad types of mating systems: resource-based territoriality and lek territoriality (Mungall 1978; Ranjitsinh 1989; Prasad 1989; Isvaran 2005a). In some populations, males defend mating territories in areas visited by females while grazing. As female groups move through these territories, males may display to and try to mate with those in estrous. In a few populations, males defend tiny display territories which do not contain resources, such as forage or water. These territories are clustered (a cluster is called a lek) and are thought to attract females in search of mates. Females may visit one or more territories, mate with a male and then leave the lek. Lekking is thought to be a high-cost high-benefit mating tactic (Hoglund and Alatalo 1995; Isvaran and Jhala 2000). Males feed less and engage in costly displays and fights when compared to males in all-male herds and males following other tactics. Male mating success is highly skewed on leks with relatively few males obtaining the majority of matings. The evolution of this mating system is, however, still not fully understood.

6.4.4 Mating System: What Maintains Variation in Mating System Among and Within Populations and When Do We See Lekking?

The categories, resource-based territoriality and lek territoriality, hide much variation. Blackbuck mating territoriality varies from solitary dispersed territories to tightly clustered classical leks (Prasad 1989; Ranjitsinh 1989; Isvaran and Jhala 2000). Thus, lekking can be viewed as one extreme in a continuum of territorial distributions. Lekking is a rare mating system seen in very few mammals, and blackbuck provide an opportunity to identify the ecological conditions that favour lekking, particularly since this antelope displays so much variation in mating system.

Across populations, the degree of clustering of territories is closely correlated with local female density, captured well by female group size (Isvaran 2005a). Leks are found in populations where typical group sizes are extremely high (e.g. 325 at Velavadar in Gujarat). One explanation for this relationship is that local female densities represent the maximum possible benefits available to a male following a particular mating tactic. Thus, costly tactics such as lekking are likely to be favoured when mating benefits are sufficiently high. Further evidence for this hypothesis comes from examining the frequency of other male mating tactics. The frequency of males holding solitary, dispersed territories (low-cost tactic) is highest in populations where females move in small groups and local female density is low, and this frequency decreases with group size. The frequency of holding territory in a loose cluster (intermediate-cost tactic) increases initially with group size and decreases when female group sizes are very high (Isvaran 2005b).

Apart from variation across populations in mating system, many blackbuck populations also show much variation in the size and clustering of territories within a population (Isvaran 2005a, b). At this smaller scale, within a population, local female distribution again appears to be important. The number of territories and territory clustering were found to be closely correlated with local female numbers and female group size across 1 km^2 spatial units within a population. This suggests that, at this scale too, similar to the inference from among-population comparisons, males adopt more costly tactics when mating benefits are relatively high. Thus, the wide variation in mating tactics within population, may be explained, in part at least, by environmental heterogeneity. In populations in which ecological conditions, particularly habitat openness and forage abundance, vary over space, female distribution is expected to vary in space (as suggested by the data on group sizes) from areas of low local female density to areas of high local density. Accordingly, male territorial tactics may also vary within the population, from solitary dispersed territories in areas with small female group sizes to classical leks in areas with high local female concentrations. Other processes might also be involved in maintaining variation in mating tactics within a population. For example, blackbuck may show condition-dependent choice of male mating tactic, where the tactic adopted by an individual varies with the health or social status of the individual; individuals in poor condition may adopt low-cost tactics while those in good condition would be expected to show lekking behaviour. There is evidence for condition dependence from another lekking antelope, topi *Damaliscus lunatus* (Brojorgensen and Durant 2003).

At even smaller spatial scales within a population, the relationship between female numbers and mating tactic is not so clear: why should males cluster their territories in a very small part of the area used by female herds and incur the costs of clustering (Isvaran 2003)? There are several hypotheses for the evolution of clustering. One of the most popular hypotheses for the evolution of lekking is that males may cluster their territories in response to a female preference for mating with clustered males, but the evidence for its generality is mixed (Isvaran and Ponkshe 2013). Other major hypotheses for lek evolution are that, leks may form as a result of lower quality males clustering their territories around those of males that attract females, or as a result of males establishing territories in areas of high female traffic, or because clusters are better able to retain females when male harassment of females is high (Hoglund and Alatalo 1995). While there is some evidence that the hot spot and male harassment hypotheses may not explain lekking in blackbuck (Isvaran 2005a), lek-evolution processes need to be investigated in greater detail in this species.

6.4.5 *Adaptive Variation in Male Behaviour at a Small Scale, Within a Lek*

Looking at male behaviour at an even smaller spatial scale, i.e. within a lek, there is evidence that males adjust their allocation towards costly lekking behaviour to match potential mating gains (Isvaran and Jhala 2000). As in some other lekking species, male mating success is higher on central territories than on peripheral lek territories (Fig. 6.10). This appears to be a result of a female mating preference for central territories. Corresponding to this pattern in mating benefits, central males spend more time on the lek and engage in more frequent fights than peripheral males (Fig. 6.11). Furthermore, during periods of low female visitation rates, both central and peripheral males appear to reduce their investment in lekking, so much so that the time they invest in maintenance activities, such as foraging, is very similar to non-breeding males in bachelor herds (Fig. 6.11).

6.4.6 *Social Behaviour May in Turn Influence Habitat Conditions*

Studies over the past five decades have clearly documented the influence of ecological conditions on behaviour (Jarman 1974; Davies et al. 2012). In contrast, only

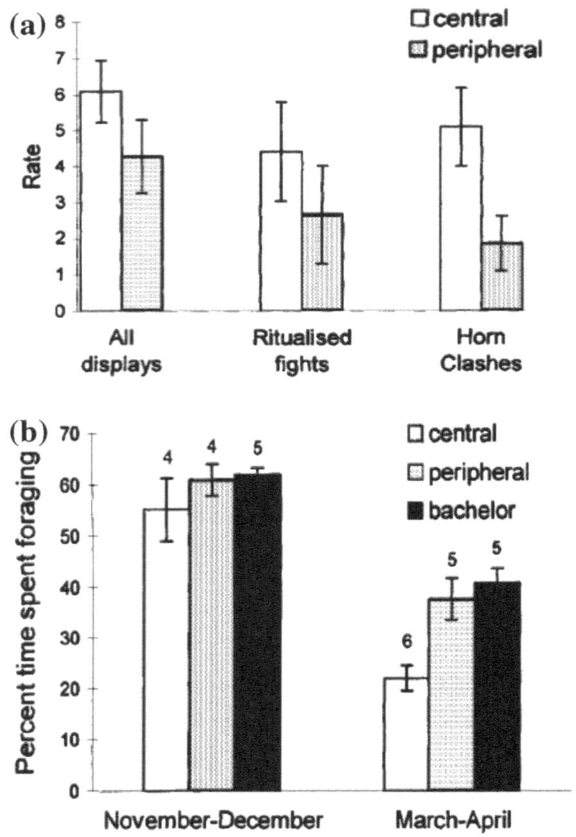

Fig. 6.11 Comparing **a** rates of displays and interactions between central and peripheral lekking males during high breeding activity and **b** time spent foraging by central and peripheral lekking males and non-breeding bachelor males in periods of low and high breeding activity (reproduced from Isvaran and Jhala 2000 with kind permission from Koninklijke Brill NV). Error bars represent standard errors

relatively recent studies have shown that, in turn, animals, through their behaviour, may modify their habitat (e.g., Jones et al. 1994; Wright et al. 2002), which may then feedback to influence their behaviour and the behaviour of other species in the community. Recent work suggests that blackbuck through their extreme mating strategies have the potential to modify the structure of their habitat and the distribution of resource, which in turn, may influence their behaviour and ecology (Jadeja et al. 2013). Mesquite *P. juliflora*, is a widespread invasive in semi-arid landscapes in India. It produces nutritious seed pods that are eaten and dispersed by many mammals. A study at a lekking blackbuck population found evidence to indicate that blackbuck may aid in the spread of *P. juliflora* and facilitate the conversion of grassland into woodland (Jadeja et al. 2013). Intriguingly, lekking males may be largely responsible for this facilitation. Lekking males spend a large part of their time on their small lekking territories and forage in the vicinity of the lek and thus, have a relatively restricted daily range. Herds on the other hand range widely. Both males and females feed on *P. juliflora* pods and disperse them through their dung. However, seeds are primarily found on territories, with very few seeds found at random sites in the grassland. On territories, seeds are further concentrated on to dung piles, scent marks that males create by repeatedly urinating and defecating at a location. Seedlings too are much more abundant on territories than on random sites in the grassland, suggesting successful recruitment of dispersed seeds on territories. Thus, territorial males could faciliate the colonisation of *P. juliflora* deep into open grasslands, relatively far from parent trees (Jadeja et al. 2013).

Such spatial concentration of *P. juliflora* recruitment suggests a particular spatial pattern in the spread of this woody invasive: recruitment onto territorial grounds leading to locally patchy and dense thickets. Since territorial male blackbuck and visiting females prefer open areas, the establishment of a thicket on a territorial ground is likely to be followed by movement of territorial males into new open areas, where, in turn, *P. juliflora* is likely to establish as a patchy dense thicket (Jadeja et al. 2013).

Blackbuck-facilitated spread of woody invasives into open grassland is likely to influence blackbuck behaviour and ecology. The correlations established by previous studies (Shankar Raman et al. 1995; Isvaran 2005a, b, 2007) between open grassland patches, high local blackbuck density, large group sizes and leks, suggest that such habitat modification will result in a reduction in group size, a shift from lekking to more dispersed resource-based territoriality, and ultimately a decrease in population density. The change in habitat is likely to affect other grassland fauna too.

6.5 Synthesis

- Blackbuck are found in a wide range of habitats but reach greatest local densities in semi-arid grasslands. The blackbuck is primarily a grazer, with adaptations to the nutritional seasonality in grasses and the limited water availability, which are characteristic of this habitat. Populations fluctuate largely in response to catastrophic events, namely floods and droughts. Blackbuck show high fecundity, which indicates that populations can respond quickly to stochastic events that cause high mortality. Selective predation by wolves appears to be important in shaping blackbuck population structure. Additionally, combined predation by jackals and wolves, especially on calves, likely regulates blackbuck populations.
- Adult sex ratio is strongly female-biased and a rare long-term data set on survivorship clearly indicates sex differences in mortality. Their mating system suggests that these differences are probably related to strong sexual selection acting on males but not on females; this pattern in sexual selection is, in turn, likely associated with high local animal densities.
- Blackbuck social organisation is highly variable and appears to be strongly influenced by habitat. Large group sizes are associated with high local abundances, which in turn, are closely associated with open grassland habitat. Their anti-predator behaviour and diet suggest that open grasslands favour large group sizes both through habitat structure (openness selecting for grouping strategies)

and resource distribution (large grass patches minimising intraspecific competition and allowing high local concentrations of animals).

- Blackbuck mating system similarly shows a close association with habitat. Like the large group sizes, the rare and spectacular lek mating system is associated with high local numbers and open grassland habitat, perhaps because the costly lek mating behaviour only pays off when the stakes (mating gains) are sufficiently high. Local female densities place a ceiling on expected mating gains.
- Taken together, the behaviour, nutritional ecology, breeding biology and demography of the blackbuck are suggestive of an antelope highly specialised to the shortgrass semi-arid biome.

References

Aiyadurai A, Jhala YV (2006) Foraging and habitat use by golden jackals (*Canis aureus*) in the Bhal region, Gujarat, India. J Bombay Nat Hist Soc 103:5–12

Bagchi S, Goyal SP, Sankar K (2008) Social organisation and population structure of ungulates in a dry tropical forest in western India (Mammalia, Artiodactyla). Mammalia 72:44–49

Barrette C (2004) Barking Deer or Muntjac. In: Sankar K, Goyal SP (eds) Ungulates of India. ENVIS Bulletin: Wildlife and Protected Areas (Vol. 7, No. 1). Wildlife Institute of India, Dehradun, pp 17–28

Biswas S, Sankar K (2002) Prey abundance and food habit of tigers (*Panthera tigris tigris*) in Pench National Park, Madhya Pradesh, India. J Zool (Lond) 256:411–420

Bouchenakkhelladi YANIS, Anthony Verboom G, Hodkinson TR, Salamin N, Francois O, Ni Chonghaile G, Savolainen V (2009) The origins and diversification of C4 grasses and adapted ungulates. Glob Change Biol 15:2397–2417

Bowyer RT, Bleich VC, Stewart KM, Whiting JC, Monteith KL (2014) Density dependence in ungulates: a review of causes and concepts with some clarifications. Calif Fish Game 100:550–572

Brashares JS, Arcese P (2002) Role of forage, habitat and predation in the behavioural plasticity of a small African antelope. J Anim Ecol 71:626–638

Brashares JS, Garland T, Arcese P (2000) Phylogenetic analysis of coadaptation in behavior, diet, and body size in the African antelope. Behav Ecol 11:452–463

Bro-Jørgensen J, Durant SM (2003) Mating strategies of topi bulls: getting in the centre of attention. Anim Behav 65:585–594

Caughley G (1977) Analysis of vertebrate populations. Wiley Interscience, London

Clutton-Brock TH, Deutsch JC, Nefdt RJC (1993) The evolution of ungulate leks. Anim Behav 46:1121–1138

Clutton-Brock TH, Isvaran K (2007) Sex differences in ageing in natural populations of vertebrates. Proc R Soc (Biol) 274:3097–3104

Dave C, Jhala YV (2011) Is competition with livestock detrimental for native wild ungulates? A case study of chital (*Axis axis*) in Gir Forest, India. J Trop Ecol 27:239–247

Davies NB, Krebs JR, West SA (2012) An introduction to behavioural ecology. Wiley

Dhungel SK, O'Gara BW (1991) Ecology of the hog deer in Royal Chitwan National Park, Nepal. Wildlife Monogr 119:3–40

Dookia S, Jakher GR (2013) Social organisation and population dynamics of Indian gazelle (*Gazella bennettii*) in Thar desert of Rajasthan, India. Tigerpaper 40:5–14 du Toit JT, Cumming DHM (1999) Functional significance of ungulate diversity in African savannas and the ecological implications of the spread of pastoralism. Biodiver Conserv 8:1643–1661

Errington PL (1946) Predation and vertebrate populations. Q Rev Biol 21:144–177

Fryxell JM (1995) Aggregation and migration by grazing ungulates in relation to resources and predators. In: Sinclair ARE, Arcese P (eds) Serengeti II: dynamics, management, and conservation of an ecosystem (vol 2). University of Chicago Press, pp 257–273

Hassanin A, Douzery EJ (1999) The tribal radiation of the family Bovidae (Artiodactyla) and the evolution of the mitochondrial cytochrome *b* gene. Mol Phylogenet Evol 13:227–243

Höglund J, Alatalo RV (1995) Leks. Princeton University Press, Princeton

Hofmann RR (1989) Evolutionary steps of ecophysiological adaptation and diversification of ruminants: a comparative view of their digestive system. Oecologia 78:443–457

Isvaran K (2003) The evolution of lekking: insights from a species with a flexible mating system. Dissertation, University of Florida

Isvaran K (2005a) Female grouping best predicts lekking in blackbuck (*Antilope cervicapra*). Behav Ecol Sociobiol 57:283–294

Isvaran K (2005b) Variation in male mating behaviour within ungulate populations: patterns and processes. Curr Sci India 89:1192–1199

Isvaran K (2007) Intraspecific variation in group size in the blackbuck antelope: the roles of habitat structure and forage at different spatial scales. Oecologia 154:435–444

Isvaran K, Jhala YV (2000) Variation in lekking costs in blackbuck (*Antilope cervicapra*): relationship to lek-territory location and female mating patterns. Behaviour 137:547–563

Isvaran K, Ponkshe A (2013) How general is a female mating preference for clustered males in lekking species? A meta-analysis. Anim Behav 86:417–425

Jadeja S, Prasad S, Quader S, Isvaran K (2013) Antelope mating strategies facilitate invasion of grasslands by a woody weed. Oikos 122:1441–1452

Janis C (2008) An evolutionary history of browsing and grazing ungulates. In: Gordon IJ, Prins H (eds) The ecology of browsing and grazing. Springer, Berlin, pp 21–45

Jarman PJ (1974) The social organization of antelope in relation to their ecology. Behaviour 48:215–267

Jethva B, Jhala YV (2004) Foraging ecology, economics and conservation of Indian wolves in the Bhal region of Gujarat, Western India. Biol Conserv 116:351–357

Jhala YV (1991) Habitat and population dynamics of wolves and blackbuck in Velavadar National Park, Gujarat, India. Dissertation. Virginia Polytechnic Institute and State University, Blacksburg

Jhala YV (1993a) Damage to *Sorghum* crop by blackbuck. Trop Pest Manage 39:23–27

Jhala YV (1993b) Predation on blackbuck by wolves in Velavadar National Park, Gujarat, India. Conserv Biol 7:874–881

Jhala YV (1997) Seasonal effects on the nutritional ecology of blackbuck *Antelope cervicapra*. J Appl Ecol 34:1348–1358

Jhala YV, Giles RH Jr, Bhagwat AM (1992) Water in the ecophysiology of blackbuck. J Arid Environ 22:261–269

Jones CG, Lawton JH, Shachak M (1994) Organisms as ecosystem engineers. Oikos 69:373–386

Karanth KU, Sunquist ME (1992) Population structure, density and biomass of large herbivores in the tropical forests of Nagarhole, India. J Trop Ecol 8:21–35

Khan JA, Chellam R, Johnsingh AJT (1995) Group size and age-sex composition of three major ungulate species in Gir lion sanctuary, Gujarat, India. J Bombay Nat Hist Soc 92:295–302

Krause J, Ruxton GD (2002) Living in groups. Oxford University Press, Oxford

Mani MS (1974) Ecology and biogeography in India. Junk Publishers, Dr. W

McNaughton SJ, Georgiadis NJ (1986) Ecology of African grazing and browsing mammals. Ann Rev Ecol Syst 17:39–65

Mech DL, Peterson RO (2003) Wolf-prey relations. In: Mech DL, Boitani L (eds) Wolves: behavior, ecology, and conservation. The University of Chicago Press, Chicago and London, pp 131–160

Mech LD (1970) The wolf: the ecology and behavior of an endangered species. Doubleday Publishing Co., New York

Mungall EC (1978) The Indian blackbuck antelope: a Texas view. Kleberg Studies in Natural Resources, College Station

Nair SS (1976) A population survey and observations on the behaviour of the blackbuck in the Point Calimere Sanctuary, Tamil Nadu. J Bombay Nat Hist Soc 73:304–310

Prasad NLNS (1989) Territoriality in the Indian blackbuck, *Antilope cervicapra* (Linnaeus). J Bombay Nat Hist Soc 86:187–193

Priyadarshini KVR (2005) Interactions between forage, recruitment and activity patterns of the Indian blackbuck. PhD Thesis. Saurashtra University, Rajkot

Rahmani AR (1990) Distribution, density, group size and conservation of the Indian gazelle or chinkara (*Gazella bennetti* Sykes 1831) in Rajasthan, India. Biol Conserv 51:177–189

Ranjitsinh MK (1989) The Indian blackbuck. Natraj Publishers, Dehradun

Robbins CR (1983) Wildlife feeding and nutrition. Academic Press, New York

Shankar Raman TR, Menon RKG, Sukumar R (1995) Decline of blackbuck (*Antilope cervicapra*) in an insular nature reserve: The Guindy National Park, Madras. Curr Sci India 68:578–580

Skalski JR, Ryding KE, Millspaugh J (2005) Wildlife demography: analysis of sex, age, and count data. Elsevier Academic Press, CA

Srinivasulu C, Srinivasulu B (1999) A population survey and observations on the herd size and structure of the blackbuck *Antilope cervicapra* Linn. in the Mahaveer Harina Vanasthali, Hyderabad, Andhra Pradesh. Ecol Env Cons 5:257–264

Stearns SC (1992) The evolution of life histories. Oxford University Press, Oxford

Terborgh J, Janson CH (1986) The socioecology of primate groups. Annu Rev Ecol Syst 17:111–135

Trivers RL (1972) Parental investment and sexual selection. In: Campbell B (ed) Sexual selection and the descent of man. Aldine-Atherton, Chicago, pp 136–179

Varman KS, Sukumar R (1995) The line transect method for estimating densities of large mammals in a tropical deciduous forest: an evaluation of models and field experiments. J Biosci 20:273–287

Wright JP, Jones CG, Flecker AS (2002) An ecosystem engineer, the beaver, increases species richness at the landscape scale. Oecologia 132:96–101

Chapter 7
Do Large Herbivores Influence Plant Allocation to Above- and Belowground Compartments?

Sumanta Bagchi

Abstract Plants allocate biomass to aboveground (shoot) and belowground (root) structures to capture light and nutrients, respectively. Plant allocation strategies are a key feature of ecosystem structure, and function, as they have broad influence over material and energy flow. Herbivores can exercise strong control over plant allocation, and thereby influence ecosystem structure and function. Theoretical models of plant allocation strategies predict that root:shoot ratio is related to leaf nitrogen (*N*) concentration; specifically root:shoot ratio declines monotonically over a broad range of *N* concentrations. Here, I assess whether grazing by large mammalian herbivores alters the inverse relationship between root:shoot ratio and leaf-*N* concentration. Data from an herbivore exclusion experiment in the arid Trans-Himalayan ecosystem were in general agreement with predictions of theoretical models. Herbivores did not affect the slope of the relationship, rather they lowered the intercept, and this change was independent of their influence over plant community composition. Reduction in root:shoot ratio with increasing leaf-*N*, and lowering biomass allocation to roots with grazing, are consistent with other studies on how herbivores influence plants.

Keywords Ecosystem structure and function · Grazer effects · Leaf nitrogen · Plant strategies · Root:shoot ratio · Trans-Himalayan

7.1 Introduction

Plant growth occurs via the acquisition of two fundamentally different types of resources: belowground resources such as nutrients and water, and aboveground ones such as light and carbon dioxide (CO_2). To meet these requirements, plants allocate biomass to shoots for capturing sunlight and CO_2, and to roots for uptake of soil nutrients such as nitrogen (*N*). Plant strategies for biomass allocation to roots

S. Bagchi (✉)
Centre for Ecological Sciences, Indian Institute of Science, 560012 Bangalore, India
e-mail: sbagchi@ces.iisc.ernet.in

© Springer Science+Business Media Dordrecht 2016

F.S. Ahrestani and M. Sankaran (eds.), *The Ecology of Large Herbivores in South and Southeast Asia*, Ecological Studies 225,
DOI 10.1007/978-94-017-7570-0_7

and shoots are a key feature of ecosystem structure and function, and have a wide range of influences on energy and nutrient dynamics. In grazing ecosystems, herbivores can exercise strong controls over plant allocation, and hence can have wide ranging indirect effects on energy and nutrient dynamics.

Over the last few decades, there has been considerable empirical and theoretical interest in understanding strategies of plant allocation (Gadgil and Gadgil 1979; Agren 1985; Agren and Ingestad 1987; Bazzaz et al. 1987; Levin et al. 1989; Hilbert 1990; Hilbert and Reynolds 1991; Sugiura and Tateno 2011). These studies have shown that plant allocation above- and belowground is related to internal *N* concentration. A principal pattern is that root:shoot ratio declines monotonically over a broad range of *N* concentrations, which allows plants to respond optimally to soil resource conditions (Box 7.1). In comparison, our understanding of how herbivory can alter plant allocation to above- and belowground components across gradients of *N* availability, remains incomplete.

Box 7.1 Theoretical formulation of optimal biomass allocation strategies follow from simplified mathematical considerations of Agren and Ingestad (1987), which were re-examined by Levin et al. (1989).

Let S be shoot biomass, and R be root biomass in a plant whose total amount of nitrogen is N. Let α be nitrogen productivity, or biomass produced per unit nitrogen per unit time, and P be net photosynthetic rate per unit shoot biomass. Now, growth rate of the plant is given by:

$$\frac{\mathrm{d}}{\mathrm{d}t}(S+R) = PS = \alpha N. \tag{7.1}$$

Now, P depends on nitrogen concentration in tissues, with $n = \frac{N}{S+R}$. So, we can write $P(n) = P_{\max} f(n)$, where $P_{\max}$ is maximal photosynthetic rate and $0 \leq f(n) \leq 1$.

If we express Eq. (7.1) in terms of relative growth rate, we get:

$$\frac{1}{(S+R)}\frac{\mathrm{d}}{\mathrm{d}t}(S+R) = \frac{\mathrm{d}}{\mathrm{d}t}\ln(S+R) = \frac{\alpha N}{S+R} = \alpha n = \frac{P(n)S}{S+R} \tag{7.2}$$

However, to account for the fact that there is a minimal N concentration (n^*) below which photosynthesis does not occur, and without photosynthesis there can be no growth, we can modify (7.2) as follows:

$$\frac{\mathrm{d}}{\mathrm{d}t}\ln(S+R) = \begin{cases} \alpha(n-n^*) & \text{for } n > n^* \\ 0 & \text{for } n \leq n^* \end{cases} \tag{7.3}$$

and

$$\frac{\mathrm{d}}{\mathrm{d}t}\ln(S+R) = \begin{cases} \frac{P(n)S}{S+R} & \text{for } n > n^* \\ 0 & \text{for } n \leq n^* \end{cases} \tag{7.4}$$

Then, from (7.3) and (7.4) it follows that

$$\frac{R}{S} = \begin{cases} \frac{P(n)}{\alpha(n-n^*)} - 1 & \text{for } n > n^* \\ \infty & \text{for } n \leq n^* \end{cases}$$

Thus, root:shoot ratio is expected to decline monotonically with increasing nitrogen concentration.

Herbivores can alter plant allocation patterns, and also simultaneously influence N content in plant tissue through nutrient-mediated feedbacks. Therefore, in addition to their direct and indirect effects on plants and nutrients (McNaughton 1976; Hobbs 1996; McNaughton et al. 1997), herbivores can also influence plant strategies in grazing ecosystems. These effects can vary across sites depending on how herbivores influence the trade-off in plant allocation to above- and belowground compartments by changing relative allocation patterns. Therefore, herbivores can be a factor in how plants allocate biomass to above versus belowground compartments in response to variation in N availability in their environment.

Our understanding of the effect of herbivore-induced defoliation on root:shoot ratio and on N content comes primarily from studies where plants are considered to be in a state of balanced exponential growth where all fractions allocated to different plant compartments are in steady state (Johnson and Thornley 1987; Hilbert and Reynolds 1991). Simulations indicate that new allocation to leaf proteins initially increases immediately after a single defoliation event, whereas new allocation to root- and shoot fractions initially declines. The time taken for each fraction to return to steady state is variable, but the root-fraction is generally the slowest to recover (Hilbert and Reynolds 1991). These inherent time-lags pose challenges in predicting the broad and general effect of herbivores on plants growing under heterogeneous soil conditions that experience repeated defoliation events.

Consideration of the theoretical relationship between root:shoot ratio and leaf-N concentration for grazed and ungrazed plants can, however, generate testable predictions. Grazing can result in positive, neutral, or even negative changes in the slope of the inverse relationship between root:shoot ratio and leaf-N concentration. Further, grazing can similarly alter the intercept of the inverse relation between root:shoot ratio and leaf-N concentration. The critical factor determining how grazers influence this relationship is the magnitude of the difference between their simultaneous effects on root:shoot ratio and leaf-N. If grazer effects on both aspects are balanced, then there may be no change in slope. But, there may still be a net effect on the intercept. In case grazer effects on root:shoot and plant nitrogen are unbalanced, and grazing disproportionately affects one aspect over the other, then grazers can induce a change in the slope of the relationship between root:shoot ratios and leaf *N* concentrations. This way, the difference between simultaneous grazer effects on root:shoot ratio and leaf-*N* can help determine whether and how plants respond to their environmental settings under grazing.

A change in the slope of the relationship under grazing can be interpreted as a physiological response in plants that may allow them to re-establish a steady-state condition between the different compartments (Johnson and Thornley 1987; Hilbert and Reynolds 1991). In fact, under nutrient stress, plants maintain relatively low growth rates. When grazed, they may accelerate their growth rate to recover from defoliation (Hilbert et al. 1981; Bagchi and Ritchie 2011). In comparison, a change in the intercept under grazing indicates that plants may alter relative allocation to either roots, or shoots, for a given internal *N* concentration (Box 7.1). This translates into grazer effects over plant allocation to above- and belowground compartments, depending on concomitant changes in *N* concentration. As this effect may differ between plant species, such a pattern may be related to grazer-induced changes in plant community composition. Therefore, it is also important to investigate whether herbivory-induced responses in internal *N*-concentration and/or biomass allocation co-vary with other factors such as grazing intensity and community composition.

Here, I evaluate whether and how herbivores alter the relationship between root: shoot ratio and leaf-*N* content using a dataset from the Trans-Himalayan ecosystem in Spiti region of northern India (Bagchi and Ritchie 2010a, b, 2011; Bagchi et al. 2012). Further, I assess whether these effects are related to plant species composition, and grazing intensity.

7.2 Herbivory, Nitrogen, and Biomass Allocation: An Experiment in the Trans-Himalaya

This dataset from Spiti region of the Trans-Himalayan ecosystem in northern India comes from an herbivore exclusion experiment initiated in 2005. Twenty-four herbivore exclosures (ungrazed), each 10 m × 10 m in area, and accompanying paired control plots (grazed) were used to obtain information on above- and belowground plant biomass, foliar *N*-concentration, grazing intensity, and change in species composition in response to grazer exclusion. Exclosures were set up in 2005, and here I use measurements obtained after 4-years of grazer-exclusion. Change in species composition was estimated with two indices—(a) Sørensen index, which indicates change in species identity, and (b) Bray–Curtis index, which indicates change in species identity as well as their relative abundance between paired plots. Grazing intensity was measured as percent biomass consumed by herbivores. Grazer effect on a parameter was measured as the log-ratio of values in paired grazed and ungrazed plots. Further details on the dataset and experimental design can be found in Bagchi and Ritchie (2010a, b, 2011) and Bagchi et al. (2012).

Using this dataset, I tested the following predictions derived from theoretical models (Box 7.1): (a) aboveground allocation (shoot fraction) should show a monotonically increasing trend with foliar *N*-concentration, (b) belowground allocation (root fraction) should show a monotonically decreasing trend with foliar *N*-concentration, (c) root:shoot ratio should show a monotonically decreasing trend with foliar *N*-concentration; the slope and/or the intercept of the relationship may be altered by grazing, and (d) change in slope and/or intercept could reflect a background influence of grazer-induced shift in plant species composition. All these expected patterns were analyzed using linear models in R 3.0 (R Development Core Team 2008).

7.3 Co-variation in Root:Shoot Ratio and Leaf-*N* Concentration

As expected, shoot fraction showed a monotonic increase with foliar *N*-concentration (Fig. 7.1a). Likewise, root fraction showed a monotonic decline with foliar *N*-concentration (Fig. 7.1b). As a result, and consistent with patterns expected from theory, root:shoot ratio showed a monotonic decline with foliar *N*-concentration (Fig. 7.1c). The relationship between root:shoot ratio and foliar *N*-concentration was better described by a semi-log function, than by a linear function (Fig. 7.1c). There was no evidence that grazing altered the slope of the relationship between

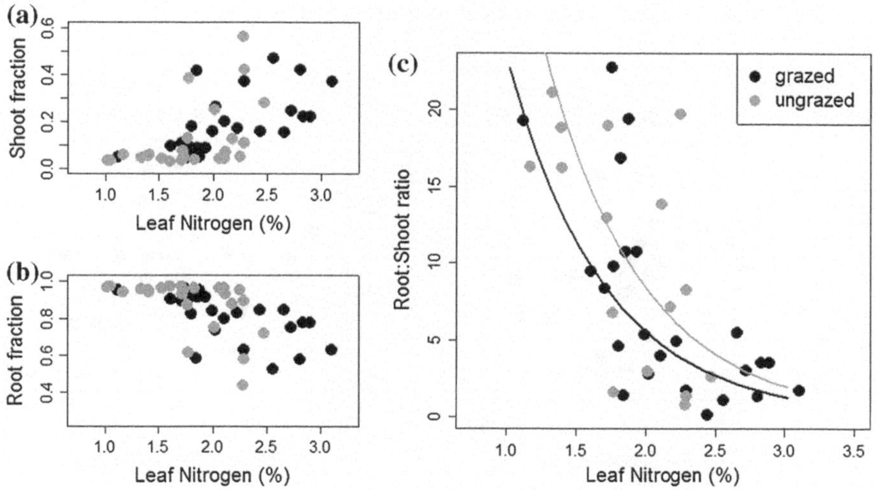

Fig. 7.1 Co-variation between plant allocation aboveground (shoot fraction, **a**) and belowground (root fraction, **b**) with foliar nitrogen concentration for grazed and ungrazed plants in the Trans-Himalayan ecosystem in Spiti region of northern India. Co-variation between root:shoot ratio and foliar nitrogen follows the theoretically expected pattern of a monotonic decline for both grazed and ungrazed plants (**c**)

Table 7.1 Summary of statistical models that attempt to account for co-variation between grazer effect on root:shoot ratio and on leaf *N*-concentration

Model type	Model AIC	Factor	df	F	P
Linear	332.91	Leaf *N*	1, 44	23.54	<0.001
		Grazing	1, 44	20.38	<0.001
		Interaction	1, 44	1.86	0.18
Linear	331.90	Leaf *N*	1, 45	23.10	<0.001
		Grazing	1, 45	20.07	<0.001
Semi-log	135.46	Leaf *N*	1, 44	23.12	0.001
		Grazing	1, 44	11.91	<0.001
		Interaction	1, 44	0.04	0.85
Semi-log	132.59	Leaf *N*	1, 45	23.62	<0.001
		Grazing	1, 45	12.16	0.001

Results indicate that a semi-log model with one slope and two intercepts is the best descriptor of the observed pattern (i.e., log(root:shoot ratio) as a function of leaf *N*)

root:shoot ratio and foliar *N*-concentration (Table 7.1). However, grazing did alter the intercept (Table 7.1). Specifically, the intercept in grazed plots was lower than that in ungrazed plot, indicating a reduced investment in roots for a given *N* concentration when grazed, relative to ungrazed conditions.

Grazer effect on root:shoot ratio, and on foliar *N*-concentration was unrelated to change in plant species composition, both as Sørensen index as well as Bray–Curtis index (Fig. 7.2). Grazer effect on root:shoot ratio showed a hump-shaped pattern with grazing intensity ($F_{2,21} = 3.25$, $P = 0.05$, $R^2 = 0.23$, Fig. 7.3). But, grazer effect on foliar *N*-concentration was unrelated to grazing intensity (Fig. 7.3).

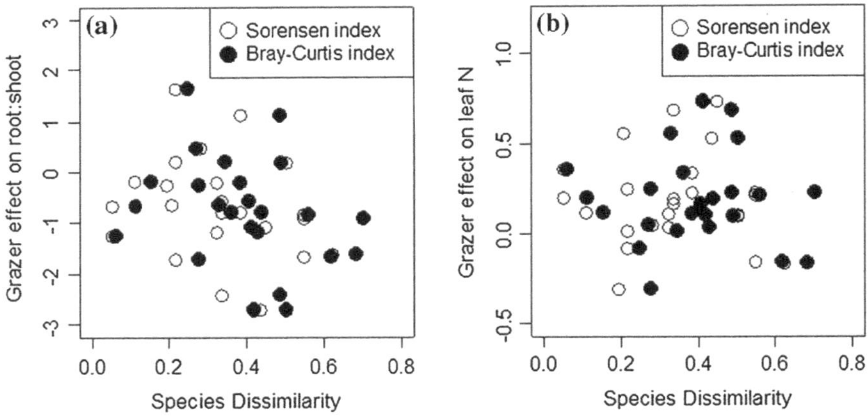

Fig. 7.2 Co-variation between grazer effect on root:shoot ratio (measured as log-ratio of paired grazed and ungrazed plots) and change in plant species composition measured as dissimilarity between grazed and ungrazed plots in (**a**). Co-variation between grazer effect on leaf nitrogen concentration (measured as log-ratio of grazed and ungrazed plots) and change in plant species composition in (**b**). Data are from an herbivore exclusion experiment in the Trans-Himalayan ecosystem in Spiti region of northern India

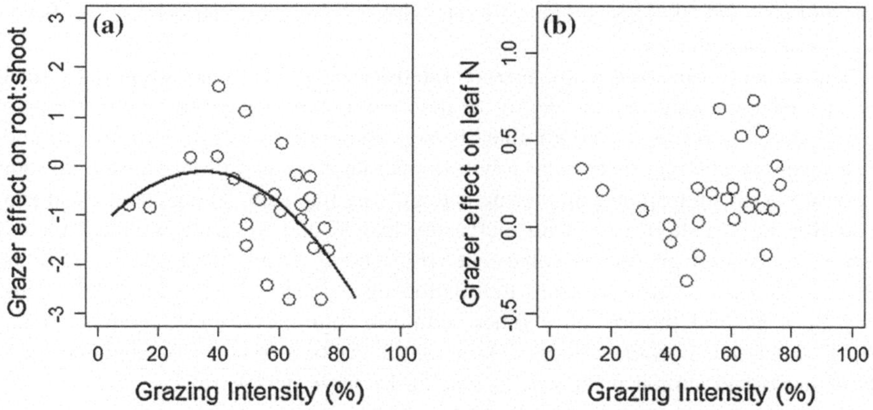

Fig. 7.3 Co-variation between grazer effect on root:shoot ratio (measured as log-ratio of grazed and ungrazed plots) and grazing intensity (measured as fraction of plant production consumed by herbivores), suggesting the presence of the hump-shaped pattern (**a**). Co-variation between grazer effect on leaf nitrogen concentration and grazing intensity (**b**). Data are from an herbivore exclusion experiment in the Trans-Himalayan ecosystem in Spiti region of northern India

The data are in general agreement with theoretical models of plant allocation, based on their nutrient status (Fig. 7.1). The shoot fraction increased, while the root fraction decreased monotonically with leaf-*N* (Gadgil and Gadgil 1979; Agren 1985; Agren and Ingestad 1987; Johnson and Thornley 1987; Levin et al. 1989; Hilbert and Reynolds 1991). Not surprisingly, both grazed and ungrazed plants showed a monotonic decline in root:shoot ratio with increasing leaf-*N* (Fig. 7.1c) (Agren 1985; Agren and Ingestad 1987; Levin et al. 1989). But, pattern of allocation, represented by the slope of the relationship, did not change with grazing (Fig. 7.1, Table 7.1). However, grazed plants had a lower intercept compared to the ungrazed plant. This implies that for a given status of internal *N*-concentration, ungrazed plants maintain a higher root:shoot ratio than grazed plants. Previous analyses of this dataset have discussed grazing-induced reduction in root biomass, and concomitant increase in shoot biomass in this ecosystem, and such patterns are consistent with results from other similar semi-arid rangelands (e.g., North American shortgrass steppe, Derner et al. 2006).

One potential explanation for the observed change in intercept associated with grazing would be plant species replacement. Since different plant species vary in their *N*-use efficiency (Agren 1985; Agren and Ingestad 1987; Levin et al. 1989); relative allocation to root and shoot compartments for a given *N*-concentration may also differ between species. But, there was no strong evidence to suggest that species replacement, or changes in the relative abundance of species under grazing, was related to the influence of grazers on root:shoot ratio and on leaf-*N* (Fig. 7.2). Thus, there was little indication of any indirect influence of herbivores on plant allocation strategies via their influence on plant community composition. Given these results, the observed change in intercept is very likely due to average community-level changes in growth rate and *N*-efficiency (Hilbert et al. 1981; Bagchi and Ritchie 2011).

The patterns observed with grazing intensity (Fig. 7.3) lend support to such direct effects mediated by the degree of defoliation experienced by plants. Change in root:shoot ratio was related nonlinearly to grazing intensity (Fig. 7.3). At low and high levels of grazing, there were low-to-moderate changes in root-shoot ratio, but there was a strong response at intermediate grazing intensity. The general trend is a reduction in root:shoot ratio with increasing leaf-*N*, and lowering biomass allocation to roots with grazing is consistent with studies documenting the conditions under which plants recover from defoliation events (Hamilton et al. 1988). This may be a general pattern since other semi-arid high-elevation ecosystems (e.g., North American shortgrass steppe, Derner et al. 2006) also show comparable root: shoot ratios and similar responses to grazing by large herbivores.

Acknowledgments The dataset used for this analysis was created with funding support from Syracuse University, US National Science Foundation (DDIG DEB-0608287), Rufford Foundation, and Wildlife Conservation Society. I am grateful to M.E. Ritchie, Y.V. Bhatnagar, and C. Mishra for numerous discussions on plant–herbivore interactions. I am also grateful to the Himachal Pradesh Department of Forest Farming and Conservation, and to all the assistants, interns, and volunteers who have worked very hard in helping me in the field. Initial drafts were greatly improved by critiques from anonymous referees and the editors.

References

Agren GI (1985) Theory for growth of plants derived from the nitrogen productivity concept. Physiol Plant 64:17–28

Agren GI, Ingestad T (1987) Root: shoot ratio as a balance between nitrogen productivity and photosynthesis. Plant, Cell Environ 10:579–586

Bagchi S, Bhatnagar YV, Ritchie ME (2012) Comparing the effects of livestock and native herbivores on plant production and vegetation composition in the Trans-Himalayas. Pastoralism Res Policy Pract 2(21)

Bagchi S, Ritchie ME (2011) Herbivory and plant tolerance: experimental tests of alternative hypotheses involving non-substitutable resources. Oikos 120:119–127

Bagchi S, Ritchie ME (2010a) Herbivore effects on above- and belowground plant production and soil nitrogen availability in the Trans-Himalayas. Oecologia 164:1075–1082

Bagchi S, Ritchie ME (2010b) Introduced grazers can restrict potential soil carbon sequestration through impacts on plant community composition. Ecol Lett 13:959–968

Bazzaz FA, Chiariello NR, Coley PD, Pitelka LF (1987) Allocating resources to reproduction and defense. Bioscience 37:58–67

Derner JD, Boutton TW, Briske DD (2006) Grazing and ecosystem carbon storage in the North American Great Plains. Plant Soil 280:90

Gadgil M, Gadgil S (1979) Adaptive significance of the relation between root and shoot growth. J Indian Inst Sci 61:25–40

Hamilton EW, Giovannini MS, Moses A, Coleman JS, McNaughton SJ (1988) Biomass and mineral element responses of a Serengeti short-grass species to nitrogen supply and defoliation: compensation requires a critical [N]. Oecologia 116:407–418

Hilbert DW (1990) Optimization of plant root: shoot ratios and internal nitrogen concentration. Ann Bot 66:91

Hilbert DW, Reynolds JF (1991) A model allocating growth among leaf proteins, shoot structure and root biomass to produce balanced activity. Ann Bot 68:417–425
Hilbert DW, Swift DM, Detling JK, Dyer MI (1981) Relative growth rates and the grazing optimization hypothesis. Oecologia 51:14–18
Hobbs NT (1996) Modification of ecosystems by ungulates. J Wildlife Manage 60:695–713
Johnson IR, Thornley JHM (1987) A model of shoot:root partitioning with optimal growth. Ann Bot 60:133–142
Levin SA, Mooney HA, Field C (1989) The dependence of plant root: shoot ratios on internal nitrogen concentration. Ann Bot 64:71
McNaughton SJ (1976) Serengeti migratory wildebeest: facilitation of energy flow by grazing. Science 191:92–94
McNaughton SJ, Banyikwa FF, McNaughton MM (1997) Promotion of the cycling of diet-enhancing nutrients by African grazers. Science 278:1798–1800
R Development Core Team (2008) R: A language and environment for statistical computing. R Foundation for Statistical Computing, Vienna, Austria. http://www.R-project.org
Sugiura D, Tateno M (2011) Optimal leaf-to-root ratio and leaf nitrogen content determined by light and nitrogen availabilities. PLoS ONE 6:e22236

Chapter 8
Grazing and Fire Effects on Community and Ecosystem Processes in a Tall-Grass Mesic Savanna Ecosystem in Southern India

Mahesh Sankaran

Abstract Grazing and fire are processes that are integral to savannas and grasslands worldwide. However, very few studies have evaluated the effects of fire and grazing on community and ecosystem processes in savannas and grasslands in the Indian context. Here, I report results from an experimental study that investigated the individual and interactive effects of grazing and fire on community and ecosystem level processes in mesic tall-grass savannas of the southern Western Ghats, India. Burning had no effect on species richness or composition of herbaceous plant communities in these grasslands, and within 2 years following burning vegetative cover in burnt plots was indistinguishable from unburned plots. Grazing treatments, on the other hand, significantly reduced the relative abundance of the dominant tall-grass species *Cymbopogon flexuosus* in plots, resulting in a concomitant increase in herbaceous plant species richness relative to unburned plots. At the ecosystem level, grazers did not enhance soil N availability and N turnover rates in this mesic tall-grass savanna, which is most likely a consequence of the low herbivore densities in the study area. Burning, on the other hand, resulted in greater soil N availability in plots at the start of the growing season, but such effects were short-lived and did not persist. Interestingly, greater soil N availability in burnt plots appeared to be a consequence of lowered N uptake by plants, associated with lower plant biomass in burned plots, rather than an increase in N mineralization rates in soils. Results from this study suggest that tall-grass communities dominated by *C. flexuosus* are quite stable with respect to perturbations by fire. However, grazers, when coupled with fire, have the potential to significantly enhance species richness and induce plant compositional shifts in this mesic tall-grass savanna, particularly when grazing intensity is high.

M. Sankaran (✉)
National Centre for Biological Sciences, Tata Institute of Fundamental Research, GKVK Campus, Bellary Road, Bangalore 560065, India
e-mail: mahesh@ncbs.res.in; m.sankaran@leeds.ac.uk

M. Sankaran
School of Biology, University of Leeds, Leeds LS2 9JT, UK

© Springer Science+Business Media Dordrecht 2016

F.S. Ahrestani and M. Sankaran (eds.), *The Ecology of Large Herbivores in South and Southeast Asia*, Ecological Studies 225,
DOI 10.1007/978-94-017-7570-0_8

Keywords *Cymbopogon flexuosus* · Ecosystem processes · Grasslands · Nitrogen cycling · Plant community composition · Soil nitrogen/N

8.1 Introduction

Grazing and fire are common occurrences in tropical savannas and grasslands worldwide (Hobbs 1996; Sankaran et al. 2004, 2005; Bond 2008). The large amounts of standing dead biomass present during the dry season, and the high surface to volume ratios and flammability of grass tissues render savannas and grasslands especially prone to fires when compared to other ecosystems in the tropics such as forests (D'Antonio and Vitousek 1992; Bond 2008; Bond and Parr 2010; Ratnam et al. 2011). Tropical savannas and grasslands, particularly in Africa and Asia, are also characterized by a high diversity and biomass of large mammalian grazers (see Chap. 1), in contrast to forest ecosystems where the mammalian herbivore community tends to be browser-dominated (Bodmer 1990). Indeed, the rise to ecological dominance of C_4 grasslands and savannas, along with the near-synchronous adaptive radiation of vertebrate grazers and the shift from low- to high-crowned teeth in ungulates (indicating a switch from browsing to grazing diets) during the late Miocene ~ 8–10 million years ago, is suggestive of a tight coevolution between grasses and vertebrate grazers (MacFadden and Ceding 1994; Beerling and Osborne 2006; Edwards et al. 2010, also see refs in Bouchenak-Khelladi et al. 2009).

Grazing and fire are analogous to some degree in that both exert top–down control on ecosystem structure by 'consuming' vegetation (Bond and Keeley 2005; Bond 2008; Sankaran et al. 2013). By consuming part, or all, of the aboveground biomass, fires initiate a secondary successional sequence that can alter the species richness, diversity, composition and structure of grasslands and savannas (Collins and Gibson 1990; Collins et al. 1995; Higgins et al. 2007). Further mediation of community change by fires can occur through the alteration of local microclimatic conditions such as temperature, light, and moisture, and by their effects on ecosystem nutrient pools and cycling (Hobbs and Schimel 1984; Hulbert 1986; Stock and Lewis 1986; Kauffman et al. 1994; Ojima et al. 1994; Turner et al. 1997; Blair 1997; Reich et al. 2001; Augustine et al. 2014). Consumption of aboveground plant biomass by fires can result in loss of accumulated nutrients to the atmosphere by volatilization or alternately, the return of nutrients to the soil as ash or incompletely combusted plant material (Seastedt and Ramundo 1990; Kauffman et al. 1994; Blair 1997). Fire effects on community and ecosystem processes are, however, contingent on fire regimes (Raison 1979; Seastedt et al. 1991; Ojima et al. 1994; Blair 1997). In the short term, burning can enhance N availability and microbial activity thereby resulting in increased above and below ground plant production (Raison 1979; Stock and Lewis 1986; Ojima et al. 1994). However, repeated burning can result in substantial losses of N from the system and result in plant production being limited by N availability (Vitousek and Howarth 1991; Seastedt et al. 1991; Ojima et al. 1994; Blair 1997).

Like fires, grazers too can have manifold effects on a range of community and ecosystem processes. Selective herbivory can result in the competitive inhibition of some species and the release of others, thereby altering the relative abundance of different plant species in a community (Augustine and McNaughton 1998). By virtue of their feeding strategies and their physical alteration of the environment, herbivores can influence plant growth form, species composition and distribution patterns, successional trajectories, and the balance between trees, bush, and open grasslands (McNaughton et al. 1988; Naiman 1988; Dublin et al. 1990; Augustine and McNaughton 1998; Sankaran et al. 2005, 2008). At the ecosystem level, herbivores can influence energy and nutrient cycling rates by: (i) altering patterns of energy capture by plants, i.e., production, and thus the quantity of plant inputs to soils, (ii) altering the quality of plant inputs to soils as a result of either grazer-induced physiological responses at the individual plant level or grazer-induced shifts in plant species composition patterns, (iii) returning a substantial portion of plant N consumed to the soil via dung and urine in forms that are more readily available for plant and microbial uptake, and (iv) by modulating soil biological processes over short-time scales through changes in non-detrital inputs such as root exudates (McNaughton 1985; Ruess and McNaughton 1987; Holland and Detling 1990; Frank et al. 1994; Hobbs 1996; Augustine and McNaughton 1998; Ritchie et al. 1998; Bardgett et al. 1998; Hamilton and Frank 2001; Sankaran and Augustine 2004).

Fire and herbivory are, however, not mutually exclusive events in grasslands and savannas, with each influencing the characteristics and occurrence probabilities of the other (Hobbs et al. 1991; Hobbs 1996). By consuming vegetation and thereby reducing fuel loads, grazers can influence the frequency, intensity, and extent of fires (Hobbs 1996; McNaughton 1992). Because grazing is typically heterogeneous across the landscape, grazing-induced spatial variation in plant biomass can also influence fire spread characteristics, resulting in a more spatially variable post fire community (Hobbs et al. 1991; McNaughton 1992; Archibald et al. 2005). Further, by incorporating plant nutrients into herbivore tissue, and by transferring nutrients from above ground plant biomass to the soil via urine and feces, grazing can also result in lowered volatile losses of N during fires when compared to ungrazed grasslands (Hobbs et al. 1991). Conversely, fires can also influence the nature and frequency of subsequent grazing events. Grazers have been observed to preferentially utilize previously burned patches when compared to unburned patches, although this is dependent on season (Vinton et al. 1993), herbivore body size (Wilsey 1996), size of the burned patch, and time since burning (Archibald and Bond 2004; Archibald et al. 2005; Allred et al. 2011). Fires can also have a homogenizing effect on grazing patterns, and promote a more uniform grazing environment (Hobbs 1996; Archibald and Bond 2004). In unburned areas, the probability of a grazed area being regrazed tends to be high, resulting in spatially patchy distribution of plant biomass (Hobbs 1996; Archibald and Bond 2004; Archibald et al. 2005). In contrast, grazing tends to be more evenly distributed in burnt patches, resulting in a relatively more homogenous distribution of plant biomass across space (Hobbs 1996; Archibald and Bond 2004). Because fire and grazing interact in a variety of ways, it is necessary to consider both their direct individual effects, as well as indirect interactive effects, in order to understand their influences on plant community structure and ecosystem processes.

Although there has been much work on the individual and interactive effects of fire and grazing in regulating community and ecosystem processes, the bulk of this work has been carried out in ecosystems in North America and elsewhere. Much less is known about the roles of fire and grazing in the Indian context, despite the fact that the Indian subcontinent supports both a diverse and abundant assemblage of large mammalian herbivores (see Table 1.1 in Chap 1) and is subject to frequent fires. Here, I focus on the effects of fire and grazing on community and ecosystem processes in tall-grass savanna communities of the southern Western Ghats, a biodiversity hotspot. Mesic savannas dominated by tall-grass species such as *Cymbopogon flexuosus* and *Themeda cymbaria* are fairly widespread across the southern Western Ghats, particularly in the rain shadow areas to the east of the Ghats and along the eastern hill slopes. They are quite prone to frequent fires, with many tall-grass savannas in the Western Ghats burning on an annual basis (Sankaran 2001, 2005, pers. obs). Although mesic savanna grasslands do not support as much herbivore biomass as their more arid short-grass counterparts, they are nevertheless important habitats for herbivores, particularly large-bodied ones, immediately following burning (pers obs, Sankaran 2001). Given their large spatial extent, the frequent fires they support, and their relevance for herbivore populations, understanding the dynamics of mesic savanna grasslands in response to fire and grazing is critical for their effective management. In this chapter, I largely draw on the results of an experimental study carried out between 1997–1999 in the Mundanthurai plateau of the Kalakad Mundanthurai Tiger Reserve (KMTR), Tamil Nadu, India, that looked at the effects of fire and grazing on the dynamics of savannas dominated by the tall-grass species *Cymbopogon flexuosus*. These savannas are representative of mesic tall-grass savannas in the wider landscape of the southern Western Ghats.

8.2 Study Area and Study Design

The Kalakad Mundanthurai Tiger Reserve (KMTR; 8°25′–8°50′, 77°15′–77°40′) is situated at the southern tip of India's Western Ghats mountain range; a region recognized as a global biodiversity "hotspot" (Myers 2000). The reserve covers an area of approximately 1000 km^2 and spans an altitudinal gradient from about 100–1867 m above m.s.l. (Kothari et al. 1989). This southernmost region of the Western Ghats including KMTR is also recognized as one of the centers of high plant diversity in India (Ganesh et al. 1996) and harbors some of the largest contiguous tracts of non-equatorial tropical evergreen forests in the country (Ramesh et al. 1997). Besides tropical forests, the reserve also supports large tracts of savannas and short-grass habitats that collectively account for about 38 % of the reserve area

(~320 km^2, Sankaran 2001). Fire is a common occurrence in the reserve, with KMTR having some of highest reported fire frequencies of any protected area in India (Kothari et al. 1989). Fires in the reserve typically occur between February and May, and are mostly restricted to savanna habitats, with some savanna grasslands burning on a nearly annual basis (Sankaran 2001). At the time of the study, between 1704–2045 ha of the park was estimated to burn annually, representing 2–2.3 % of the total area of the park and 6–6.6 % of the area covered by savanna grasslands (Sankaran 2001). Fires tend to burn over larger areas along the eastern boundary of the reserve where savanna grassland habitats are contiguous. Average burn size at the time was estimated to be 28.4 ha (range: 0.7–203.8 ha), with most burned patches tending to be small (<10 ha; Sankaran 2001).

The Mundanthurai plateau, the focal area for the study, covers an area of approximately 60 km^2 at an elevation of about 200 m.s.l. in the east-central section of the park. Annual temperatures in the plateau fluctuate between 17 and 37 °C (Katti 1997), and annual rainfall averages around 1200 mm, the bulk of which falls between October and December during the northeast monsoon (10-year average of 963 mm), and a lesser amount between June and August during the southwest monsoons (307 mm, Joshua and Johnsingh 1987; Johnsingh and Joshua 1994). Six mammalian terrestrial herbivore species including spotted deer or chital *Axis axis*, sambar *Rusa unicolor*, chevrotain or mouse deer *Moschiola indica*, black-naped hare *Lepus nigricollis*, cattle *Bos indicus* and domestic buffalo *Bubalis bubalis* occur in the plateau. Gaur *Bos gaurus*, although not present in the area at the time of study, have been observed to increasingly frequent the plateau in recent years (A.J. T. Johnsingh *pers comm*). In the two decades prior to the start of the study, the Mundanthurai plateau witnessed a strong decline in herbivore populations (Johnsingh and Sankar 1991; Viswanathan 1996, Sankaran unpubl.). Declines were most pronounced in the case of spotted deer or chital *Axis axis*, from ~200 in 1984 (Johnsingh and Sankar 1991) to levels too low in 1995 to permit accurate density estimation (Viswanathan 1996). At the time of the study, wild herbivore biomass density in the plateau was estimated to be 498 kg/Km^2, while that of domestic livestock 566 kg/km^2 (Sankaran 2001).

To evaluate the community and ecosystem level consequences of altered herbivore densities, and the utility of prescribed burning as a management tool to enhance herbivore habitat quality, an experiment was initiated in the Mundanthurai plateau in 1997 to look at the interactive effects of fires and grazing on understory plant communities and nutrient cycling patterns in tall-grass mesic savannas dominated by *C. flexuosus*. Communities dominated by *C. flexuosus* are the most widespread grassland type in the plateau. They occur on acidic sandy loam soils (pH: 4.9–5.9) with soil %C and %N contents of ~0.93 % and 0.08 %, respectively (Sankaran 2001). Vegetation heights average about a meter, and plant production estimated over a 7.5-month period using moveable exclosures (McNaughton 1985) averaged 3.3 g m^{-2} d^{-1}, the bulk of which was produced during the major wet season between October and December (Sankaran 2001). Peak standing biomass in unburned areas is ~580 g m^{-2}, of which ~75 % comprises standing dead material.

As a result, fires are fairly common in these communities. Grasses and sedges account for about 85 % of green cover, with forbs and shrubs/trees contributing 5 and 10 %, respectively. *C. flexuosus* individuals account for ~77 % of the live cover in these communities. These communities typically sustain low grazing pressure when mature, although stands tend to be grazed immediately following fires.

The experimental design was a 3 × 3 factorial experiment involving three burning treatments (unburned, burnt once a year, burnt once in 2 years) and three grazing treatments (ungrazed, naturally grazed and experimentally clipped to simulate high-intensity grazing). At each of three different sites, a 15 × 15 m enclosure was set up consisting of a square grid of nine 4 × 4 m^2 plots with a 1 m gap between plots and a 50 cm walkway around the edge. Each of the nine plots at a site was randomly assigned to one of nine burning × grazing treatments. All experimental burns were conducted at the start of the dry season in March 1997. A further subset of these was burned again in March 1998. High-intensity grazing was simulated by clipping vegetation in plots down to ~5 cm every fortnight. Within each 16 m^2 plot, species richness and cover of all species was quantified in eight 1 m^2 plots using a systematic subsampling scheme. Plots were sampled prior to the initiation of the experiment in March 1997, and subsequently at the end of the first and second year following the start of the experiment. Data were analyzed as a factorial experiment blocked by site. Although there was no data available on the long-term fire regimes in the region, none of the study sites had burned for at least 3 years prior to the start of the experiment.

In addition to responses at the plant community level, the effect of burning and grazing on patterns of N availability and transformation in soils was also quantified for a subset of plots and treatments in the second year of the study. At two of the three experimental sites, N availability and N mineralization rates were determined for unburned and annually burned plots for all three grazing treatments for five different time periods: ~11–12 months following the first experimental burn and just prior to the second set of burns (Feb–Mar 1998), and then subsequently 2 (May–June 1998), 4 (July–August 1998), 11 (January–February 1999), and 12 months (February–March 1999) following burning. At each sampling period, soil cores (three subsamples from each plot) were collected to determine initial standing stocks of available nitrogen (T0). At the same time, soils from paired adjacent cores were loaded into mineralization tubes (20 cm long PVC tubes with an inside diameter of 2.5 cm), which were then incubated in the field and collected 30 days later (T30). Following collection in the field, 50 ml of KCl was added to soil samples (~10 g), which were vigorously shaken, allowed to sit for 24 h, and then filtered. Filtrates were analyzed at a later date for ammonium (NH_4^+), and nitrate N (NO_3^-) using a Lachat Quikchem Autoanalyzer (Milwaukee, Wisc.). Estimates based on soil samples collected from the initial cores (T0) were used to quantify total available N (NH_4^+ and NO_3^-) at each time period. Nitrogen mineralization rate was quantified as the difference in extractable nitrogen between the two sampling periods (T30–T0), and expressed on a per day basis. Due to logistical constraints N availability and mineralization rates were not quantified between

August and December, which included the peak wet season. Data were analyzed separately for each time period as a factorial experiment blocked by site, with replicate cores from each plot treated as subsamples. All analyses were carried out using the open source statistical package R (R Core Team 2012).

8.3 Community Level Effects of Grazing and Fire

There were no significant differences in species richness of plots assigned different burning and grazing treatments at the start of the study. However, species richness of all plots, irrespective of experimental treatment, was higher at the end of year-1 and year-2 compared to the start of the study as a result of above-average rainfall conditions during the intervening period (1997: 2168 mm and 1998: 1544 mm relative to the long-term mean of 1200 mm). None of the burning treatments had any effect on species richness of plots ($F_{2,17} = 1.2$, $P = 0.32$). Grazing treatments, on the other hand, had a significant impact on species richness of plots ($F_{2,17} = 4.0$, $P = 0.03$; Fig. 8.1). Across both years, plots that were experimentally clipped

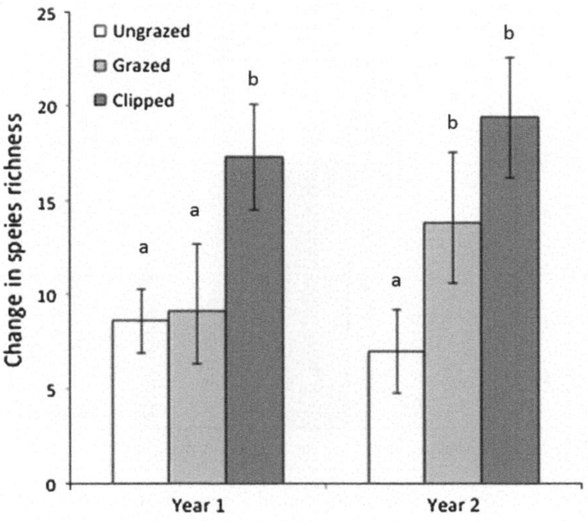

Fig. 8.1 Change in species richness relative to initial conditions over a 2-year period in tall-grass *C. flexuosus* dominated communities in response to different grazing treatments. There was no effect of burning treatments on change in species richness and so data are pooled across different burning treatments. *Error bars* represent ±1 S. E, and letters above *bars* denote treatments that were significantly different from one another based on Tukey's HSD post hoc test. All plots typically showed an increase in species richness relative to the start as a result of wetter than average conditions during the course of the study. However, across both years, the magnitude of increase in species richness was consistently higher in *clipped plots* when compared to *ungrazed plots*. Naturally *grazed plots* were more similar to *ungrazed plots* in year-1, and to *clipped plots* in year-2

showed a consistently greater increase in species richness when compared to ungrazed plots (Fig. 8.1). Response of plots that were naturally grazed varied between years and was similar to ungrazed plots in year-1 and to experimentally clipped plots in year-2 (Fig. 8.1).

Burning and grazing treatments had independent, but not interactive, effects on total vegetative cover in plots at the end of 2 years (Fig. 8.2; Burning: $F_{2,17} = 14.6$, $P < 0.001$, Grazing: $F_{2,17} = 26.8$, $P < 0.001$, Burning × Grazing: NS). Total cover in plots that had been burnt one year previously (Fig. 8.2, treatment B1 pooled across all grazing treatments) was significantly lower than plots that were unburned or burned two years previously. However, two years following burning, total vegetative cover in plots was indistinguishable from unburned plots (Fig. 8.2). When pooled across all burning treatments, total vegetative cover at the end of 2 years was the highest in ungrazed plots followed by plots that were naturally grazed, and lowest in experimentally clipped plots (Fig. 8.2).

Burning had no effect on the relative cover of the dominant tall-grass species *C. flexuosus* in plots at the end of 2 years ($F_{2,17} = 0.98$, $P = 0.40$). However, the

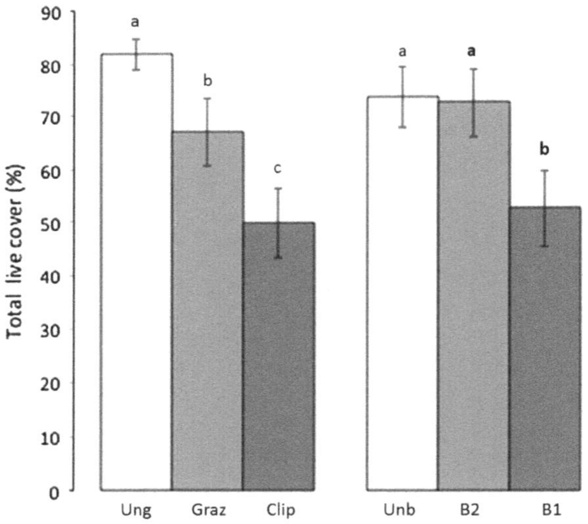

Fig. 8.2 Total vegetative cover in plots at the end of 2 years in response to different burning and grazing treatments. Both burning and grazing had independent, but not interactive, effects on total cover in plots. The *bars* on the *left* depict cover in plots subject to different grazing treatments (*Ung* ungrazed, *Graz* naturally grazed, *Clip* experimentally clipped; data pooled across different burning treatments), while the *bars* on the *right* depict total cover in plots with different burning treatments (*Unb* Unburned, *B2* burned once in 2 years, i.e., values are for 2 years post-burn, *B1* burnt annually, i.e., data are for 1 year post burn; data pooled across grazing treatments). *Error bars* represent ±1 S. E, and letters above *bars* denote significant differences based on Tukey's HSD post hoc test. In terms of grazing treatments, cover was highest in ungrazed plots followed by grazed plots, and lowest in experimentally clipped plots. In terms of burning treatments, annually burned plots (B1) had significantly lower total cover when compared to plots burned 2 years previously or unburned plots; the latter two were not significantly different from one another

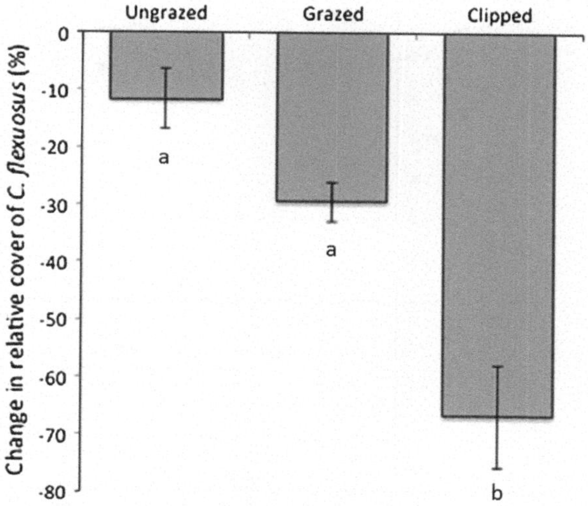

Fig. 8.3 Change in the relative abundance of the tall-grass species *C. flexuosus* in plots in response to different grazing treatments. There was no effect of burning on the relative abundance of *C. flexuosus* and so data are pooled across different burning treatments. Clipping resulted in a significant and substantial decline in the relative abundance of *C. flexuosus* in plots. Although relative abundance of *C. flexuosus* was lower at the end of the experiment relative to the start in both *grazed* and *ungrazed plots*, the magnitude of the decline was not significantly different between these two treatments. *Error bars* represent ±1 S. E, and letters above *bars* denote significant differences based on Tukey's HSD post hoc test

relative dominance of *C. flexuosus* was significantly impacted by different grazing treatments ($F_{2,17} = 26.2$, $P < 0.001$; Fig. 8.3). At the start of the study, *C. flexuosus* individuals on average accounted for ~78 % of total live cover in plots. Over the course of the study, the relative abundance of *C. flexuosus* in plots declined on average across all treatments as a result of the greater representation of annual and ephemeral species in response to the above-average rainfall conditions during the course of the study. However, clipping in particular resulted in a large and significant reduction in the relative dominance of *C. flexuosus* individuals in plots (Fig. 8.3), from ~78 % to ~14 % at the end of 2 years. At the end of the experiment, *C. flexuosus* accounted for ~50 % of live cover in the grazed plots and ~62 % in ungrazed plots, although the magnitude of the change relative to the start in the case of these two treatments was not significantly different from one another (Fig. 8.3).

Compositional shifts at the whole-plot level (i.e., 4 × 4 m plots) in response to different burning and grazing treatments were assessed following ordination of communities using detrended correspondence analysis. There was no evidence to suggest that any of the burning treatments elicited any shifts in the community composition of plots (no segregation of points in ordination space, results not shown). However, results suggest compositional shifts induced by grazing treatments (Fig. 8.4). Clipped plots, in particular, occupied an ordination space that was

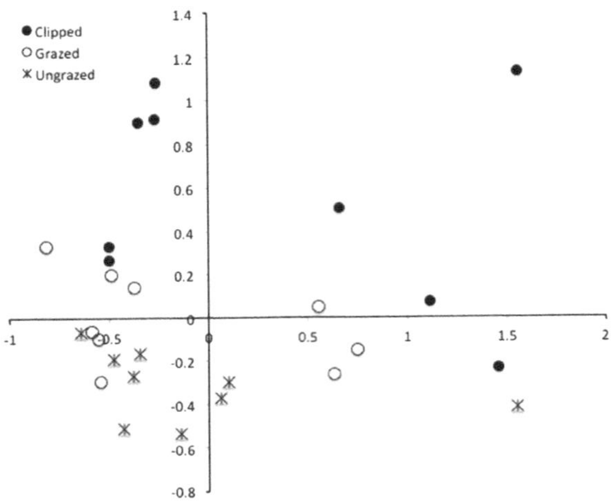

Fig. 8.4 Plot of the detrended correspondence analysis scores depicting differences in community composition between plots that were *ungrazed*, *grazed*, and experimentally *clipped*. There was no segregation of plots based on different burning treatments, and so data are pooled across burning treatments for ease of interpretation. The first two ordination axes accounted for 66 % of the variance in the data. All analyses were carried out using the package *vegan* in the statistical software R (R Core Team 2012)

distinct and separate from ungrazed plots (Fig. 8.4). Species composition of grazed plots also appeared to differ somewhat from ungrazed plots, although the magnitude of the separation was not as great in this case (Fig. 8.4). Overall, at the end of 2 years, species composition of ungrazed plots tended to be more similar to one another (as evidenced by the clustering of points in ordination space) when compared to grazed or experimentally clipped plots (Fig. 8.4). The greater spread of points in the case of clipped plots (Fig. 8.4) suggests that high-intensity grazing, as simulated by experimental clipping, can potentially induce divergent responses in terms of community level shifts in composition depending on site-specific characteristics and burning regimes.

These results are in agreement with both theory and previously published empirical work which suggests that grazers are capable of significantly enhancing species diversity and inducing plant compositional shifts, particularly in more mesic and productive sites (Milchunas et al. 1988; Belsky 1992; Milchunas and Lauenroth 1993; Proulx and Mazumder 1998; Olff and Ritchie 1998). Increases in herbaceous plant species richness in response to grazing treatments, both natural and simulated (experimental clipping), have been reported in mesic grassland and savanna systems from a range of different sites including the Serengeti National Park, Tanzania (Belsky 1992), Konza Prairie, USA (Collins et al. 1998; Koerner and Collins 2014), Kruger National Park, South Africa (Koerner and Collins 2014), secondary savannas of the Vindhyan Plateau, India (Pandey and Singh 1991), and the Royal Bardia National Park, Nepal (Karki et al. 2000), amongst others. Burning, on the other hand, did not appear to have any effect on plant species diversity in this mesic savanna grassland site. Although burning has been shown to negatively impact plant species richness in some grassland ecosystems (e.g., tall-grass prairie, Collins et al. 1998), plant diversity in other savanna grassland ecosystems, particularly mesic ones, appear to be independent of fire regimes (Belsky 1992; Uys et al. 2004). Plant species in this mesic savanna grassland, both grasses and forbs, thus appear to have vegetative traits that allow them to persist in the face of frequent fires (Bond and Parr 2010) such that the community as a whole is relatively stable to perturbations in the form of fire.

8.4 Fire and Grazing Effects on Nitrogen Availability and Transformation in Soils

In general, soil available N in plots was low early in the dry season (Fig. 8.5; February), increased as the dry season progressed to a peak in April, and then started declining during May and June following the onset of the rains. Soil N

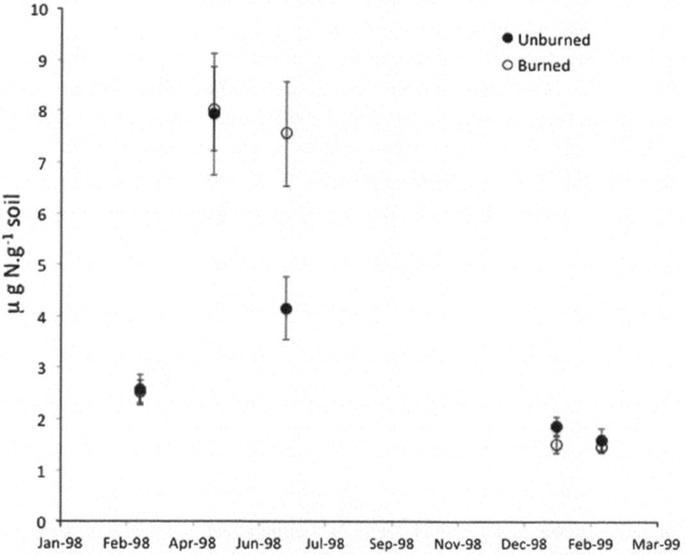

Fig. 8.5 Temporal patterns in total available soil N (NH_4^+ and NO_3^-) as a function of burning. There was no effect of grazing on N availability in soils at any time period and so data are pooled across different grazing treatments. *Error bars* represent ± 1 S. E calculated based on all replicate cores within each burning treatment. Soil N availability was low at the end of the wet season and early in the dry season (December–February) and increased over the course of the dry season to reach a peak in April. There was no significant effect of burning treatments at any time period except July following the onset of the rains, at which time total available soil N was significantly higher in *burned plots* when compared to *unburned plots*

availability was similarly low at the end of the wet season/start of the dry season the following year (January and February 1999). Although estimates are not available from the peak wet season (October–December), available data seem to suggest a pattern of low inorganic N availability in soils during the wet season, potentially as a result of plant uptake of N, with inorganic N availability in soils increasing during the course of the dry season, following the cessation of plant growth, to peak at the end of the dry season just prior to the start of the rains.

None of the grazing treatments had any effect on total soil available N at any time period ($P > 0.05$ all cases). Burning significantly influenced soil available N, but only during the sampling period immediately following the start of the rains in June (Fig. 8.5; June: $F_{1,5} = 16.5$, $P = 0.01$; NS for all other time periods). The effect of burning on soil available N was similar across all grazing treatments (burning x grazing interaction: NS for all time periods). Following the onset of the rains, soil N availability declined more strongly, and was significantly lower in unburned plots when compared to burned plots (Fig. 8.5 June; data pooled across grazing treatments). It is likely that the lowered amounts of live aboveground biomass in burned plots at the start of the growing season resulted in lowered plant uptake of soil N relative to unburned plots, resulting in higher levels of available soil N in burned plots. However, such effects were not long-lived, and 9–10 months following burning (January), burned, and unburned plots were indistinguishable in terms of soil inorganic N concentrations (Fig. 8.5).

N mineralization showed a contrasting pattern to soil N availability overtime (Fig. 8.6). In general, N mineralization rates tended to be highest during the early part of the dry season (February–March 1998 and 1999), and lowest at the start of the short wet season (June) and at the end of the long wet season (Jan; Fig. 8.6). As was the case with soil available N, no significant effects of either burning or grazing on N mineralization rates were observed at any time periods except during June–July,

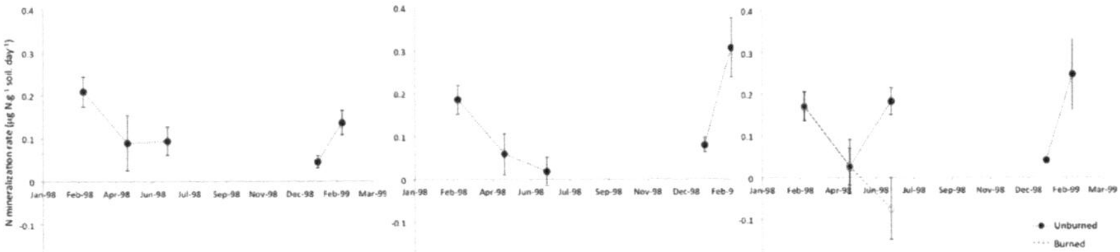

Fig. 8.6 Temporal patterns in net N ($NH_4^+ + NO_3^-$) mineralization rates as a function of different burning and grazing treatments. Panel (a) shows data for ungrazed plots, (b) clipped plots, and (c) grazed plots. There was no difference in net N mineralization rates between burned and unburned treatments for plots that were ungrazed (a) or clipped (b) and so data are pooled across both burning treatments in these cases. In grazed plots, burning had no effect on net N mineralization rates at any time period except during June–July immediately following the onset of the rains (c) when there was an increase in net mineralization rates in *unburned plots*, but net immobilization of N in *burned plots*. *Error bars* represent ±1 S. E calculated based on all replicate cores pooled within treatments as appropriate

immediately following the start of the rains. However, in this case, there was a significant interactive effect of burning and grazing on net N mineralization rate (Burning × Grazing: $F_{2,5} = 8.87$, $P = 0.02$; Fig. 6). Burning had no effect on net N mineralization rates in plots that were either ungrazed or clipped (Fig. 8.6a, b). However, under natural levels of grazing, burning resulted in the immobilization of N in soils while unburned plots showed a net positive mineralization of N (Fig. 8.6c).

Grazers are well recognized as important regulators of energy and nutrient flows in ecosystems (McNaughton et al. 1985, 1997; Hobbs 1996), capable of enhancing N availability and increasing N turnover rates in soils, particularly in more arid and eutrophic systems (McNaughton 1985, 1997; Hobbs 1996; Frank et al. 2000; Augustine and McNaughton 2006). However, in this mesic savanna ecosystem, grazer effects on N availability and turnover in soils were, for the most part, largely negligible. This is not surprising considering that herbivore biomass and consumption in the study area were low, and that grazer stimulation of N cycling rates and primary production is typically highest at intermediate levels of grazing (McNaughton 1985; Frank and McNaughton 1993; Augustine and McNaughton 1998). Furthermore, the clipping treatments used to simulate high-intensity grazing in this study did not encompass the increased return of N by herbivores in more available forms such as dung and urine, which is a critical pathway through which herbivore effects on N-cycling rates are manifested (McNaughton 1985; McNaughton et al. 1997; Augustine and McNaughton 1998). To the best of my knowledge, there have been no other studies that have looked at grazer effects on N-cycling rates in the grasslands and savannas of the Western Ghats and peninsular India , and the extent to which grazers regulate nutrient cycling rates in these systems, particularly in areas where they occur at high densities, remains currently unknown.

Fires are also an important driver of the nutrient economy of ecosystems, capable of increasing N availability in soils in the short term through the return of nutrients to soils as ash or incompletely combusted plant material, and also influencing N transformation rates in soils by altering microclimatic conditions such as temperature, light, and water availability (Raison 1979; Hobbs et al. 1984; Stock and Lewis 1986; Blair 1997; Augustine et al. 2014). Burning has been previously shown to significantly increase soil inorganic N pools in short term, particularly NH_4^+–N, with such effects in some cases persisting through the growing season (Stock and Lewis 1986; Singh et al. 1991; Augustine et al. 2014). Although fires did have an effect on N availability in this tall-grass savanna, such effects were largely limited to the period immediately following the onset of the rains, and did not persist. In contrast to previously reported results from other ecosystems, fires did not appear to increase N availability in soils in this mesic savanna (Fig. 8.5). The higher amounts of total available soil N in burned plots following the onset of the rains (and the marginally higher [NH_4^+–N]; $P = 0.07$; data not shown) appear to be a consequence of lowered uptake of N by vegetation in burned plots relative to unburned plots rather than an increase in N mineralization rates or N availability following burning.

In contrast to previous studies (Singh et al. 1991), burning did not also appear to have any significant effects on N turnover rates in soils except for the period immediately following the rains when net immobilization was observed in plots that were burnt and grazed (Fig. 8.6c). The reasons underlying these trends are unclear. Greater root death and turnover in burned relative to unburned plots under grazing could have increased C availability for microbes, resulting in microbial growth and subsequent mobilization of N in microbial tissue. Why similar patterns were not observed in experimentally clipped plots is, however, unclear. Given the small sample sizes used in this experiment, these results must be interpreted with caution. A comprehensive understanding of the seasonal dynamics of N in this system, and how it is impacted by fire and grazing, will require not only larger sample sizes than those utilized in this study, but also the elucidation of complete seasonal N stocks and fluxes both within and between different components of the plant-soil-microbe system.

In many temperate and tropical ecosystems, plant growth and microbial activity appear to be temporally synchronized such that rainfall initiates a pulse of plant growth and a synchronous increase in N mineralization rates (Pastor et al. 1984; Singh et al. 1989, 1991; McNaughton et al. 1997; Frank and Groffman 1998). In contrast, in some other systems, plant growth and N mineralization patterns appear to be decoupled, with significant N mineralization occurring when plants are senescing or otherwise dormant (Jackson et al. 1988; Brooks et al. 1998; Augustine and McNaughton 2004). Although estimates of N mineralization during the peak wet season are unavailable, patterns observed during the sampled time periods in this study suggest that plant growth and soil mineralization are asynchronous in this tall-grass savanna. N mineralization rates were high early in the dry season when the plants were senescing, declined as the dry season progressed, and tended to be low immediately following the onset of the rains and at the end of the wet season. In contrast, soil available N increased over the course of the dry season to reach a peak just before the onset of the rains, and it is likely that plant growth early in wet season was primarily supported by this available N pool rather than wet season N mineralization (Augustine and McNaughton 2004). One possible reason for the high levels of mineralization observed early in the dry season could be the result of enhanced microbial activity during this period. Microbes are typically believed to be less sensitive to water stress when compared to plants (Bernhard-Reversat 1982; Singh et al. 1989). Low soil moisture levels early in the dry season, while limiting for plant growth could have been sufficient for microbes, and when coupled with the high temperatures could have resulted in increased microbial activity and N mineralization at this time. However, as the dry season progressed, microbes would have become increasingly water-stressed with subsequent reduction in microbial activity and N mineralization rates.

8.5 Conclusions and Implications for Management

Grazing and fire, although they both exert top–down forces in ecosystems by consuming aboveground plant matter, appear to have contrasting effects on community versus ecosystem processes in this tall-grass savanna ecosystem. Community level variables including species richness, composition, and dominance were strongly influenced by grazing treatments, but were largely unaffected by burning. In contrast, the only effects of burning detected in this mesic tall-grass savanna were on N availability in soils, but such effects were short-lived and restricted to the period immediately following the onset of the rains.

Tall-grass mesic savannas dominated by *C. flexuosus* appear to be quite stable to perturbations by fire, with species richness and cover of burnt plots indistinguishable from unburned plots within 2 years following burning. Conversely, and as is often the case with more 'productive' ecosystems (Proulx and Mazumder 1998), grazers in this tall-grass savanna appear to have the potential to regulate and increase herbaceous species richness at the community level by reducing the competitive dominance of *C. flexuous* individuals (see results of from the clipping treatments). However, communities dominated by *Cymbopogon* species, both in India and elsewhere (e.g., Serengeti National Park, Tanzania; McNaughton 1983), are typically not 'preferred' by herbivores, except following burning (Vessey-Fitzgerald 1973). Herbivores have been previously shown to successfully suppress *Cymbopogon* species when burnt (Vessey-Fitzgerald 1973), but similar results were not observed in this study under natural levels of grazing (Fig. 8.3). Although burning resulted in a five-fold increase in the number of individuals of *C. flexuosus* grazed in plots over the short term (up to 4 months following burning; Fig. 8.7; Sankaran 2005), such effects were ephemeral, with levels of grazing on *C. flexuosus* returning to negligible levels within a year following burning (Fig. 8.7; Sankaran 2005). This was most likely a consequence of the low herbivore densities in the study area such that grazing pressure on *C. flexuosus* individuals were too low to prevent the species from quickly reattaining competitive dominance (in contrast to the clipping treatments; see Fig. 8.3, Sankaran 2001, 2005). Burning as a management tool to improve herbivore habitat quality by reducing the dominance of *C. flexuosus* is thus likely to only be effective when the densities of herbivores, particularly large bodied bulk feeders (See Chap. 4 by Ahrestani et al.), are above a critical threshold.

Savannas are fairly widespread across semiarid and mesic regions of peninsular India . Although large herbivores and fires are both integral component of savannas, few studies thus far have investigated their roles in modulating community and ecosystem processes in these ecosystems in the Indian context. Managers in particular have traditionally viewed fire as a 'bad' and 'unnatural' disturbance, a view that persists to the present day. As a result, there has been a policy of active fire suppression in these systems, with attendant implications for herbivore populations, as well as herbaceous and woody plant dynamics. The virtual lack of any detailed studies on the top–down roles of herbivory and fire in these systems means that

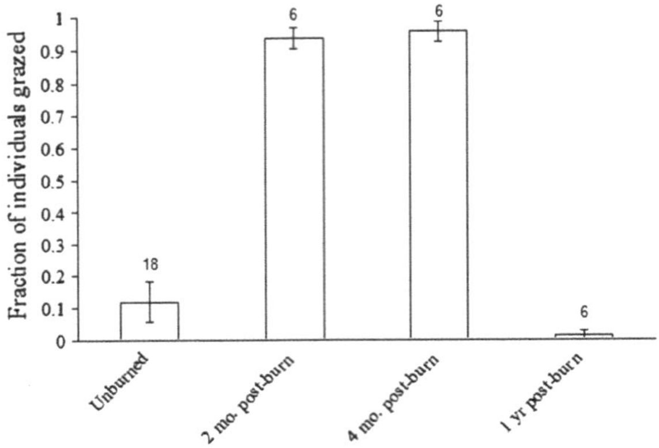

Fig. 8.7 The fraction of C. flexuosus individuals grazed in unburned plots and in burned plots at different time periods following burning. *Error bars* represent ±1 S. E. Values for unburned plots are based on data from all plots assigned the 'grazed' treatment prior to experimental manipulation (N = 9), as well as three unburned plots sampled at the three different sampling periods (2 months, 4 months, and 1 year post-burn) during the first year of the study. Values for the burned plots were estimated using data from six replicate treatment plots at each time period from the first year of the study (N = 3 each from the 'burned once' treatment and 'burned twice treatments) (Graph reproduced from Sankaran (2005) Conservation and Society 3:4–25)

most management decisions are currently being made in the absence of any underlying scientific knowledge of the dynamics of these ecosystems. Clearly, there is a critical and urgent need for detailed long-term studies investigating the role of large herbivores and fire in regulating the dynamics of these key ecosystems.

References

Allred BW, Fuhlendorf SD, Engle DM et al (2011) Ungulate preference for burned patches reveals strength of fire–grazing. Ecol Evol 1:132–144

Archibald S, Bond WJ (2004) Grazer movements: spatial and temporal responses to burning in a tall-grass African savanna. Int J Wildland Fire 13:377–385

Archibald S, Bond WJ, Stock WD et al (2005) Shaping the landscape: fire-grazer interactions in an African savanna. Ecol Appl 15:96–109

Augustine DJ, McNaughton SJ (1998) Ungulate effects on the functional species composition of plant communities: herbivore selectivity and plant tolerance. J Wildl Manag 62:1165–1183

Augustine DJ, McNaughton SJ (2004) Temporal asynchrony in soil nutrient dynamics and plant production in a semiarid ecosystem. Ecosystems 7:829–840

Augustine DJ, McNaughton SJ (2006) Interactive effects of ungulate herbivores, soil fertility, and variable rainfall on ecosystem processes in a semi-arid savanna. Ecosystems 9:1242–1256

Augustine DJ, Brewer P, Blumenthal DM et al (2014) Prescribed fire, soil inorganic nitrogen dynamics, and plant responses in a semiarid grassland. J Arid Environ 104:59–66

Bardgett RD, Wardle DA, Yeates GW (1998) Linking above-ground and below-ground interactions: how plant responses to foliar herbivory influence soil organisms. Soil Biol Biochem 30:1867–1878

Beerling DJ, Osborne CP (2006) The origin of the savanna biome. Glob Change Biol 12:2023–2031
Belsky AJ (1992) Effects of grazing, competition, disturbance and fire on species composition and diversity in grassland communities. J Veg Sci 3:187–200
Bernhard-Reversat F (1982) Biogeochemical cycle of nitrogen in a semi-arid savanna. Oikos 38:321–332
Blair JM (1997) Fire, N availability, and plant response in grasslands: a test of the transient maxima hypothesis. Ecology 78:2359–2368
Bodmer RE (1990) Ungulate frugivores and the browser-grazer continuum. Oikos 57:319–325
Bond WJ (2008) What limits trees in C_4 grasslands and savannas? Annu Rev Ecol Evol Syst 39:641–659
Bond WJ, Keeley JE (2005) Fire as a global 'herbivore': the ecology and evolution of flammable ecosystems. Trends Ecol Evol 20:387–394
Bond WJ, Parr CL (2010) Beyond the forest edge: ecology, diversity and conservation of the grassy biomes. Biol Conserv 143:2395–2404
Bouchenak-Khelladi Y, Verboom GA, Hodkinson TR et al (2009) The origins and diversification of C_4 grasses and savanna-adapted ungulates. Glob Change Biol 15:2397–2417
Brooks PD, Williams MW, Schmidt SK (1998) Inorganic nitrogen and microbial biomass dynamics before and during spring snowmelt. Biogeochemistry 43:1–15
Collin SL, Gibson DJ (1990) Effects of fire on community structure in tallgrass and mixed-grass prairie. In: Collins SL, Wallace LL (eds) Fire in North American tallgrass prairies. University of Oklahoma Press, Norman, pp 81–98
Collins SL, Glenn SM, Gibson DJ (1995) Experimental analysis of intermediate disturbance and initial floristic composition: decoupling cause and effect. Ecology 76:486–492
Collins SL, Knapp AK, Briggs JM et al (1998) Modulation of diversity by grazing and mowing in native tallgrass prairie. Science 280:745–747
D'Antonio CM, Vitousek PM (1992) Biological invasions by exotic grasses, the grass/fire cycle, and global change. Annu Rev Ecol Syst 23:63–87
Dublin HT, Sinclair ARE, McGlade J (1990) Elephants and fire as causes of multiple stable states in the Serengeti-Mara woodlands. J Anim Ecol 59:1147–1164
Edwards EJ, Osborne CP, Strömberg CA et al (2010) The origins of C_4 grasslands: integrating evolutionary and ecosystem science. Science 328:587–591
Frank DA, Groffman PM (1998) Ungulate versus topographic control of soil carbon and nitrogen processes in grasslands of Yellowstone National Park. Ecology 79:2229–2241
Frank DA, McNaughton SJ (1993) Evidence for the promotion of aboveground grassland production by native large herbivores in Yellowstone National Park. Oecologia 96:157–161
Frank DA, Inouye RS, Huntly N et al (1994) The biogeochemistry of a north-temperate grassland with native ungulates: nitrogen dynamics in Yellowstone National Park. Biogeochemistry 26:163–188
Frank DA, Groffman PM, Evans RD et al (2000) Ungulate stimulation of nitrogen cycling and retention in Yellowstone Park grasslands. Oecologia 123:116–121
Ganesh T, Ganesan R, Soubadra Devy M et al (1996) Assessment of plant biodiversity at a mid-elevational evergreen forest of Kalakad-Mundanthurai Tiger Reserve, Western Ghats, India. Curr Sci 71:379–392
Hamilton EW, Frank DA (2001) Can plants stimulate soil microbes and their own nutrient supply? Evidence from a grazing tolerant grass. Ecology 82:2397–2402
Higgins SI, Bond WJ, February EC et al (2007) Effects of four decades of fire manipulation on woody vegetation structure in savanna. Ecology 88:1119–1125
Hobbs NT (1996) Modification of ecosystems by ungulates. J Wildl Manag 60:695–713
Hobbs NT, Schimel DS (1984) Fire effects on nitrogen mineralization and fixation in mountain shrub and grassland communities. J Range Manag 37:402–405
Hobbs NT, Schimel DS, Owensby CE et al (1991) Fire and grazing in the tallgrass prairie: contingent effects on nitrogen budgets. Ecology 72:1374–1382

Holland EA, Detling JK (1990) Plant response to herbivory and below-ground nitrogen cycling. Ecology 71:1040–1049
Hulbert LC (1986) Fire effects on tallgrass prairie. In: Clambey GK, Pemble RH (eds) Proceedings of the Ninth North American Prairie Conference. Tri-College University Center for Environmental Studies, Fargo, North Dakota, pp 138–142
Jackson L, Strauss RB, Firestone M et al (1988) Plant and soil nitrogen dynamics in California annual grassland. Plant Soil 110:9–17
Johnsingh AJT, Joshua J (1994) Avifauna in three vegetation types of Mundanthurai Plateau, South India. J Trop Ecol 10:323–335
Johnsingh AJT, Sankar K (1991) Food plants of chital, sambar and cattle on Mundanthurai Plateau, Tamil Nadu, south India. Mammalia 55:57–63
Joshua J, Johnsingh AJT (1987) Observations on birds on Mundanthurai Plateau, Tamil Nadu. J Bombay Nat Hist Soc 85:565–577
Karki JB, Jhala YV, Khanna PP (2000) Grazing lawns in Terai Grasslands, Royal Bardia National Park, Nepal1. Biotropica 32:423–429
Katti M (1997) Ecology and evolution of non-breeding distributions in the old world leaf warblers. Dissertation, University of California, San Diego
Kauffman JB, Cummings DL, Ward DE (1994) Relationships of fire, biomass and nutrient dynamics along a vegetation gradient in the Brazilian cerrado. J Ecol 82:519–531
Koerner SE, Collins SL (2014) Interactive effects of grazing, drought, and fire on grassland plant communities in North America and South Africa. Ecology 95:98–109
Kothari A, Pande P, Singh S et al (1989) Management of National Parks and Sanctuaries in India: a status report. Indian Institute of Public Administration, New Delhi, India
MacFadden BJ, Ceding TE (1994) Fossil horses, carbon isotopes and global change. Trends Ecol Evol 9:481–486
McNaughton SJ (1983) Serengeti grassland ecology: the role of composite environmental factors and contingency in community organization. Ecol Monogr 53:291–320
McNaughton SJ (1985) Ecology of a grazing ecosystem: the Serengeti. Ecol Monogr 55:259–294
McNaughton SJ (1992) The propagation of disturbance in savannas through food webs. J Veg Sci 3:301–314
McNaughton SJ, Ruess RW, Seagle SW (1988) Large mammals and process dynamics in African ecosystems. Bioscience 38:794–800
McNaughton SJ, Banyikwa FF, McNaughton MM (1997) Promotion of the cycling of diet-enhancing nutrients by African grazers. Science 278:1798–1800
Milchunas DG, Lauenroth WK (1993) Quantitative effects of grazing on vegetation and soils over a global range of environments. Ecol Monogr 63:327–366
Milchunas DG, Sala OE, Lauenroth WK (1988) A generalized model of the effects of grazing by large herbivores on grassland community structure. Am Nat 132:87–106
Myers N, Mittermeier RA, Mittermeier CG et al (2000) Biodiversity hotspots for conservation priorities. Nature 403:853–858
Naiman RJ (1988) Animal influences on ecosystem dynamics. Bioscience 38:750–752
Ojima DS, Schimel DS, Parton WJ et al (1994) Long-and short-term effects of fire on nitrogen cycling in tallgrass prairie. Biogeochemistry 24:67–84
Olff H, Ritchie ME (1998) Effects of herbivores on grassland plant diversity. Trends Ecol Evol 13:261–265
Pandey CB, Singh JS (1991) Influence of grazing and soil conditions on secondary savanna vegetation in India. J Veg Sci 2:95–102
Pastor J, Aber J, McClaugherty C et al (1984) Aboveground production and N and P cycling along a nitrogen mineralization gradient on Blackhawk Island, Wisconsin. Ecology 65:256–268
Proulx M, Mazumder A (1998) Reversal of grazing impact on plant species richness in nutrient-poor vs. nutrient-rich ecosystems. Ecology 79:2581–2592
Raison RJ (1979) Modification of the soil environment by vegetation fires, with particular reference to nitrogen transformations: a review. Plant Soil 51:73–108

Ramesh BR, Menon S, Bawa KS (1997) A vegetation based approach to biodiversity gap analysis in the Agastyamalai Region, Western Ghats, India. Ambio 26:529–536

Ratnam J, Bond WJ, Fensham RJ et al (2011) When is a 'forest' a savanna, and why does it matter? Glob Ecol Biogeogr 20:653–660

Reich PB, Peterson DW, Wedin DA et al (2001) Fire and vegetation effects on productivity and nitrogen cycling across a forest-grassland continuum. Ecology 82:1703–1719

Ritchie ME, Tilman D, Knops JMH (1998) Herbivore effects on plant and nitrogen dynamics in oak savanna. Ecology 79:165–177

Ruess RW, McNaughton SJ (1987) Grazing and the dynamics of nutrient and energy regulated microbial processes in the Serengeti grasslands. Oikos 49:101–110

Sankaran M (2001) Disturbance, diversity and community dynamics in a Southern Indian savanna-grassland ecosystem. Dissertation, Syracuse University

Sankaran M (2005) Fire, grazing and the dynamics of tall-grass savannas in the Kalakad-Mundanthurai Tiger Reserve, South India. Conserv Soc 3:4–25

Sankaran M, Augustine DJ (2004) Large herbivores suppress decomposer abundance in a semiarid grazing ecosystem. Ecology 85:1052–1061

Sankaran M, Ratnam J, Hanan NP (2004) Tree–grass coexistence in savannas revisited—insights from an examination of assumptions and mechanisms invoked in existing models. Ecol Lett 7:480–490

Sankaran M, Hanan NP, Scholes RJ et al (2005) Determinants of woody cover in African savannas. Nature 438:846–849

Sankaran M, Ratnam J, Hanan NP (2008) Woody cover in African savannas: the role of resources, fire and herbivory. Glob Ecol Biogeogr 17:236–245

Sankaran M, Augustine DJ, Ratnam J (2013) Native ungulates of diverse body sizes collectively regulate long-term woody plant demography and structure of a semi-arid savanna. J Ecol 101:1389–1399

Seastedt TR, Ramundo RA (1990) The influence of fire on belowground processes of tallgrass prairie. In: Collins SL, Wallace LL (eds) Fire in North American tallgrass prairies. University of Oklahoma Press, Norman, pp 99–117

Seastedt TR, Briggs JM, Gibson DJ (1991) Controls of nitrogen limitation in tallgrass prairie. Oecologia 87:72–79

Singh JS, Raghubanshi AS, Singh RS et al (1989) Microbial biomass acts as a source of plant nutrients in dry tropical forest and savanna. Nature 338:499–500

Singh RS, Raghubanshi AS, Singh JS (1991) Nitrogen-mineralization in dry tropical savanna: effects of burning and grazing. Soil Biol Biochem 23:269–273

Stock WD, Lewis OAM (1986) Soil nitrogen and the role of fire as a mineralizing agent in a South African coastal fynbos ecosystem. The Journal of Ecology 74:317–328

Turner CL, Blair JM, Schartz RJ et al (1997) Soil N and plant responses to fire, topography, and supplemental N in tallgrass prairie. Ecology 78:1832–1843

Uys RG, Bond WL, Everson TM (2004) The effect of different fire regimes on plant diversity in southern African grasslands. Biol Conserv 118:489–499

Vesey-Fitzgerald DR (1973) East African Grasslands. East African Publishing House, Nairobi, Kenya

Vinton MA, Hartnett DC, Finck EJ et al (1993) Interactive effects of fire, Bison (Bison bison) grazing and plant community composition in tallgrass Prairie. Am Midl Nat 129:10–18

Viswanathan V (1996) Estimation of density of some herbivores in Mundanthurai using the line transect method. MSc Dissertation, Pondicherry University

Vitousek PM, Howarth RW (1991) Nitrogen limitation on land and in the sea: how can it occur? Biogeochemistry 13:87–115

Wilsey B (1996) Variation in the use of green flushes following burns among African Ungulate species: the importance of body size. Afr J Ecol 34:32–38

Chapter 9
Density and Activity Patterns of the Globally Significant Large Herbivore Populations of Cambodia's Eastern Plains Landscape

Thomas N.E. Gray, Sovanna Prum and Channa Phan

Abstract The northern and eastern plains of Cambodia support the largest extent of lowland deciduous forest remaining in Southeast Asia. This landscape has also been identified as the highest priority site for tiger *Panthera tigris* recovery in Indochina. We estimated ungulate tiger prey densities using distance-based line transect sampling from two protected areas in the Eastern Plains Landscape between 2009 and 2011. Densities for large ungulates ranged from 1.1 ± SE 0.2 individuals/km^2 for banteng *Bos javanicus* to 2.2 ± SE 0.2 individuals km^2 for red muntjac *Muntiacus muntjak*. The ungulate activity patterns were correlated with activity patterns of extant large carnivores in the landscape with leopard *Panthera pardus* and dhole *Cuon alpinus* showing substantial activity pattern overlap with wild pig *Sus scrofa* and red muntjac, respectively. Overall tiger prey biomass was more than 540 kg/km^2 of which the endangered banteng comprised greater than 80 %. However, ungulate densities were much lower than in ecologically similar sites in South Asia. This was mainly due to the absence of large deer species like *Cervus* deer, which have historically been extensively hunted. Nevertheless, the Eastern Plains landscape likely supports 50–60 % of the global banteng population and remains a high priority area for the conservation of large herbivores in Southeast Asia.

T.N.E. Gray (✉)
WWF Greater Mekong, #21, Street 322, Sangkat Boeun Keng Kang 1,
Phnom Penh, Cambodia
e-mail: tomnegray@hotmail.com

S. Prum · C. Phan
WWF Cambodia Country Program, #21, Street 322, Sangkat Boeun
Keng Kang 1, Phnom Penh, Cambodia

S. Prum
Forestry Administration, Ministry of Agriculture Forestry and Fisheries,
40 Preah Norodom Blvd. Phsar Kandal 2, Khann Daun Penh,
Phnom Penh, Cambodia

© Springer Science+Business Media Dordrecht 2016

F.S. Ahrestani and M. Sankaran (eds.), *The Ecology of Large Herbivores in South and Southeast Asia*, Ecological Studies 225,
DOI 10.1007/978-94-017-7570-0_9

Keywords Banteng · Cambodia · Distance-based line transect sampling · Tiger prey

9.1 Introduction

The northern and eastern Cambodian provinces of Mondulkiri, Ratanakiri, Stung Treng, and Preah Vihear support some of the largest extents of lowland deciduous forest in Southeast Asia (Todorff et al. 2005). These forests had once been described as one of the "great game-lands of the world; a Serengeti of Asia" and had historically supported a diverse and abundant megafaunal assemblage of ungulates, predators and scavengers (Wharton 1957; Todorff et al. 2005). The ungulate component included at one time an assemblage of four sympatric wild cattle species (gaur *Bos gaurus*, banteng *Bos javanicus*, kouprey *Bos sauvelli* and wild water buffalo *Bubalus arnee*); a unique distinction as few regions in the world have supported more than two sympatric wild cattle species.

The Eastern Plains Landscape (EPL), Mondulkiri province, in eastern Cambodia (Fig. 9.1) forms part of the Lower Mekong Dry Forest Ecoregion (Olson and Dinerstein 1998; Tordoff et al. 2005). This landscape of rolling lowlands, generally under 150 m asl, is characterized by extensive deciduous dipterocarp forest, including open grassland areas (*veal*) with a high frequency fire regime, and smaller patches of mixed deciduous and semi-evergreen forests on higher ground and along watercourses. The EPL deciduous forest is bisected by a number of rivers including one major tributary of the Mekong, the Srepok River, and is studded with small seasonal wetlands (*trapeang*).

The wider landscape in which the EPL is situated suffered considerable political instability and conflict throughout the latter half of the 20th century, which intensified during the Lon Nol (1970–75) and Pol Pot (1975–79) regimes, and persisted into the late 1990s. During this time there was evidence of large declines in the population and distribution of large mammal species including tiger *Panthera tigris*, leopard *P. pardus*, Asian elephant *Elephas maximus*, banteng, gaur and Eld's deer *Cervus eldii* (Loucks et al. 2008). These declines were associated with a proliferation of firearms, the development of an external market for wildlife products and, particularly during the Khmer Rouge era, government sponsored hunting (Loucks et al. 2008). This hunting pressure led to the global extinction of one large herbivore species endemic to Indochina, the kouprey (Timmins et al. 2008a, b; Timmins 2011), and more recently the extinction of the tiger from the landscape due to targeted poaching of the remaining individuals (Gray et al. 2012a; O'Kelly et al. 2012).

Almost four–fifths of the EPL (13,730 km^2) lie within protected areas; four wildlife sanctuaries managed by the General Department for Administration of

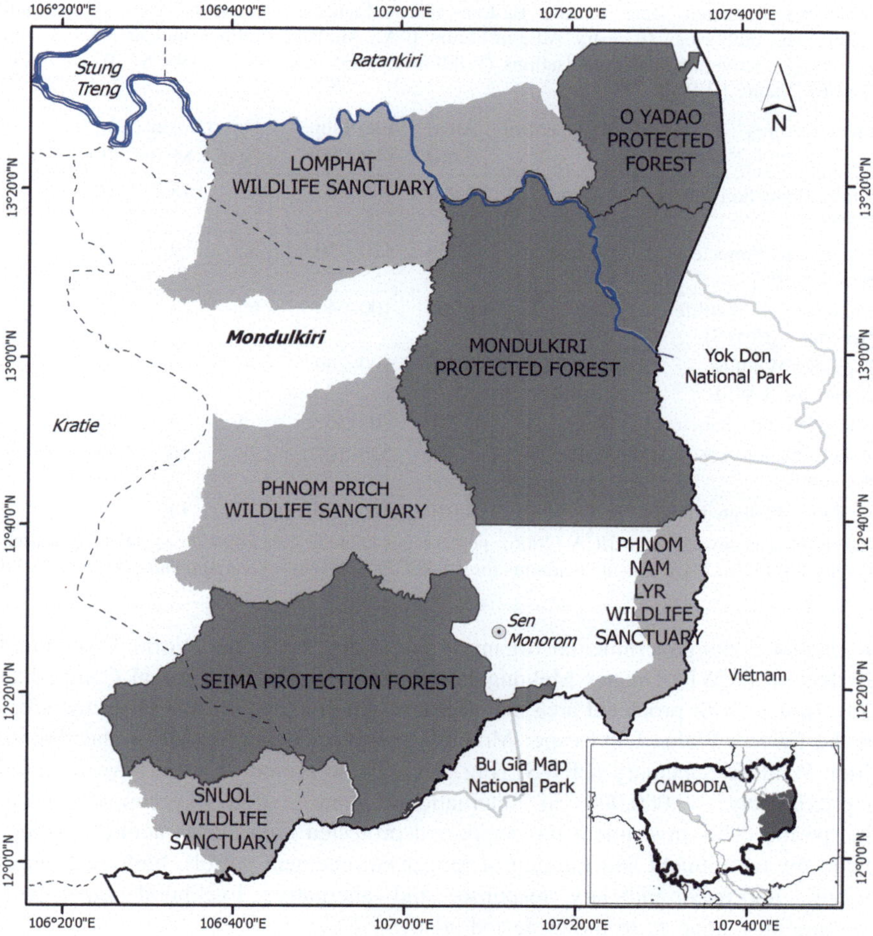

Fig. 9.1 Protected areas within the Eastern Plains Landscape Cambodia and Vietnam

Nature Conservation and Protection of the Ministry of Environment, and three protection forests managed by the Forestry Administration of the Ministry of Agriculture, Forestry and Fisheries (Table 9.1). Furthermore, two areas in Vietnam adjacent to the EPL are gazetted as national parks; Bu Gia Map (12.7°N 107.1°E; 260 km^2), South of Seima Protection Forest, and Yok Don (12.5°N 107.4°E; 1,155 km^2), the largest national park in Vietnam, to the east of Mondulkiri Protected Forest (13.0°N 107.3°E). With improved security resulting from a stable political climate the EPL region has, since the early 2000s, been the focus of conservation activity by government conservation departments supported by

Table 9.1 Protected areas of the Eastern Plains Landscape province with government management authorities (Forestry Administration [FA], Ministry of Environment [MoE]), and supporting nonovernmental organisations (Wildlife Conservation Society [WCS], World Wide Fund for Nature [WWF].

Protected area	Management	Area (km^2)	Elevation range (m)	Forest cover (%)		
				DDF	MDF	SEGF/EGF
Seima Protection Forest [SPF]	FA and WCS	2,990	70–650	42	15	31
Mondulkiri Protected Forest [MPF]	FA and WWF	3,730	120–450	82	9	1
Phnom Prich Wildlife Sanctuary [PPWS]	MoE and WWF	2,200	100–640	69	23	5
Lumphat Wildlife Sanctuary [LWS]	MoE and BirdLife	2,510	100–260	78	6	6
Snoul Wildlife Sanctuary	MoE	750	70–130	3	8	65
Phnom Nam Lyr Wildlife Sanctuary	MoE	540	320–1070	28	16	44
O'Yadao Protected Forest	FA	1,010	100–320	72	14	7

Forest cover estimated from JICA (2006); percentages exclude non-forest areas. DDF deciduous dipterocarp forest; MDF mixed deciduous forest; SEGF/EGF semi-evergreen, and evergreen forest

international nongovernmental organisations. Since 2002, the World Wide Fund for Nature (WWF) Greater Mekong Program has assisted the Royal Cambodian Government with protected area management activities within two protected areas in the Eastern Plains Landscape: Mondulkiri Protected Forest (MPF) and Phnom Prich Wildlife Sanctuary (PPWS) (Fig. 9.1). These protected areas have received approximately $9,000,000 in international conservation investment during this period. This investment has supported protected area infrastructure development, the recruitment and training of ranger enforcement patrols, biological monitoring, improved judiciary response, and alternative livelihood work with communities adjacent to the protected area.

Mondulkiri Protected Forest (MPF) and Phnom Prich Wildlife Sanctuary (PPWS) cover approximately 7,000 km^2 in the heart of the Eastern Plains Landscape and comprise a mosaic of deciduous dipterocarp, mixed deciduous, and semi-evergreen forest (Table 9.1). The two protected areas support at least 19 critically endangered or endangered mammal, bird, and reptile species (Table 9.2). Since 2008 we have been studying large herbivore and predator communities within the core areas of MPF and PPWS (Phan and Gray 2010; Gray and Phan 2011; Gray and Prum 2012; Gray 2012; Gray et al. 2012b, 2013; Gray et al. 2014), largely though distance-based line transect sampling and camera trapping.

Table 9.2 Critically endangered (CR) and endangered (EN) species of mammal, bird, and reptile recorded from Phnom Prich Wildlife Sanctuary (PPWS) and Mondulkiri Protected Forest (MPF) and the conservation significance of the landscape population

Species		IUCN	PPWS	MPF	Significance
Wild Water Buffalo	*Bubalus arnee*	CR			Global?
Tiger	*Panthera tigris*	CR			Regional?
Silvered Langur	*Trachypithecus germaini*	EN			Global
Black-shanked Douc	*Pygathrix nigripes*	EN			Global
Yellow-cheeked Crested Gibbon	*Nomascus gabriellae*	EN			Global
Eld's Deer	*Cervus eldii*	EN			Global
Banteng	*Bos javanicus*	EN			Global
Dhole	*Cuon alpinus*	EN			Regional
Asian Elephant	*Elephas maximus*	EN			Regional
White-rumped Vulture	*Gyps bengalensis*	CR			Global
Slender-billed Vulture	*Gyps. tenuirostris*	CR			Global
Red-headed Vulture	*Sarcogyps calvus*	CR			Global
Giant Ibis	*Pseudibis gigantea*	CR			Global
White-shouldered Ibis	*Pseudibis davisoni*	CR			Global
Green Peafowl	*Pavo munticus*	EN			Global
White-winged Duck	*Cairina scutulata*	EN		??	Regional
Masked Finfoot	*Heliopais personata*	EN			Regional?
Siamese Crocodile	*Crocodylus siamensis*	CR			Global
Elongated Tortoise	*Indotestudo elongata*	EN			Regional?

Shaded cells indicate confirmed presence in each of the protected areas

9.2 Large Herbivore Densities and Their Conservation Significance in Cambodia's Deciduous Dipterocarp Forests

We surveyed 110 line transects that were 1–4 km long and randomly distributed within the 3,400 km^2 core areas of Mondulkiri Protected Forest and Phnom Prich Wildlife Sanctuary during the dry seasons of 2009/10 and 2010/11. Surveys followed the protocols of Karanth and Nichols (2002) for line transect sampling of large herbivores. Total survey effort was 1,310 km resulting in 325 encounters with large herbivores (Table 9.3). We used the conventional distance sampling (CDS) engine in software DISTANCE 6.0 (Thomas et al. 2010) to estimate densities following the protocols of Thomas et al. (2010). Despite the extensive survey effort the numbers of encounters of three large ungulate species (gaur, sambar *Cervus unicolor*, and Eld's deer) were too low for density estimation.

Table 9.3 Number of encounters, mean (and range) cluster size, and encounter rates (encounters per 10 km surveyed) of large ungulates during line transect surveys in Mondulkiri Protected Forest and Phnom Prich Wildlife Sanctuary in 2010 and 2011

Species	Number of encounters	Cluster size	Encounter rate (encounters/10 km)
Gaur	3	1.7 (1–3)	0.02
Banteng	63	4.7 (1–16)	0.48
Sambar	5	1.2 (1–2)	0.04
Eld's Deer	2	2 (1–3)	0.02
Red Muntjac	198	1.1 (1–4)	1.51
Wild Pig	54	4.7 (1–25)	0.41

Species with sufficient observations for density estimation are highlighted

Table 9.4 Sample size of observations used in models (N), densities per km^2(± SEM), 95 % confidence interval range of density estimates and associated population size (number of individuals) (± SEM) and 95 % confidence interval range of population size for Banteng, Red Muntjac, and Wild Pig in the core areas of Mondulkiri Protected Forest and Phnom Prich Wildlife Sanctuary based on estimates from distance-based line-transect sampling

Species	N	Density km^{-2} ± SE	Density % CV	Density 95 % CI range	Population size ± SE	Population size 95 % CI range
Banteng	51	1.1 ± 0.2	19	0.8–1.7	3920 ± 750	2700–5690
Red muntjac	182	2.2 ± 0.2	10	1.8–2.6	7400 ± 750	6060–9030
Wild pig	48	1.4 ± 0.4	25	0.9–2.3	4900 ± 1220	3040–7970

Eastern Plains Landscape wide densities for large ungulates ranged from 1.1 ± SE 0.2 individuals/km^2 for banteng to 2.2 ± SE 0.2 individuals/km^2 for red muntjac *Muntiacus muntjak* (Table 9.4). Using estimated species weights from Karanth and Sunquist (1992), the densities suggest a prey biomass of approximately 540 kg/km^2, of which banteng comprise more than 80 %. Estimated population sizes of species across the entire study area were 2700–5700 banteng, 6000–9000 red muntjac, and 3000–8000 wild pig *Sus scrofa* (Table 9.4).

As far as we are aware the only other site with similarly robust estimates for ungulates in Indochina (sensu Laos, Cambodia, and Vietnam) is the adjacent Seima Protected Forest. In 2010 density estimates, from the 1,800 km^2 core area of Seima, based on a 1,600 km survey effort of 40 transects lines, were 0.29 wild cattle (both banteng and gaur)/km^2, 1.8 red muntjac/km^2, 2.0 wild pig/km^2, and 0.1 sambar/km^2 (O'Kelly and Nut 2010). This corresponds to an overall large ungulate density of 4.2 individuals/km^2 providing a biomass of approximately 260 kg/km^2.

A number of other studies have attempted to quantify tiger prey densities and biomass, largely in semi-evergreen and evergreen forests, in Southeast Asia. However, only a few of these studies employed the scientifically robust distance-based line transect sampling methodology that had been used to survey the Eastern Plains Landscape (O'Brien et al. 2003; WCS-Thailand 2008). For example, in the mosaic of semi-evergreen, mixed deciduous, and deciduous dipterocarp forests in Huai Kha Khaeng Wildlife Sanctuary, western Thailand, overall large ungulate densities are estimated at 6.2 individuals per km^2 (2008 data based on 45+ line transects and 1,000 km+ survey effort; WCS-Thailand 2008). Red muntjac was the most abundant species (2.2 individuals/km^2) followed by sambar (1.9 individuals/km^2) and wild pig (1.6 individuals/km^2). Wild cattle densities were estimated at 0.2 banteng/km^2 and 0.4 gaur/km^2 (WCS-Thailand 2008). Steinmetz and Mather (1996) estimated *Muntiacus* spp. density in semi-evergreen and mixed deciduous forest in Thung Yai Naresuan Wildlife Sanctuary, western Thailand, to be 1–2 individuals/km^2 based on 360 km of line transect sampling.

Large ungulate density estimates have also been published from lowland rainforest in Bukit Barisan National Park, Sumatra where sambar density was 0.6–1.4 individuals/km^2, red muntjac 1.8–4.4 individuals/km^2, and wild pig 4.4–6.0 individuals/km^2 (O'Brien et al. 2003). Srikosamatara (1993) estimated densities and biomass of wild cattle, sambar, and red muntjac from Huai Kha Khaeng Wildlife Sanctuary based on distance sampling of droppings from line transects calibrated against defecation and decomposition rates. Density estimates were approximately 1.8 wild cattle/km^2, 1.9–4.2 sambar/km^2, and 3.1 red muntjac/km^2 giving an overall biomass of 1,250 kg/km^2. By extrapolating of camera-trap encounter rates, Kawanishi and Sunquist (2004) estimated densities of wild pig (3.4–4.6 individuals/km^2) and red muntjac (3.2–4.1 individuals/km^2), which provided a biomass of 200–400 kg/km^2 of tiger prey species in Taman Negara, peninsular Malaysia. Using the Royle-Nichols occupancy heterogeneity model, Vongkhamheng (2011) estimated an ungulate density of 5.3 individuals/km^2 (primarily *Muntiacus* spp. and wild pig with a few sambar) in largely evergreen forest in Nam Et-Phou Louey NPA, northern Laos. However, calculating abundance from such occupancy models based on sign-encounter rates is dependent upon a large number of assumptions and is unlikely to be as robust, or accurate, as estimates generated from distance-based line transect sampling. In contrast to the paucity of reliable data from South–East Asia, there is, however, a wealth of robust data on large ungulate densities from South Asia, where published densities, based on distance-based line transect sampling, in protected areas range from 7 (Wang 2010) to 250+ individuals per km^2 (Wegge and Storaas 2000) with 50+ individual ungulates per km^2 being the norm in most Indian tiger reserves (Karanth and Nichols 2000). The EPL estimate of <5 individuals per km^2 is clearly at the low end of the ungulate density spectrum for deciduous dipterocarp forests; for example, in ecologically similar lowland sal deciduous forest in Ranthambore Tiger Reserve ungulate density is approximately 75 animals/km^2 (Bagchi et al. 2003). Despite this disparity densities of smaller ungulates (red muntjac and wild pig) within the Eastern Plains Landscape are similar to many South and Southeast Asian protected areas (Table 9.5). Steinmetz et al. (2010) found that in areas with minimal hunting in western Thailand, muntjac densities were generally 1–3 individuals/km^2.

Although wild pig have been recorded to reach densities of 40+ individuals per km^2 following mass seeding of dipterocarps in lowland evergreen dipterocarp forest in peninsular Malaysia (Ickes 2001), densities of 1–4 individuals/km^2 are typical throughout the species' cosmopolitan world range (Melis et al. 2006). Although O'Kelly and Nut (2010) suggested red muntjac and wild pig densities are lower in

Table 9.5 Densities of selected large ungulates from distance-based line transect surveys from South and South–East Asian Tiger landscape

Location	<45 kg		>45 kg	
	Red muntjac	Wild pig	Sambar	Chital
Bardia, Nepal[a]	1.5	1.5	n/a	267.0
Rajaji, India[b]	n/a	1.9	14.6	49.9
Pench, India[c]	n/a	n/a	9.6	51.3
Nagarahole, India[d]	4.2		50.6	5.5
Ranthambhore, India[e]	n/a	9.8	17.2	31.0
Kanha, India[c]	0.6	n/a	1.5	49.7
Bandipur, India[c]	0.7	n/a	5.6	20.1
Bhadra, India[f]	3.6	n/a	0.9	4.5
Jigme, Bhutan[g]	2.2	3.7	1.2	n/a
South Asian mean	***2.1***	***4.2***	***12.7***	***59.9***
Huai Kha Khaeng, Thailand[h]	2.2	1.6	1.9	n/a
Bukit Barisan, Indonesia[i]	1.8–4.4	4.4–6.0	0.6–1.4	n/a
PPWS-MPF, Cambodia[j]	2.2	1.4	n/a	n/a
SPF, Cambodia[k]	1.8	2.0	0.1	n/a
South–East Asian mean	***2.3***	***2.6***	***1***	

Data sources [a] Wegge and Storaas (2000); [b] Harihar et al. (2008); [c] Karanth and Nichols (2000); [d] Karanth and Sunquist (1992); [e] Bagchi et al. (2003); [f] Jathana et al. (2003); [g] Wang (2010); [h] WCS-Thailand (2008); [i] O'Brien et al. (2003); [j] This study; [k] O'Kelly and Nut (2010)

Seima Protected Forest than would be expected in sites without hunting, we believe that populations of red muntjac and wild pig are relatively healthy across the Eastern Plains Landscape. Although there was evidence that both species were still being hunted in all protected areas surveyed, the density estimates of both species from the Eastern Plains Landscape were similar to those from ecologically similar protected areas in South and Southeast Asia (Table 9.5), suggesting that both species may be fairly resilient to current levels of hunting pressure in the landscape.

The shortfall in prey densities within the Eastern Plains Landscape in comparison with ecologically similar sites in South Asia and Southeast Asia appears to be mainly due to the low abundances of larger (>45 kg) deer species, like sambar and Eld's deer (Table 9.5). This is worrying as large cervids make up more than three-quarters of prey consumed by tiger across most of its range (Karanth and Nichols 2002). The reasons for the low densities of sambar and Eld's deer in the Eastern Plains are unclear, but similar patterns of low densities and slow recoveries of *Cervus* deer, even when other ungulate species are increasing, have been noted elsewhere in Southeast Asia. For example, in Thung Yai Wildlife Sanctuary, western Thailand, Steinmetz et al. (2010) suggested that the sambar's mating system, in which females select for prime males, made the sambar less resilient to recovery following hunting when compared to other ungulates (such as red muntjac, wild pig and gaur) when prime sambar males had been selectively targeted as hunting trophies. Similarly Aung et al. (2001), from a radio tracking study of Eld's deer in deciduous dipterocarp forest in Myanmar, suggested that the species' social organization and life history traits differed from other tropical cervids. Eld deer showed tightly synchronized seasonal breeding and low reproductive output by females, which could contribute to low levels of population recovery following hunting. Given the importance of large cervids in tiger diets, additional research into natural history and the role of behavior in these species' recovery patterns is essential.

9.3 Significance of Banteng Population

The Eastern Plains Landscape supports two globally endangered large ungulates: banteng and Eld's deer. While we obtained insufficient observations for density estimation of Eld's deer, our data, also published in Gray et al. (2012b), provides the first robust density and population estimate of banteng from anywhere within the species' global range. This suggests that Mondulkiri Protected Forest and Phnom Prich Wildlife Sanctuary support the majority of the global population of banteng, and that previous estimates of the species' Cambodian and global population sizes are low. The IUCN Red List suggests the global population of banteng is '*unlikely to be more than 8,000 and is quite possibly fewer than 5,000 animals… no subpopulation is believed to exceed 500 individuals and only 6–8 subpopulations of more than 50 animals, are known*' (Timmins et al. 2008a, 2008b). If our estimate of 2,700–5,700 banteng (with a mean of 3,900 individuals) in Mondulkiri Protected Forest and Phnom Prich Wildlife Sanctuary is accurate, then the IUCN information would need to be substantially revised. This study and that of O'Kelly and Nut (2010) in the adjacent Siema Proteced Forest, which suggested a population 100–700 (mean 250) individuals in Siema, are the only banteng population estimates we could find to be derived from robust sampling methodologies accounting for imperfect detection. A review of our data indicates that the assumptions of distance sampling were largely met, and therefore we strongly believe that there is little reason to doubt the accuracy of our estimates. Indeed, the majority of sampling errors which are likely to affect distance sampling of large ungulates, for example evasive movement away from transect lines prior to detection and failing to observe all individuals in groups, will lead to underestimating rather than overestimating density (Wegge and Storaas 2000).

Assessing the global significance of the banteng population in the Eastern Plains Landscape is hindered by the lack of any accurate, scientific population estimates from elsewhere in the species' range. Table 9.6 summarizes the species' status from

Table 9.6 Banteng population estimates from range states

Range State (subspecies)	Population size	Data source
B. j. javanicus		
Java, Indonesia	700–1300	Pudyatmoko (2004)
B. j. lowi		
Kallimantan, Indonesia	? < 50?	IUCN-red list
Sabah, Malaysia	?300–500?	IUCN-red list
B. j. birmanicus		
Myanmar	?<500?	Estimate
Laos	?<50?	Estimate
Vietnam	74–103	Pedrono et al. (2009)
Thailand	470	Srikosamatara and Suteethorn (1995)
MPF-PPWS-SPF	2800–6400	This study and O' Kelly and Nut (2010)
Cambodia-rest	?1000–1500?	Estimate

Estimates are marked with? not based on published sources

other range countries, though numbers for Myanmar, Laos, Borneo, and the remainder of Cambodia are essentially optimistic estimates of maximum populations based on limited fieldwork. A better global population estimate, therefore, might be 5,600–11,000 individuals, though there is great uncertainty in the population estimates besides those from the Eastern Plains Landscape. We strongly recommend robust surveying for Banteng elsewhere in the species' range, particularly at sites such as Huai Kha Khaeng Wildlife Sanctuary, Thailand (estimated population 290 individuals; Srikosamatara and Suteethorn 1995), and Ujong Kulon (estimated population 300–800 individuals; Pudyatmoko 2004) and Baluran (estimated population 206 individuals; Pudyatmoko 2004) National Parks, Java, where distance-based line transect surveys appear practical. As in our study areas it is possible that these may result in substantial upwards revisions of known population estimates.

Considering the maximum possible population sizes elsewhere in the species' range, the protected area complex of Mondulkiri Protected Forest-Phnom Prich Wildlife Sanctuary-Siema Protected Forest supports 50–60 % (upper and lower 95 % confidence intervals respectively) of the global banteng population (Table 9.6). These protected areas are therefore clearly the global stronghold and are irreplaceable for the conservation of this enigmatic and beautiful species. Given the presence of banteng in at least three other protected areas in eastern Cambodia (Lumphat Wildlife Sanctuary and O'Yadao Protected Forest) and adjacent Vietnam (Yok Don National Park—population estimate 30–44 individuals; Pedrono et al. 2009), it is possible that the banteng population in EPL may be even higher than our estimates. However hunting, for wild meat and trophy horns, remains a major threat in these protected areas where enforcement levels are much lower than in our study areas.

9.4 Large Herbivore Activity Patterns and Interactions with Carnivores

Understanding the control that extant large predator populations exert on large herbivore populations is important for conservation management focussed at reviving populations of both threatened large herbivores and their predators. The deciduous dipterocarp forests of the Eastern Plains Landscape, similar to much of tropical Asia, historically supported two *Panthera* species (tiger and leopard) together with dhole *Cuon alpinus* and a suite of smaller cat species (Gray et al. 2014). Where the three species co-occur today, inter-specific competition appears to affect behavior, movement patterns , and prey selection, particularly where large ungulate prey densities are reduced (Stoen and Wegge 1996; Karanth and Sunquist 2000; Odden et al. 2010). The tiger, however, is most likely extinct in the landscape (O'Kelly et al. 2012), but the leopard remains relatively abundant with an estimated density of 3.8 (±SE 1.9) individuals/100 km^2 (Gray and Prum 2012), which is comparable with estimates in well protected national parks in Thailand.

We wanted to test, like foraging theory suggests, if whether carnivores synchronize their activity patterns with those of their principal prey species to optimize foraging behavior (Stephens and Krebs 1987; Linkie and Ridout 2011). We, therefore, deployed 50 camera-trap pairs (Reconyx RapidFire Professional PC90; Reconyx, Inc., Holmen, WI) within approximately 210 km^2 of the core area of Mondulkiri Protected Forest during the late dry season (March–July 2009) for a total of 3,711 camera-trap pair nights (mean 77.5 per location). Camera-traps were located either side of routes (i.e., motorbike trails, dry-river beds, and ridgelines) designed to maximize encounters with large carnivores (for more details of camera-trap study design see Gray and Prum 2012). All independent encounters (defined as successive photographs separated by >20 min; Phan et al. 2010) of leopard, dhole, and large herbivores were recorded.

Overlap between the activity patterns of leopard, dhole, and the two most frequently photographed large herbivores (red muntjac and wild pig) were assessed using a statistical model developed by Ridout and Linkie (2009). The Ridout and Linke (2009) model calculates probability density functions of each species' activity pattern using kernel density estimates (Taylor 2008). Overlap between the density distributions of any two species is estimated using the coefficient of overlapping Δ_4 (Weitzman 1970), which ranges from 0 (no overlap) to 1 (complete overlap; Ridout and Linkie 2009). Linkie and Ridout (2011) suggested that $\Delta_4 > 0.8$ indicated strongly overlapping activity patterns.

The range of independent encounters of the four target species captured by the camera-traps was between 33 (dhole) and 442 (red muntjac) (Table 9.7). Although active throughout the day and night (i.e., cathemeral activity pattern), leopard displayed clear crepuscular activity with 26 % of encounters between 0401–0659 h and 19 % between 1701–1959 h. Dhole were predominantly diurnal (Table 9.7), and wild pig and red muntjac were cathemeral with encounters throughout the 24-h cycle. However, wild pig was distinctly more nocturnal than red muntjac

Table 9.7 Number of independent encounters (sensu Phan et al. 2010), and proportion of nocturnal (i.e., 18h01–05h59) encounters, of leopard, dhole, red muntjac, and Eurasian wild pig, during camera-trapping in Mondulkiri Protected Forest, March–June 2009

Species	# independent encounters	# (%) of nocturnal (18h01–05h59) encounters
Leopard	141	81 (57)
Dhole	33	4 (12)
Red muntjac	442	96 (21)
Wild pig	307	158 (51)

(Table 9.7); red muntjac activity peaked after dawn with 36 % of encounters between 0601–0859 h. The activity pattern of leopard overlapped with wild pig more than it did with red muntjac, while the activity pattern of dhole overlapped more with red muntjac than it did with wild pig (Figs. 9.2 and 9.3).

The close temporal overlap between the daily activities of leopard and wild pig, and between dhole and red muntjac suggest that there may be specialization in prey selection by the two large carnivores in Mondulkiri Protected Forest. Although considered a generalist predator, a detailed meta-analysis found that leopard preferentially prey on ungulate species that have a body mass of 10–40 kg and occur in small herds in dense forest habitats (Hayward et al. 2006). However, inter-specific competition with tiger may result in leopard targeting suboptimal prey, for example langurs *Semnopithecus* spp., particularly when ungulate densities are reduced (Ramakrishnan et al. 1999; Karanth and Sunquist 2000; Steinmetz et al. 2013). In sites such as Mondulkiri Protected Forest, where tigers are absent, wild pig may be an ideal prey for leopard given their relative abundance (approximate density 1.9 individuals per km^2), optimal body mass (~38 kg; Karanth and Sunquist 1992), and their herding behavior.

Dhole preferentially prey upon medium to large cervids in South Asia, particularly chital and sambar, and it has been suggested that the primarily diurnal activity patterns of the dhole reflects those of its prey (Karanth and Sunquist 2000;

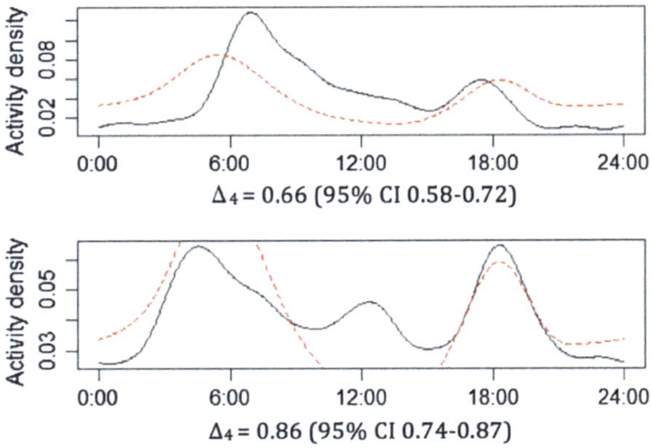

Fig. 9.2 Activity overlap between leopard and red muntjac (*top*), and leopard and wild pig (*below*) based on camera-trapping in Mondulkiri Protected Forest, Cambodia. Δ_4, a measure of activity overlap, was calculated based on Linkie and Ridout (2011); $\Delta_4 > 0.8$ indicates strongly overlapping activity patterns

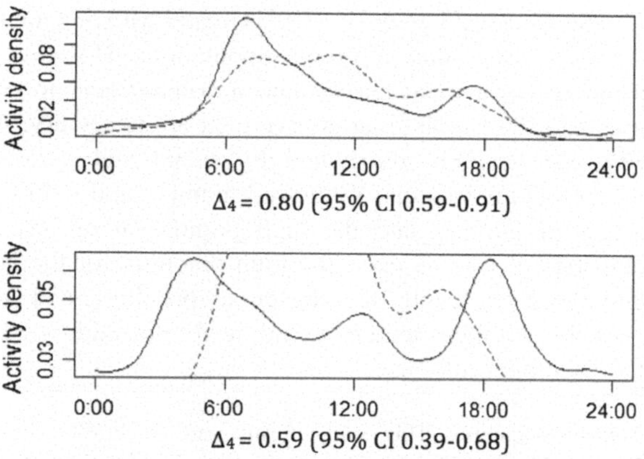

Fig. 9.3 Activity overlap between dhole and red muntjac (*top*) and dhole and wild pig (*below*) based on camera-trapping in Mondulkiri Protected Forest, Cambodia. Δ_4, a measure of activity overlap, was calculated based on Linkie and Ridout (2011); $\Delta_4 > 0.8$ indicates strongly overlapping activity patterns

Andheria et al. 2007; Borah et al. 2009). However, in tropical Southeast Asia, where ungulate densities are reduced, muntjac spp., and even mouse-deer *Tragulus* spp., predominate in dhole scat analysis (Grassman et al. 2005; Kawanishi and Sunquist 2008; Kamler et al. 2012; Steinmetz et al. 2013). The strong temporal overlap found in this study between the activity patterns of the muntjac and dhole in Mondulkiri Protected Forest provides further support for the importance of muntjac in dhole diets in Southeast Asia. Sambar, which dominates dhole diet in other sites, occurs at very low densities in Mondulkiri Protected Forest (4 camera-trap encounters during this study) probably due to hunting pressure. Recovery of sambar populations, therefore, appears to be important for large carnivore conservation in Mondulkiri Protected Forest. Diet studies, involving scat analysis, are recommended for further understanding carnivore–prey interactions in eastern Cambodia.

9.5 Conclusions

The data that our co-workers and us have presented over the past 6 years (Phan and Gray 2010; Gray and Phan 2011; Clements et al. 2012; Gray 2012; Gray and Prum 2012; Gray et al. 2012b; O'Kelly et al. 2012; Wright et al. 2012; Gray et al. 2013) have confirmed the global significance of Mondulkiri Protected Forest, Seima Protection Forest, and Phnom Prich Wildlife Sanctuary for biodiversity conservation. This particularly applies for the long-term conservation of large herbivores in the Southeast Asian region. The documented banteng population is substantially higher than previously estimated and is likely to represent >50 % of the global population of this endangered species. However overall large ungulate densities, <5 individuals per km^2, are much lower than the intrinsic carrying capacity of deciduous dipterocarp forest. This appears to be largely due to the scarcity of large deer (sambar and Eld's deer).

The landscape goals of WWF Greater Mekong are to recover large ungulate, and thus tiger and other carnivore populations. However, for this to happen, stronger protected area management and a commitment to conservation from the highest levels in the Cambodian government are necessary. All protected areas within the EPL are severely threatened by social and agricultural land concessions and infrastructure projects. While red muntjac and wild pig may be resilient to current levels of hunting and the high population of banteng is encouraging, the low densities of *Cervus* deer, the high potential for the banteng population to decline, and the high levels of widespread poaching and hunting are major causes for concern. All these factors would need to be addressed for the EPL to be once again the 'Serengeti of Asia'.

Acknowledgments This study forms part of WWF Greater Mekong Cambodia Country Program's Eastern Plains Landscape Project. Major funding comes from WWF-US, WWF-Sweden, WWF-Germany, and Humanscale. Work in Mondulkiri Protected Forest is with permission of the Forestry Administration of the Ministry of Agriculture Forestry and Fisheries with support from His Excellency Cheng Kimsun, Men Phymean, Song Keang, and Keo Sopheak. Work in Phnom Prich Wildlife Sanctuary is with permission of the Ministry of the Environment and support from His Excellency Chay Samith, Sanrangdy Vicheth, and Han Sakhan. This project would not have been possible without the hard work and dedication of our field teams: Khaev Oudom, Ing Seangnirithy, Lien Nor, and Van Sonny (MPF) and Sary Tre, Sin Somoan, Se Noun, and Chan Touy (PPWS). Craig Bruce, Pin Chanrattanak, Nick Cox, Keith Metzner, Amy Malling, Pem Maly, Michelle Owen, and Seng Teak provided logistical and administrative support whilst Hannah O'Kelly, Bivash Pandav, Abishek Harihar, Paul Doggerty, Shannon Barber-Meyer, and Barney Long assisted in study design, survey objectives, and provided statistical advice.

References

Aung M, McShea WJ, Htung S et al (2001) Ecology and social organization of a tropical deer (*Cervus eldi thamin*). J Mammal 82:836–847

Andheria AP, Karanth KU, Kumar NS (2007) Diet and prey profiles of three sympatric large carnivores in Bandipur Tiger Reserve. India. J. Zool. 273:169–175

Bagchi S, Goyal SP, Sankar K (2003) Prey abundance and prey selection by tigers (*Panthera tigris*) in a semi-arid, dry deciduous forest in western India. J Zol Soc Lon 260:285–290

Borah JK, Deka S, Gupta RP (2009) Food habits of dhole (*Cuon alpinus*) in Satpura Tiger Reserve, Madhya Pradesh, India. Mammalia 73:85–88

Clements T, Gilbert M, Rainey HJ et al. (2012) Vultures in Cambodia: populations, threats and conservation. Bird Cons Int. doi:10.1017/S0959270912000093

Grassman LI, Tewes ME, Silvy NJ et al (2005) Spatial ecology and diet of the dhole *Cuon alpinus* (Canidae, Carnivora) in North Central Thailand. Mammalia 69:11–20

Gray TNE (2012) Studying large mammals with imperfect detection: status and habitat preferences of wild cattle and large carnivores in eastern Cambodia. Biotropica 44:531–536

Gray TNE, Prum S (2012) Leopard density in post-conflict Cambodia: evidence from spatially explicit capture recapture. J Wil Man 76:163–169

Gray TNE, Phan C (2011) Habitat preferences and activity patterns of the larger mammal community in Phnom Prich Wildlife Sanctuary, Cambodia. R Bull Zool 59:311–318

Gray TNE, Ou R, Huy K et al (2012a) The status of large mammals in eastern Cambodia: a review of camera-trapping data 1999-2007. Cam J Nat His 2012:42–55

Gray TNE, Prum S, Pin C et al (2012b) Distance sampling reveals Cambodia's Eastern plains landscape supports largest global population of the endangered banteng *Bos javanicus*. Oryx 46:563–566

Gray TNE, Phan C, Pin C et al (2013) Establishing a monitoring baseline for threatened large ungulates in eastern Cambodia. Wil Biol 18:406–413

Gray TNE, Phan C, Pin C et al. (2014) The status of jungle cat and sympatric small cats in Cambodia's Eastern Plains Landscape. Cat News Special Issue 8:19-23

Harihar A, Pandav B, Goyal SP (2008) Responses of tiger (*Panthera tigris*) and their prey to removal of anthropogenic influences in Rajaji National Park, India. Eur J Wil Res 55:97–105

Hayward MW, Henschel P, O'Brien J et al (2006) Prey preferences of the leopard (*Panthera pardus*). J Zool 270:298–313

Ickes K (2001) Hyper-abundance of Native Wild Pigs (*Sus Scorfa*) in a lowland dipterocarp rain forest in Penisular Malaysia. Biotropica 33:682–690

Jathanna D, Karanth KU, Johnsingh AJT (2003) Estimation of large herbivore densities in the tropical forests of southern India using distance sampling. J of the Zol Soc Lon 261:285–290

Kamler JF, Johnson A, Vongkhamheng C et al (2012) The diet, prey selection, and activity of dholes (*Cuon alpinus*) in northern Laos. J Mammal 93:627–633

Karanth KU, Nichols JD (2000) Ecological status and conservation of tigers in India. In: Final technical report to the Division of International Conservation, US Fish and Wildlife Service, Washington DC and Wildlife Conservation Society, New York. Centre for Wildlife Studies, Bangalore, India

Karanth KU, Nichols JD (eds) (2002) Monitoring tigers and their prey: a manual for researchers, managers and conservationistsin tropical Asia. Centre for Wildlife Studies, Bangalore

Karanth KU, Sunquist ME (1992) Population structure, density and biomass of large herbivores in the tropical forests of Nagarhole. India J Trop Trop Ecol 8:21–35

Karanth KU, Sunquist ME (2000) Behavioral correlates of predation by tiger (*Panthera tigris*), leopard (*Panthera pardus*) and dhole (*Cuon alpinus*) in Nagarahole. India J Zool 250:255–265

Kawanishi K, Sunquist ME (2004). Conservation status of tigers in a primary rainforest of Peninsular Malaysia. Biol Cons 120:329–344

Kawanishi K, Sunquist ME (2008) Food habitats and activity patterns of the Asiatic Golden Cat (*Catopuma temminckii*) and Dhole (*Cuon alpinus*) in a primary rainforest of Penisular Malaysia. Mam St 33:173–177

Linkie M, Ridout MS (2011) Assessing tiger-prey interactions in Sumatran rainforests. J Zool 284:224–229

Loucks L, Mascia MB, Maxwell A et al (2008) Wildlife decline in Cambodia, 1953–2005: exploring the legacy of armed conflict. Con Let 2:82–92

Melis C, Szafranska PA, Jedrzejewska B et al (2006) Biogeographical variation in the population density of Wild Boar in western Eurasia. J Biogeog 33:803–811

O'Brien TG, Kinnaird MF, Wibisono HT (2003) Crouching tigers, hidden prey: Sumatran tiger and prey populations in a tropical forest landscape. Anim Cons 6:131–139

O'Kelly H, Nut MH (2010) Monitoring of ungulate, primate and peafowl populations using line transect surveys in Seima Protection Forest, Cambodia 2005–2010. Wildlife Conservation Society Cambodia, Phnom Penh

O'Kelly HJ, Evans TDE, Stokes EJ et al (2012) Identify conservation successes, failures and future opportunities; assessing recovery potential of wild ungulates and tiger in eastern Cambodia. PLoS ONE 7(10):e40482. doi:10.1371/journal.pone.0040482

Odden M, Wegge P, Fredriksen T (2010) Do tigers displace leopards? If so, why? Ecol Res 25:875–881

Olson DM, Dinerstein E (1998) The Global 200: a representation approach to conserving the Earth's most biologically valuable ecoregions. Cons Biol 12:502–515

Pedrono M, Ha MT, Chouteau P et al (2009) Status and distribution of the endangered banteng *Bos javanicus birmanicus* in Vietnam: a conservation tragedy. Oryx 43:618–625

Phan C, Gray TNE (2010) Ecology and natural history of Banteng in eastern Cambodia: evidence from camera-trapping in Mondulkiri Protected Forest and Phnom Prich Wildlife Sanctuary. Cam J Nat His 2:118–126

Phan C, Prum S, Gray TNE (2010) Recent camera-trap records of globally threatened species from the Eastern Plains Landscape, Cambodia. Cam J Nat His 2:89–93

Pudyatmoko S (2004) Does the banteng (Bos javanicus) have a future in Java? Challenges of the conservation of a large herbivore in a densely populated island. In: Report of the 3rd IUCN World Conservation Congress. Bangkok

Ramakrishnan U, Coss RG, Pelkey NW (1999) Tiger decline caused by the reduction of large ungulate prey: evidence from a study of leopard diets in southern India. Biol Cons 89:113–120

Ridout MS, Linkie M (2009) Estimating overlap of daily activity patterns from camera-trap data. J Agri Biol Env Stat 14:322–337

Srikosamatara S (1993) Density and Biomass of Large Herbivores and other mammals in a Dry Tropical Forest, Western Thailand. J Trop Eco 9:33–43

Srikosamatara S, Suteethorn V (1995) Populations of Gaur and Banteng and their management in Thailand. Nat His Bul Siam Soc 43:55–83

Steinmetz R, Mather R (1996) Impact of Karen villages on the fauna of Thung Yai Nareseum Wildlife Sanctuary: a participatory research project. Nat His Bul Siam Soc 44:23–30

Steinmetz R, Chutipong W, Seuaturien N et al (2010) Population recovery patterns of Southeast Asian ungulates after poaching. Biol Cons 143:42–51

Steinmetz R, Seuaturien N, Chutipong W (2013) Tigers, leopards, and dholes in a half-empty forest: assessing species interactions in a guild of threatened carnivores. Biol Cons 163:68–78

Stephens DW, Krebs JR (1987) Foraging theory. Princeton University Press, Princeton, USA

Stoen OG, Wegge P (1996) Prey selection and prey removal by tiger (*Panthera tigris*) during the dry season in lowland Nepal. Mammalia 60:363–373

Taylor CC (2008) Automatic bandwidth selection for circular density estimation. Com Stat Data Anal 52:3493–3500

Thomas L, Buckland ST, Rexstad EA et al (2010) Distance software: design and analysis of distance sampling surveys for estimating population size. J App Eco 47:5–14

Timmins RJ (2011) Searching for the Last Kouprey: report to CEPF. Global Wildlife Conservation, Austin, Texas

Timmins RJ, Hedges S, Duckworth JW (2008) *Bos sauveli*. IUCN 2009. IUCN red list of threatened species. Version 2009.1. <www.iucnredlist.org>

Timmins RJ, Duckworth WJ, Hedges S, et al. (2008) *Bos javanicus* IUCN 2010. IUCN red list of threatened species. Version 2010.2. www.iucnredlist.org

Tordoff AW, Timmins RJ, Maxwell A et al (2005) Biological assessment of the Lower Mekong dry forests eco region. WWF-Cambodia, Phnom Penh

Vongkhamheng C (2011) Abundance and distribution of tiger ad prey in montane tropical forest in northern Lao PDR. PhD thesis. University of Florida

Wang SW (2010) Estimating population densities and biomass of ungulates in the temperate ecosystem of Bhutan. Oryx 44:376–382

Thailand WCS (2008) Strengthening conservation and monitoring of Tigers in Haui Kha Khaeng Wildlife Sanctuary and the Western Forest Complex, Thailand: Thailand Tigers Forever annual report 2007–2008. WCS Thailand, Bangkok

Wegge P, Storaas T (2000) Sampling tiger ungulate prey by the distance method: lessons learned in Bardia National Park. Nepal. Anim Cons 12:78–84

Weitzman MS (1970) Measure of the overlap of income distribution of white and begro families in the United States. Technical Report No. 22, US Department of Commerse, Bureau of the Census, Washington DC

Wharton CH (1957) An ecological study of the kouprey, *Novibos sauveli* (Urbain), vol 5. Monographs of the Institute of Science and Technology, Manila, pp 1–107

Wright HL, Collar NJ, Lake IR et al (2012) First census of white-shouldered ibis *Pseudibis davisoni* reveals roost-site mismatch with Cambodia's protected areas. Oryx 46:236–239

Chapter 10
Persistence of Tropical Asian Ungulates in the Face of Hunting and Climate Change

Alys Granados and Jedediah F. Brodie

Abstract Understanding how climate change and hunting influence population persistence is crucial for the conservation of animals such as ungulates that are economically important and also serve important ecological roles. Though the effects of hunting and climate change on wildlife demography and abundance have been studied in isolation, how these impacts interact is not well understood, particularly for tropical Asian ungulates, many of which are threatened with extinction. We used simulation modeling to assess how hunting and climate change could individually and jointly affect the persistence of three tropical Asian forest ungulates spanning a range of body sizes (muntjac, sambar deer, and Asian elephant). We found that the negative effects of climate change on these species could be reduced by altering the level of allowable harvest. Species with lower intrinsic growth rates (elephant and sambar) were vulnerable to even low levels of hunting but were less affected by climate change. We identified hunting thresholds beyond which further increases in harvest level greatly reduced persistence probabilities. Importantly, climate change can lower these thresholds and make species more vulnerable to overhunting, particularly those with relatively higher intrinsic growth rates such as muntjac. Much more information is needed on tropical Asian ungulate population ecology. In the meantime, however, our analysis suggests that wildlife managers focus on monitoring carrying capacity of ungulate populations and its variability over time, as this factor may be critical for assessing population responses to climatic changes, hunting, and their synergistic interactions.

Keywords Carrying capacity · Harvest thresholds · Intrinsic growth rate · Population response · Wildlife demography and abundance

---

A. Granados (✉) · J.F. Brodie
Biodiversity Research Centre, University of British Columbia,
Vancouver, BC, Canada
e-mail: granados@zoology.ubc.ca

A. Granados
Department of Zoology, University of British Columbia, #4200-6270 University Blvd,
Vancouver, BC V6T 1Z4, Canada

© Springer Science+Business Media Dordrecht 2016

F.S. Ahrestani and M. Sankaran (eds.), *The Ecology of Large Herbivores in South and Southeast Asia*, Ecological Studies 225,
DOI 10.1007/978-94-017-7570-0_10

10.1 Introduction

Understanding how human disturbance and climatic variability affect wildlife populations is important for the development of effective management strategies (Weladji and Holand 2003; Linke et al. 2013), particularly for threatened species. Many wild ungulate species around the world are harvested by humans through sport hunting, for markets (Bodmer et al. 1988) or for subsistence (Kofinas et al. 2000). Subsistence hunting for ungulates may be especially important in developing nations, where many people depend on wild meat for protein (Fa et al. 2005, 2008; Albrechtsen et al. 2006). Harvest rates are sometimes sustainable (Bodmer et al. 1994), and indeed hunting in certain areas has persisted for millennia without appearing to compromise ungulate persistence (Brashares 2003). Elsewhere, over-harvesting has resulted in the decline and extirpation of ungulate populations over wide areas (Peres 2000; Fischer and Linsenmair 2001; Corlett 2007). Several species have even been driven to extinction in the wild by excessive hunting (e.g., Brook et al. 2012).

Another emergent threat for animal populations in many parts of the world is climate change. There is growing evidence that climatic changes can affect ungulate survival (Brodie et al. 2013) and abundance (Creel and Creel 2009). Though the effects of hunting (cf. Milner et al. 2007; Setsaas et al. 2007) and climate change (cf. Weladji and Holand 2003; Morris et al. 2008) on wildlife vital rates or population dynamics have been studied in isolation, how these factors interact is not well understood. Climate change can increase atmospheric CO_2 and variability in temperature and precipitation, and can lead to altered composition of soil nutrients. Ultimately, this can affect vegetation productivity, leading to a reduction in forage quality (Nemani et al. 2003). These effects are often predicted to lead to an increased variability in vital rates (Drake 2005), but little remains known of what this increased variability will mean for population persistence of harvested species. Understanding how climate change and hunting interact to affect population persistence is particularly crucial for animals such as ungulates that are economically important and also serve important ecological roles (cf. Goheen et al. 2010).

Our lack of knowledge about ungulate population ecology is especially acute for tropical Asian species, particularly forest-dwelling taxa (Gaillard et al. 2000; Steinmetz et al. 2010). Although they are important commercially and for subsistence (Bodmer et al. 1994; Carrillo et al. 2000), overharvest is a major threat to ungulate populations in tropical Asia, where the distributions of several species have been dramatically reduced by unsustainable hunting (Madhusudan and Karanth2002; Kumara and Singh 2004; Corlett 2007; Velho et al. 2012). This, in turn, can alter forest dynamics by compromising the regeneration of trees that depend on ungulates for seed dispersal (Brodie et al. 2009; Harrison et al. 2013, also see Chap. 5).

In order to improve management of these species, additional information on the demographic responses of tropical Asian ungulates to climate change and hunting is needed. In this chapter, we have two goals: first, to survey the literature and compile what is known about the demography, vital rates, and population dynamics of tropical forest ungulates in Asia. Second, we use simulation modeling to assess how hunting and climate change could individually and jointly affect the persistence of ungulate species. We also use these models to explore the sensitivity of ungulate persistence to uncertainty in the underlying parameters, thereby asking, in effect, "where is our uncertainty regarding ungulate vital rates hurting us the most?"

10.2 Tropical Asian Ungulate Vital Rates

We used the ISI Web of Science database to search for studies on population growth and demographics of tropical Asian ungulates. The search string we used was to = (ungulate* OR elephas maximus OR Asia* elephant* OR gaur OR bos gaurus OR muntjac OR muntiacus OR sambar OR rusa unicolor OR bearded pig* OR sus barbatus OR mousedeer* OR tragulus OR banteng OR bos javanicus OR cervid*) AND (asia) AND (reproductive output OR population structure OR population grow* rate* OR sex structure OR age structure OR population regulat* OR density dependen* OR colonization OR dispersal OR inbreed* OR carrying capacit* OR intrinsic rate of grow* OR intrinsic rate of increase* OR per capita grow* OR population densit* OR population demograph* OR recruitment rate*). We did not limit our search to studies from any specific time period. We focused on obtaining data from wild ungulate populations, but also included some information from captive or nonnative populations as these were the only data available for some species.

Overall, few studies on the population ecology of tropical ungulates in Asia exist (Gaillard et al. 1998, 2000), and even fewer provide estimates of vital rates, demography, or population characteristics. Our initial literature review yielded 73 results, though most were not relevant. This search resulted in estimates of r_{obs} for only seven species: bearded pig (*Sus barbatus*), wild pig (*S. scrofa*), red muntjac (*Muntiacus muntjak*), sambar (*Rusa unicolor*), gaur (*Bos gaurus*), banteng (*B. javanicus*), and Asian elephant (*Elephas maximus*). Estimates of spatial variance in r_{obs} were available for three of these seven species: *M. muntjak*, *R. unicolor*, and *E. maximus*. Annual variations in rates of growth were generally unavailable. We aimed to estimate the temporal variance in r_{obs} as well, but such information for Asian ungulates was completely lacking; hence, all estimations of variance in r_{obs} obtained from the literature represent spatial variance. The results from our literature search are summarized in Table 10.1.

We found no estimates of carrying capacityfor Asian ungulates. Several studies provided estimates of population density from multiple sites, but only three studies provided multiple values over different time periods (Ickes 2001; O'Brien et al. 2003; Sukumasuang 2009). This prevented us from estimating temporal changes in population density at each site and, hence, from directly ascertaining how human disturbance affects ungulate population density over time.

Table 10.1 Estimates of vital rates and population demographic information compiled from a literature search of Asian ungulate species

	Species							
	Sus barbatus	*Sus scrofa*	*Tragulus spp.*	***Muntiacus muntjak***	***Rusa unicolor***	*Bos gaurus*	*Bos javanicus*	***Elephas maximus***
Observed intrinsic rate of increase (r_{obs})	0.16[d]	0.14[a]		0.44[a]; 0.04[b]; 0.02[b]	0.22[a]	0.11[c]; 0.12[c]; 0.31[a]	0.02[f]	0.02[e]; 0.01[e]
$r_{average}$ (from r_{obs})				0.17		0.11		0.01
Carrying capacity (K)	8.9*[d]							
(individuals/km^2)	12.8**[d]							
Sex ratio (M:F)		53:100[g]		76:100[g]	42:100[g]	18:100[g]		
Average population density (D_{avg}) (individuals/km^2)	5	12.68	18.83	2.47	3.54	3.11	1.78	0.72
Average body mass (kg)	79	137.5	3.3	20	193	667	700	2088

Species used in our simulation models are in bold
[a]Steinmetz et al. (2010); [b]Hameed et al. (2009); [c]Foose (1980); [d]Lee et al. (2011); [e]Sukumar et al. (1997); [f]Bradshaw (2007); [g]Karanth and Sunquist (1992);
*exploited population; **unexploited population

Data on sex ratios, reproductive rates, and fecundity for Asian forest ungulates was also lacking. We found information on sex ratios for five species (*E. maximus, B. gaurus, R. unicolor, S. scrofa,* and *M. muntjak*), but with the exception of estimates for *M. muntjak* and *B. gaurus* (Hameed et al. 2009; Ahrestani et al. 2011), most were from a single study (Karanth and Sunquist 1992). Estimates of reproductive rate and fecundity are widespread in the literature for temperate and arctic ungulates; fecundity of young females is usually highly variable within and among populations while fecundity and survival of prime-aged females exhibits less temporal variation (Gaillard et al. 2000). There is also an association of reduced fecundity with increasing population density in some ungulates (Coulson et al. 2000). We found only one study that provided information about fecundity in a tropical Asian ungulate, *E. maximus* (Sukumar et al. 1997).

Multiple estimates of vital rates for the same species but from different sources showed substantial differences. For example, two studies that provided estimates of r_{obs} for *M. muntjak* showed a great deal of variation: 0.44 (Steinmetz et al. 2010), 0.04, and 0.02 (Hameed et al. 2009). These differences may be due to differences in data collection: Steinmetz et al. (2010) used thirty 400-m transects and an index of abundance based on animal signs to measure annual changes in population size from 2000 to 2005. They also considered the detection probability of animals in their analysis. In contrast, detectability was not included in analyses by Hameed et al. (2009) where two estimates of population growth rate (λ) (which we converted into intrinsic rates of increase) were based on multiple population counts between 1971 and 1990, and 1991 and 2005, respectively. Site-related differences may have also played a role here, in addition to human disturbances: Steinmetz et al. (2010) determined population growth rates following poaching, whereas the Hameed et al. (2009) study provided estimates from more long-term data with no information given on the level of anthropogenic presence in the area. No baseline estimates of wildlife abundance were provided by Steinmetz et al. (2010), but muntjac occurred at low density prior to the end of hunting in the area. Therefore, the relatively recent release from poaching (following increased ranger presence) may have contributed to the higher r_{obs} found by Steinmetz et al. (2010).

10.3 Demographic Simulations of Hunting and Climate Change

In the absence of data on how climate change and hunting affect tropical Asian ungulates, we used simulations to explore the effects of increasing harvest rate on persistence probability. We performed simulations across a range of demographic parameters, potential climate change scenarios, and harvest regimes. Previous studies have suggested that climate change can alter carrying capacity (Boyce et al. 2012). We note here that it could also affect intrinsic rates of increase. Climatic changes that alter vegetation productivity and/or quality can affect ungulate carrying capacities as well as population growth rate, while changes in abiotic conditions that directly alter metabolism, growth rates, or reproduction will affect a population's maximum intrinsic rate of increase, r_{max}. In all cases, our response variable was the probability of the population persisting for a century ("persistence probability," P_p), as this provides a useful metric to compare different impact and management scenarios (Bakker and Doak 2008). We used stochastic simulations to model theta logistic growth of unstructured populations while accounting for environmental and demographic stochasticity. We also included a proportional harvest term, whereby a fixed proportion of the population was hunted each year, as an approximation of a hunting scenario where hunter effort remained constant (Ludwig 2001). Population growth in year t was given by

$$\frac{dN_t}{dt} = r_{max} N_t \left(1 - \left(\frac{N_t}{K}\right)^{\theta}\right) - EN_t$$

where N_t is the population density in year t, K is the environmental carrying capacity, θ is a shape parameter controlling the expression of density dependence, and E is the annual exploitation rate, or the proportion of the population hunted each year; an E value of 0 means that no individuals were hunted, while a value of 1 refers to 100 % of the population being harvested that year.

For each simulation we used the above equation to estimate population size over 100 years, starting at $N = K$ for three species of interest, *M. muntjak, R. unicolor,* and *E. maximus*, representing a gradient of body sizes. For each time step, we randomly drew r_{max} and K values from normal distributions based on estimated values and coefficients of variation for our target species (see Fig. 10.1 legend for more details). We then calculated the proportion of simulation runs where the population remained extant as P_p. We first estimated P_p for a baseline scenario with no climate change (but still a background level of variation in r_{max} and K) and no hunting, and then assessed how various hunting and climate change scenarios affected the proportional change in persistence probability, relative to the baseline. For each species, we assessed nine different climate change scenarios: a "no change" scenario, four levels of progressively stronger negative impacts of climate change on ungulate demographic rates, and four levels of progressively stronger positive climate change impacts. Negative impacts of climate change refer to a decrease in r_{max} and K values and increases in the coefficient of variation, CV, by 10, 20, 30, and 40 % of each parameter. Positive impacts of climate change are similarly quantified as increases in r_{max} and K, and a decrease in CV for each. We ran 1000 population simulations to assess P_p for each of 50 levels of E, at each of the nine climate change scenarios, and for each of the three species.

Several patterns emerge from these simulations (Fig. 10.1). First, exploitation rate has a nonlinear relationship with P_p. After reaching a threshold, further increases in E cause P_p to drop precipitously. The position of these thresholds (i.e., where they occur on the X-axis in Fig. 10.1) varies across climate change scenarios. Negative climate change impacts reduce the threshold, while positive climate change impacts

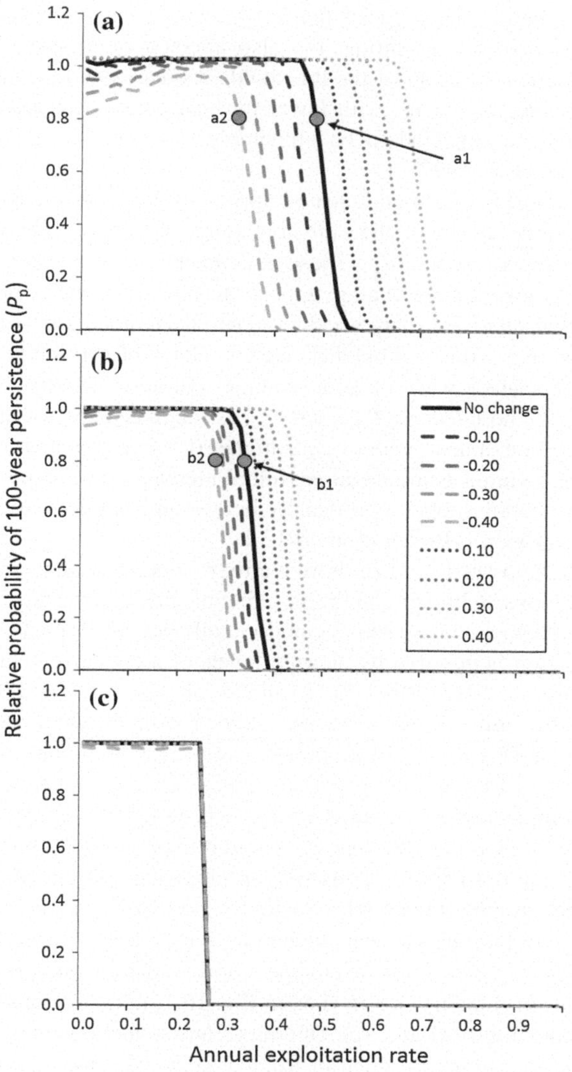

Fig. 10.1 Effects of hunting intensity and climate change on persistence probability relative to a scenario with no hunting or climate change for three species varying in size: **a** *Muntiacus muntjak* ($\bar{r}_{max}$ = 0.44, $\overline{K}$ = 600 individuals per 100 km^2, CV(r_{max}) = CV(K) = 0.2). **b** *Rusa unicolor* ($\bar{r}_{max}$ = 0.20, $\overline{K}$ = 300 individuals per 100 km^2, CV(r_{max}) = CV(K) = 0.2). **c** *Elephas maximus* ($\bar{r}_{max}$ = 0.01, $\overline{K}$ = 100 individuals per 100 km^2, CV(r_{max}) = CV(K) = 0.2). Negative (−) climate change scenarios involve a proportional decrease in $\bar{r}_{max}$ and $\overline{K}$ and a proportional increase in CV (r_{max}) and CV(K). Positive (+) scenarios involve a proportional increase in $\bar{r}_{max}$ and $\overline{K}$ and a proportional decrease in CV(r_{max}) and CV(K). *CV* coefficient of variation. *Points* a1, a2, b1, and b2 show selected intersection points with persistence probability = 0.8 and are referred to in the text

increase the threshold. This suggests that in general, species with high r_{max} values should be more resilient to hunting, but also affected more strongly by climatic changes. The average position of the thresholds decreases as r_{max} and K decrease. This is unsurprising, as species with lower intrinsic rates of increase and average abundance would be expected to be vulnerable to extinction at lower levels of hunting (Bodmer et al. 1997).

Second, the impacts of climate change appear to diminish substantially as r_{max} and K decline. In *M. muntjak*, the climate change scenarios differ clearly in their effects on persistence probability. These differences are reduced in *R. unicolor,* while in *E. maximus*, climate change appears to have essentially no effect on P_p. This suggests that relative to large-bodied generalists, small forest ungulates with more specialized (e.g., frugivorous) diets are less adaptable to changes in vegetation productivity associated with climate change (Bodmer 1990). Finally, initial increases in E have no effect on P_p when r_{max} is relatively low for all three species in our analyses. For example, when r_{max} is higher (0.44, approximating *M. muntjak*) and climate change impacts are negative, initial increases in E actually increase P_p, possibly by maintaining population density below carrying capacity where growth rates are higher (also see Boyce et al. 2012).

These results help shed light on how we may improve management of tropical ungulate populations. Climate change is affecting myriad species, but it is often unclear what managers on the ground can do to alleviate its impacts. We show here that such alleviation is possible by altering levels of a concurrent stressor, hunting (also see Brodie et al. 2012, 2013). For example, say that we would like to maintain a persistence probability of 80 % for *M. muntjak* over the next century. Without climate change, such a P_p would be possible at a 24.5 % annual exploitation rate (point a1 in Fig. 10.1). Strong negative climate change impacts would greatly reduce P_p if hunting levels remained unchanged, but if the exploitation rate was reduced to 16.5 %, then the 100-year P_p would remain at 80 % (point a2). In other words, by adjusting our harvest levels we can offset the impacts of climate change on ungulate demography. Offsetting climate impacts in larger species may require even less change to hunting patterns; maintaining $P_p = 0.8$ for *Rusa unicolor* in the face of dramatic negative climate change impacts would only require reducing exploitation rates from 35 to 28.5 % (i.e., in Fig. 10.1, note how much closer points b1 and b2 are than points a1 and a2). Of course, any reductions in harvest are much more easily said than done in much of tropical Asia. Yet, maintaining sustainable harvesting strategies is an important focus of many conservation plans (see Bennett and Robsinson 2000 and references therein) and provides a clear win-win scenario for wildlife and the livelihoods of local people that depend on the hunted animals for food. Moreover, reductions in hunting pressure may be facilitated in some areas by decreasing rural populations due to urbanization.

We also used a form of sensitivity analysis to ask how persistence probability would be affected by variation in the demographic parameters r_{max}, K, CV(r_{max}), and CV(K). As demonstrated in Table 10.1, we have extremely limited knowledge about demographic parameter values for tropical Asian ungulates; this sensitivity analysis essentially asks how strongly do errors in estimation of different demographic parameters affect our predictions of persistence probability, P_p. We determined an average P_p with no hunting or climate change impacts and $r_{max} = 0.25$, $K = 200$ individuals per 100 km^2, CV(r_{max}) = CV(K) = 0.25, and $\theta = 1$. We then varied each parameter and assessed the proportional change in P_p. For each of 50 values of r_{max} (ranging from 0 to 0.8), K (ranging from 50 to 1000 individuals per 100 km^2), CV(r_{max}) (ranging from 0 to 1), CV(K) (ranging from 0 to 1), and θ (ranging from −2 to 5) we ran 1000 simulations to assess changes in P_p.

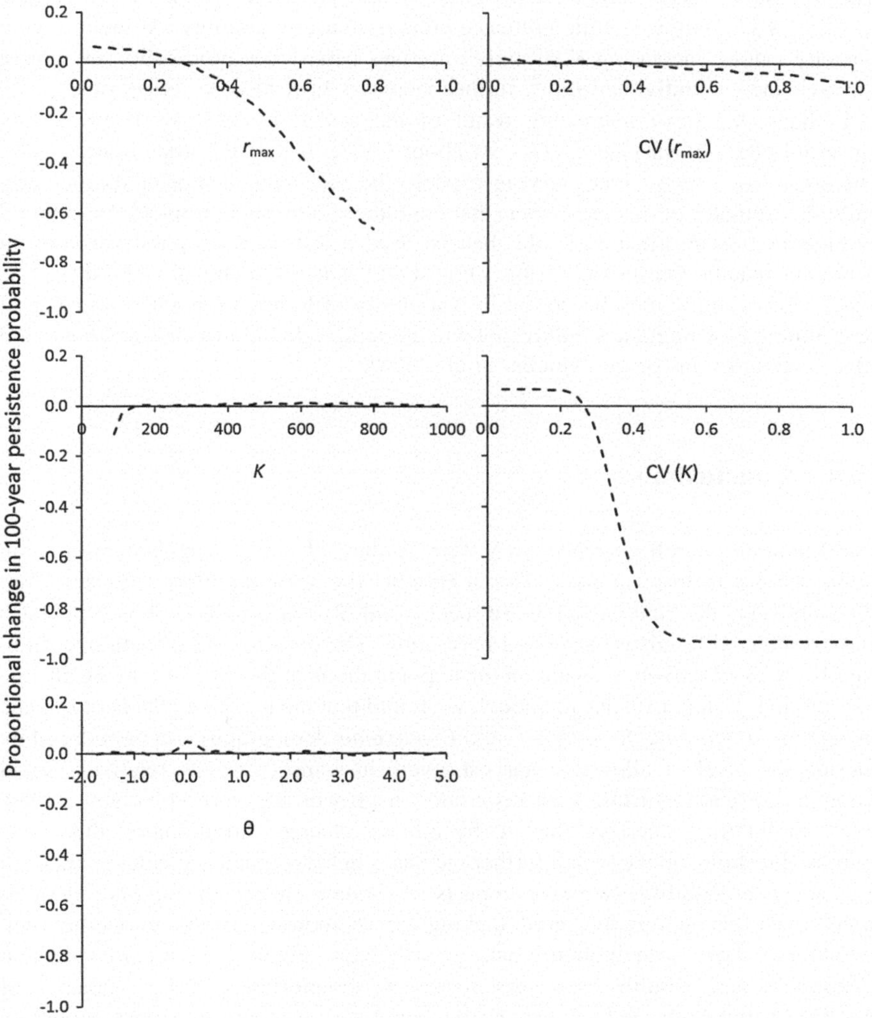

Fig. 10.2 Sensitivity of persistence probability to changes in demographic parameters. Proportional change in persistence probability for $\bar{r}_{max}$, CV(r_{max}), $\overline{K}$, CV(K), and θ from starting values of $\bar{r}_{max} = 0.25$, CV(r_{max}) = 0.25, $\overline{K} = 200$ individuals per 100 km^2, CV(K) = 0.25, and $\theta = 1$

These sensitivity analyses reveal that changes in r_{max} (or errors in estimating r_{max}) have relatively little influence on persistence probability (Fig. 10.2). From very low values up to about $r_{max} = 0.4$, P_p remains unchanged but is increased when r_{max} is under estimated. At higher r_{max} values, P_p actually decreases, likely because extremely high growth rates can cause dramatic "overshoot-collapse" population dynamics and "deterministic chaos" in population trajectories (Mills 2013). Variation in CV(r_{max}) has negligible influence on P_p; we can err grossly in our estimation of this parameter with no ill effect when it comes to estimating extinction risk. Likewise, large variations in θ (the shape parameter of the theta logistic equation) have only very little influence on persistence probability. At low carrying capacity values, increasing K strongly improves persistence probability, but above approximately 2 individuals/km^2, further increases in K have no effect on P_p.

Perhaps, the most interesting result of the sensitivity analysis is the strong threshold effect of varying CV(K). At about CV(K) = 0.3, P_p drops dramatically. Therefore, we suggest that carrying capacity be the main parameter (particularly multiple estimates of this parameter) that ungulate ecologists in tropical Asia should monitor as this information could then be used to estimate temporal variation in carrying capacity. Our model results suggest that changes in annual variability in K, which, for example may be driven by variability in forage production, are key to determining how ungulates will respond to climatic and land use changes that could affect variability in forage (Mueller et al. 2008).

10.4 Conclusions

Our literature search returned very few studies of forest ungulate population demographics in tropical Asia. This highlights the need for further research into (1) quantifying the vital rates of these species and (2) assessing how those vital rates may be affected by disturbance and harvesting. The present lack of data may limit our ability to effectively manage ungulate populations in the face of climate change and hunting. Using a stochastic model, we found that the negative effects of climate change on *M. muntjak*, *R. unicolor*, and *E. maximus* demography can be reduced by altering the level of allowable harvest, even, in some cases, by relatively small amounts. Species with lower intrinsic rates of growth are more vulnerable to low levels of hunting, but less affected by climate change. Additionally, there is a hunting threshold, above which further increases in harvest cause greater declines in persistence probability. Negative impacts of climate change on ungulate demography lower this hunting threshold, making species more vulnerable to overhunting, particularly those with high intrinsic growth rates. Finally, we recommend that researchers and wildlife managers focus on monitoring carrying capacity of ungulate populations and its variability over time, as our preliminary modeling results suggest that this factor may be critical for assessing the responses of ungulate populations to climatic change, hunting, and their cumulative effects.

Acknowledgments Both authors were funded for this work by the National Science and Engineering Research Council, Canada, and the University of British Columbia.

Literature Cited

Ahrestani F (2011) Life-history traits of gaur *Bos gaurus*: a first analysis. Mamm Rev 41:75–84

Albrechtsen L, Fa JE, Barry B et al (2006) Contrasts in availability and consumption of animal protein in Bioko Island, West Africa: the role of bushmeat. Environ Conserv 32:340–348

Bakker VJ, Doak DF (2008) Population viability management: ecological standards to guide adaptive management for rare species. Front Ecol Environ 7:158–165

Bennett E, Robsinson J (eds) (2000) Hunting for sustainability in tropical forests. Columbia University Press, New York, NY

Bodmer RE (1990) Ungulate frugivores and the browser-grazer continuum. Oikos 57:319–325

Bodmer RE, Fang TG, Ibanez LM (1988) Ungulate management and conservation in the Peruvian Amazon. Biol Conserv 45:303–310

Bodmer RE, Fang TC, Moya L et al (1994) Managing wildlife to conserve Amazonian forests: population biology and economic considerations of game hunting. Biol Conserv 67:29–35

Bodmer RE, Eisenberg JF, Redford KH (1997) Hunting and the likelihood extinction of Amazonian mammals. Conserv Biol 11:460–466

Boyce MS, Knopff K, Northrup J et al (2012) Harvest models for changing environments. In: Brodie JF, Post E, Doak DF (eds) Wildlife conservation in a changing climate. University of Chicago Press Ltd, Chicago, IL, pp 293–306

Bradshaw CJA., Isagi Y, Kaneko S, Brook BW, Bowman DMJS, Frankham R (2007) Low genetic diversity in the bottlenecked population of endangered non-native banteng in northern Australia. Mol Ecol 16:2998–3008

Brashares JS (2003) Ecological, behavioral, and life-history correlates of mammal extinctions in West Africa. Conserv Biol 17:733–743

Brodie JF, Helmy OE, Brockelman WY et al (2009) Bushmeat poaching reduces the seed dispersal and population growth rate of a mammal-dispersed tree. Ecol Appl 19:854–863

Brodie JF, Post E, Doak DF (eds) (2012) Wildlife conservation in a changing climate. University of Chicago Press Ltd, Chicago, IL

Brodie JF, Johnson H, Mitchell M et al (2013) Relative influence of human harvest, carnivores, and weather on adult female elk survival across western North America. J Appl Ecol 50:295–305

Brook SM, van Coeverden de Groot P, Scott C et al (2012) Integrated and novel survey methods for rhinoceros populations confirm the extinction of *Rhinoceros sondaicus annamiticus* from Vietnam. Biol Conserv 155:59–67

Carrillo E, Wong G, Cuarón AD et al (2000) Monitoring mammal populations in Costa Rican protected areas under different hunting restrictions. Conserv Biol 14:1580–1591

Corlett RT (2007) The impact of hunting on the mammalian fauna of tropical Asian forests. Biotropica 39:292–303

Coulson T, Milner-Gulland EJ, Clutton-Brock T (2000) The relative roles of density and climatic variation on population dynamics and fecundity rates in three contrasting ungulate species. Proc R Soc B 267:1771–1779

Creel S, Creel M (2009) Density dependence and climate effects in Rocky Mountain elk: an application of regression with instrumental variables for population time series with sampling error. J Anim Ecol 78:1291–1297

Drake JM (2005) Population effects of increased climate variation. Proc R Soc B 272:1823–1827

Fa JE, Ryan SF, Bell DJ (2005) Hunting vulnerability, ecological characteristics and harvest rates of bushmeat species in Afrotropical forests. Biol Conserv 121:167–176

Fa JE, Yuste JEG, Castelo R (2008) Bushmeat markets on Bioko Island as a measure of hunting pressure. Conserv Biol 14:1602–1613
Foose TJ (1980) Demographic management of endangered species in captivity. Inter Zoo Yearbook 20:154–166
Fischer F, Linsenmair KE (2001) Decreases in ungulate population densities. Examples from the Comoe National Park, Ivory Coast. Biol Conserv 101:131–135
Gaillard JM, Festa-Bianchet M, Yoccoz NG (1998) Population dynamics of large herbivores: variable recruitment with constant adult survival. Trends Ecol Evol 13:58–63
Gaillard JM, Festa-Bianchet M, Yoccoz NG et al (2000) Temporal variation in fitness components and popuation dynamics of large herbivores. Annu Rev Ecol Syst 31:367–393
Goheen JR, Palmer TM, Keesing F et al (2010) Large herbivores facilitate savanna tree establishment via diverse and indirect pathways. J Anim Ecol 79:372–382
Hameed W, Abbas FI, Mian A (2009) Population features of barking deer (*Muntiacus muntjak*) in Margalla Hills National Park, Pakistan. Pak J Zool 41:137–142
Harrison RD, Tan S, Plotkin JB et al (2013) Consequences of defaunation for a tropical tree community. Ecol Lett 16:687–694
Ickes K (2001) Hyper-abundance of native wild pigs (*Sus scrofa*) in a lowland dipterocarp rain forest of Penninsular Malaysia. Biotropica 33:682–690
Karanth KU, Sunquist ME (1992) Population structure, density and biomass of large herbivores in the tropical forests of Nagarahole, India. J Trop Ecol 8:21–35
Kofinas G, Osherenko G, Klein D et al (2000) Research planning in the face of change: the human role in reindeer/caribou systems. Polar Res 19:3–21
Kumara HN, Singh M (2004) The influence of differing hunting practices on the relative abundance of mammals in two rainforest areas of the Western Ghats. India Oryx 38:321–327
Lee CS, Diong CH, Goh BS, Stuebing RB, Junau DJ (2011) Towards achieving maximum sustainable yield management of the bearded pig (Sus barbatus) in a logged-over planted forest. J Biolog Dyn 5:268–276
Linke J, McDermid GJ, Fortin MJ et al (2013) Relationships between grizzly bears and human disturbances in a rapidly changing multi-use forest landscape. Biol Conserv 166:54–63
Ludwig D (2001) Can we exploit sustainably? In: Reynolds J, Mace GM, Redford KH, Robinson JG (eds) Conservation of exploited species. Cambridge University Press, Cambridge, UK, pp 16–38
Madhusudan M, Karanth KU (2002) Local hunting and the conservation of large mammals in India. AMBIO 31:49–54
Mills LS (2013) Conservation of wildlife populations: demography, genetics, and management, 2nd edn. Wiley-Blackwell, West Sussex, UK
Milner JOSM, Nilsen EB, Andreassen HP (2007) Demographic side effects of selective hunting in ungulates and carnivores. Conserv Biol 21:36–47
Morris W, Pfister C, Tuljapurkar S et al (2008) Longevity can buffer plant and animal populations against changing climate variability. Ecology 89:19–25
Mueller T, Olson K, Fuller TK et al (2008) In search of forage: predicting dynamic habitats of Mongolian gazelles using satellite-based estimates of vegetation productivity. J Appl Ecol 45:649–658
Nemani RR, Keeling CD, Hashimoto H et al (2003) Climate-driven increases in global terrestrial net primary production from 1982 to 1999. Science 300:1560–1563
O'Brien TG, Kinnaird MF, Wibisono HT (2003) Crouching tigers, hidden prey: Sumatran tiger and prey populations in a tropical forest landscape. Anim Conserv 6:131–139
Peres C (2000) Effects of subsistence hunting on vertebrate community structure in Amazonian forests. Conserv Biol 14:240–253
Setsaas TH, Holmern T, Mwakalebe G et al (2007) How does human exploitation affect impala populations in protected and partially protected areas?—A case study from the Serengeti ecosystem, Tanzania. Biol Conserv 136:563–570

Steinmetz R, Chutipong W, Seuaturien N et al (2010) Population recovery patterns of Southeast Asian ungulates after poaching. Biol Conserv 143:42–51
Sukumar R, Krishnamurthy V, Wemmer C et al (1997) Demography of captive Asian elephants (*Elephas maximus*) in Southern India. Zoo Biology 272:263–272
Sukumasuang R (2009) Population density of Asian elephants in Hua Kha Khaeng Wildlife Sanctuary. Thai J For 28:40–50
Velho N, Karanth KK, Laurance WF (2012) Hunting: a serious and understudied threat in India, a globally significant conservation region. Biol Conserv 148:210–215
Weladji RB, Holand Ø (2003) Global climate change and reindeer: effects of winter weather on the autumn weight and growth of calves. Oecologia 136:317–323

Chapter 11
The Ecology of Large Herbivores of South and Southeast Asia: Synthesis and Future Directions

Mahesh Sankaran and Farshid S. Ahrestani

Abstract The countries of South and Southeast Asia (SSEA) are home to a diverse array of large native herbivores, but the majority of these species are currently threatened with extinction. Ensuring the future survival of these species and the integrity of the ecological services they provide will require concerted management efforts, but these need to be built on a strong scientific foundation, which is currently lacking. In particular, there is an urgent need for research efforts to: (i) generate baseline data on the current status and distribution of large herbivores species in South and Southeast Asia, (ii) quantify vital rates of species and identify factors that regulate the population dynamics of different species across their ranges, (iii) understand the role of large herbivores in regulating community and ecosystem processes, and how their losses are likely to affect ecosystems, and (iv) characterize the ecological and socioeconomic drivers of human–herbivore conflicts to identify the most effective ways of reducing conflict and thereby sustain large herbivore populations across the landscape. The large herbivores of South and Southeast Asia also offer unlimited opportunities for addressing a diverse array of other basic, as well as applied, scientific questions ranging from evolution and behavior to disease dynamics and the responses of herbivore populations to changing climates. Besides establishing and sustaining research initiatives that will generate much-needed long-term scientific data on large herbivores, there is also an urgent need for greater cooperation between ecologists, sociologists, economists, politicians, land managers, and the public if we are to ensure the long-term survival of large herbivores in the region.

M. Sankaran (✉)
Ecology & Evolution Group, National Centre for Biological Sciences,
Tata Institute of Fundamental Research, GKVK Campus, Bellary Road,
Bangalore 560065, India
e-mail: mahesh@ncbs.res.in

M. Sankaran
School of Biology, University of Leeds, Leeds LS2 9JT, UK

F.S. Ahrestani
Department of Ecology, Evolution and Environmental Biology, Columbia University, New York, NY 10027, USA
e-mail: farshid.ahrestani@gmail.com

© Springer Science+Business Media Dordrecht 2016
F.S. Ahrestani and M. Sankaran (eds.), *The Ecology of Large Herbivores in South and Southeast Asia*, Ecological Studies 225,
DOI 10.1007/978-94-017-7570-0_11

Keywords Community and ecosystem processes · Conservation · Long-term monitoring · Population dynamics and trends · Research priorities · Ungulates

Large herbivores in SSEA currently face unprecedented threats from habitat loss, land-use change, and direct persecution through hunting as a result of growing human populations in the region (Schipper et al. 2008; Hoffmann et al. 2011; Ripple et al. 2015). Of the 83 species found here, nearly 37 % (31 species) are under imminent threat of extinction, with 8 species (10 %) listed as critically endangered and another 23 species (28 %) classified as endangered by the IUCN (Table 1.1 in Chap. 1, also see Fig. 2.21). An additional 35 % of the species are listed as either vulnerable (21 species) or near threatened (8 species, Table 1.1), and there is insufficient data for 9 species (11 %, classified as data deficient by the IUCN, Table 1.1 in Chap. 1, also see Fig. 2.21). Furthermore, nearly a quarter of the large herbivores in SSEA (~22 species) are potentially 'political endemics' (i.e., restricted to one country), which considerably increases their extinction risk depending on the cultural and political willingness and ability of nations to protect them (Ceballos et al. 2005). The mammal fauna of Southeast Asia, including large herbivores, also show some of the greatest range reductions since the nineteenth century, and thus local population extinctions, of all mammals globally (Ceballos and Ehrlich 2002; Morrison et al. 2007; Ripple et al. 2015). Clearly, the need for immediate conservation efforts to secure the long-term future of large herbivores in SSEA cannot be overstressed.

Ultimately, for conservation efforts in the region to be successful, they must be built on a strong foundation of in-depth scientific knowledge of the natural history and ecology of the species or guild in question, and the ecological roles they play in ecosystems. In putting this book together, we have tried to highlight and showcase some of the ongoing and past work on large herbivore ecology in SSEA. While this synthesis is by no means exhaustive, what is abundantly clear from this effort is that research on large herbivore ecology in SSEA remains largely inadequate (see Chap. 1). Neither the quantum of research being carried out in the region, nor its focus in terms of patterns and underlying processes, and spatial and temporal scales of enquiry, is sufficient to make scientifically informed management decisions for the conservation of large herbivores in SSEA (see Fig. 1.2). In this scenario, any rigorous piece of research on large herbivore ecology, irrespective of the specific question being addressed, is likely to be of significant value. However, in the following sections, we highlight some minimum data requirements and broad synthetic research themes which we believe are particularly critical and urgent, and which we hope will benefit management by filling in existing knowledge gaps and also serve to stimulate future scientific research.

11.1 Filling in the Blanks: Status and Distribution of Large Herbivores

Despite a long tradition of natural history in the region dating back to colonial times and even earlier (Jerdon 1874; Brander 1923; Peacock 1933), much remains unknown about the basic ecology of many herbivores in SSEA. This is particularly true of small-bodied forest dwelling ungulates such as muntjacs and chevrotains that inhabit the Southeast Asian countries of Myanmar, Thailand, Laos, Vietnam and Philippines, most of which are classified as data deficient by the IUCN (Table 1.2). While it is understandable that detailed information about these forest dwelling species, which are typically rare and occur at low densities, is lacking, what is more concerning is that our knowledge of even the larger, more widespread herbivore species in SSEA is fairly limited. For example, a search on Web of Science with the topic 'sambar deer'—possibly the most widespread large herbivore in SSEA—yielded 127 citations, in contrast to 'white tailed deer', which produced 5875 hits. Importantly, only a very small fraction of these papers directly addressed the ecology of the sambar, or provided insights for their conservation; a situation that is likely to hold true for most large herbivores in SSEA.

As a starting point, we need more detailed information about the current distribution and status of most of the large herbivores of SSEA; a minimum requirement for both managers and scientists alike. At present, coarse-scale distribution maps are available for many large herbivore species in the region, and these have proved extremely valuable for global and regional status assessments (Ceballos and Ehrlich 2002, 2006; Ceballos et al. 2005; Schipper et al. 2008; Morrison et al. 2007). However, their utility for conservation programs, as well as for understanding factors that regulate herbivore distribution patterns at smaller scales, is limited. Generating fine-scale distribution and status maps for the diverse array of herbivores in SSEA is not a trivial exercise, and the efforts required to establish rigorous local-scale monitoring efforts will undoubtedly be substantial. However, relatively rapid approaches such as key informant surveys within an occupancy-modeling framework have proven effective as a first cut, not only for generating finer-scale maps of the current distribution of animals, but also for uncovering spatiotemporal patterns of change in their distribution and abundance (Karanth et al. 2010; Pillay et al. 2011, 2014). While such 'rapid' assessments can be particularly useful in assessing trends over large geographic areas, they are not a substitute for more detailed and sustained monitoring of 'priority' populations of herbivores in the long-term, both inside and outside protected areas.

Long-term monitoring of herbivore populations and habitats is critical for assessing the efficacy of existing management programs, or alternately the need for new management initiatives, as they provide information about population trends and can serve as early-warning indicators of population declines (Beever and Woodward 2011). Given the logistical, financial, and manpower constraints involved with setting up and sustaining such initiatives over the long term, it is amply clear that they must ultimately be implemented and coordinated by land managers and forest department officials. Annual monitoring of wildlife populations by land managers is already in place in many protected areas in SSEA, but these efforts have often been criticized for a lack of scientific rigor (Karanth et al. 2003). A monitoring program that is poorly designed or poorly executed, or both, neither answers the question it was designed to answer, nor provides data that is scientifically useful, and is essentially a waste of valuable resources (financial, manpower, time, etc.) that could have been more effectively utilized elsewhere (Reynolds et al. 2011; Beever and Woodhouse 2011).

Given the economic, political, social, and time constraints that land managers in SSEA constantly face, it is becoming increasingly clear that collaborations between the scientific community and managers can benefit the design, execution, and evaluation of scientifically rigorous monitoring programs. Thoughtfully conceived, well-designed monitoring efforts, when replicated and coordinated across larger scales, can uncover broad-scale patterns and identify drivers of herbivore population dynamics across environmental gradients, help detect phenomena such as thresholds and nonlinear dynamics, while also allowing for more robust inferences by increasing sample size and statistical power (Beever and Woodward 2011).

Besides serving as a basic template for conservation planning, distribution maps and long-term data on population trends of species across their geographic ranges will also serve as a critical resource for scientists, both from the region and elsewhere. At present, even the little data that is available is not easily accessible to the scientific community at large. The unfortunate consequence is that neither the scientists benefit from access to the data, nor do the managers benefit from the insights that the scientific community can bring to management and conservation planning from the analysis of such data. The need for greater data transparency and public sharing of data cannot be overstated at the current time, especially given the critical status of many large herbivore species in the region. In particular, we envision an increasing role in the future for open-science initiatives and data-sharing portals (e.g., India Biodiversity Portal; http://indiabiodiversity.org) in providing open and free access to biodiversity information from the region.

11.2 Regulators of Large Herbivore Population Dynamics

Besides providing information about population trends, long-term data on herbivore populations are also critical for understanding the factors that regulate and limit herbivore population numbers; a prerequisite to their effective management (Gordon et al. 2004). Our knowledge of what limits herbivore numbers, as well as the factors that cause them to fluctuate over time, has increased substantially in recent decades, thanks largely to a few long-term ecological studies of population and community dynamics in North America and Europe (see Gaillard et al. 1998, 2000; Gordon et al. 2004 and references therein). Seminal analyses of long-term population data from multiple herbivore species (see Gaillard et al. 1998, 2000) have provided us with critical insights into the relative importance of density-dependent and density-independent factors that influence population dynamics, and how these in turn are influenced by factors such as the age-structure (Coulson et al. 2001), breeding system (Milner-Gulland et al. 2003), age- and sex-specific survival and emigration rates (Clutton-Brock et al. 2002), and harvesting (Clutton-Brock et al. 2002) of the population. The contribution of these studies to both our understanding of herbivore population dynamics, as well as their management cannot be overstated. However, the fact still remains that the majority of these studies come from temperate ecosystems, while most ungulates are tropical (Gaillard et al. 1998, 2000).

Because of differences between tropical and temperate ecosystems in terms of their seasonality, temperature versus water limitation during the lean season, disease prevalence and predation, drawing inferences about the population dynamics of tropical herbivores based on studies largely conducted in temperate ecosystems may not be straightforward (Gaillard et al. 1998, 2000). Furthermore, long-term term studies of large herbivore population dynamics from systems that still retain their complete complement of native predators, as is still the case in many areas in SSEA, is limited (Gaillard et al. 2000). Classic work from North America has highlighted the extent to which apex predators such as wolves and cougars are capable of limiting herbivore populations (Beschta and Ripple 2009; Ripple and Beschta 2012), and how extirpation or recolonization of predators can have effects that cascade through the ecosystem, influencing not only herbivore and plant trophic levels, but also other taxa such as beavers, song birds, and fish (Hebblewhite et al. 2005; Ripple and Beschta 2006), as well as ecosystem processes such as nutrient cycling (Frank 2008), river morphology, and hydrological flows (Beschta and Ripple 2012). In tropical ecosystems too, predation has, likewise, been argued to be an important regulator of population dynamics for some herbivore species and populations, but not for others (Sinclair 1985; Gasaway et al. 1996; Grange et al. 2004; Grange and Duncan 2006; Mduma et al. 1999).

At present, there is an almost complete lack of information about the long-term population dynamics and vital rates for most species of large herbivores in SSEA (see Chap. 10; Gaillard et al. 2000), and about the relative importance of 'bottom-up' versus 'top-down' forces in regulating herbivore population dynamics in the region. To our knowledge, the ongoing study of blackbuck at Velavadar National Park (See Chap. 6; Jhala and Isvaran) represents one of the only long-term studies of a species from SSEA, one that has provided us with critical insights into the factors regulating blackbuck population dynamics, vital rates, and behavior in this semi-arid system. To similarly generate accurate vital rates of other species, we require studies involving marked individuals that are monitored over time (Gaillard et al. 1998). Where it is not possible to mark individuals for one reason or another, long-term data on the abundance of the sexes, different age-classes, or even just total population size, can nevertheless provide us with insights into factors regulating herbivore populations. In this regard, ongoing long-term efforts to monitor prey populations as part of tiger conservation efforts across SSEA (see Chap. 9; Karanth and Nichols 1998; Karanth et al. 2004) are a potential source of valuable data that can contribute to a better understanding of herbivore population dynamics in the region.

The diverse array of large herbivores found in SSEA, spanning a range of body sizes and feeding strategies, provides unlimited opportunities to address basic and applied questions of both scientific and management relevance. Further, the range of habitats and environments occupied by large herbivores in SSEA, from high-altitude cold deserts to tropical rainforests, as well as the range in predator diversity and biomass, from sites with full to progressively depauperate complements of predators, allows us to address fundamental questions about the relative importance of predation versus resource-limitation in regulating herbivore populations, and how they vary as a function of herbivore body size and along resource gradients (Hopcraft et al. 2010; Fritz et al. 2011). Understanding predator-control of herbivore populations is particularly important for management in light of current poaching-driven declines and local extinctions of predators in SSEA (e.g., the local extinction of tigers in Sariska National Park, India). From a conservation perspective, there is also an urgent need for studies that investigate how hunting of large herbivores, which is both prevalent and widespread in SSEA (Madhusudhan and Karanth 2002; Velho et al. 2012), impacts vital rates and population trends of different large herbivore species.

11.3 Community and Ecosystem Consequences of Herbivory

Large herbivores are well recognized as important regulators of community and ecosystem processes. They can induce significant shifts in plant community composition and diversity, with effects that diverge widely between different ecosystems depending on site productivity and precipitation, evolutionary history of herbivory, herbivore body size, feeding selectivity of herbivores, temporal patterns of herbivory, and differences between plant species in their ability to tolerate or recover from herbivory (Milchunas et al. 1988; Augustine and McNaughton 1998; Olff and Ritchie 1998; Proulx and Mazumder 1998; Bakker et al. 2006; Diaz et al. 2007). Large herbivores also play critical roles in regulating tree community composition and the balance between trees and grasses in ecosystems by acting as agents of seed dispersal (see Chap. 5; Miller 1996), and through their effects on the growth, reproduction, recruitment and mortality of trees (Prins and van der Jeugd 1993; Augustine and McNaughton 2004; Sankaran et al. 2004, 2008, 2013; Goheen et al. 2007; Guldemond and van Aarde 2008), with such effects further modulated through interactions with fire regimes (Dublin et al. 1990; Holdo 2007; Staver et al. 2009). Large herbivore effects can also extend beyond the primary producer level, percolating through to influence the abundance of multiple other taxa including arthropods, reptiles, birds, and rodents (Pringle et al. 2007; Greenwald et al. 2008; Banks et al. 2010; Goheen et al. 2010; Foster et al. 2014). Finally, large herbivores also play critical roles in regulating the carbon and nutrient economy of ecosystems, both directly through consumption of plant material, and indirectly by redistributing nutrients across the landscape and by returning them to soils in more readily available forms via dung and urine (McNaughton 1985; Augustine and McNaughton 1998; Ritchie et al. 1998). Herbivores stimulate primary production and enhance nutrient cycling rates in fertile ecosystems when herbivory is nonselective and occurs early in the growing season, while depressing them in nutrient-poor ecosystems, and under conditions of chronic and selective herbivory (Augustine and McNaughton 1998; Ritchie et al. 1998).

Our understanding of the ecology of large herbivores in SSEA is perhaps nowhere as lacking as when it comes to understanding their roles in influencing community and ecosystem processes (also see Fig. 1.2). There have only been a handful of studies in SSEA that have used exclosures to experimentally assess herbivore effects on plant community composition and richness (see Chaps. 7 and 8; Sankaran and McNaughton 1999; Sankaran 2005; Bagchi et al. 2012), and on ecosystem processes such as carbon and nutrient cycling (see Chaps. 7 and 8; Bagchi and Ritchie 2010a, b, 2012; also see Pandey and Singh 1992; Moe and Wegge 2008). These studies have largely been restricted to herbaceous communities in grassland and savanna ecosystems, and we are unaware of any studies that have looked at herbivore effects on tree dynamics, or ecosystem processes in mixed tree-grass ecosystems and forests of SSEA. Thus, while we have some information about the roles that large herbivores play as seed dispersers (Chap. 5), we know very little about how large herbivores influence tree community composition and patterns of tree recruitment, growth, and mortality. Given the widespread declines in herbivore numbers across large parts of SSEA, there is an urgent need for research programs that focus on understanding herbivore effects at the community and ecosystem level across different ecosystem types in the region. This is critical if we are to predict how the effects of such herbivore losses are likely to cascade through ecosystems influencing not only plant communities but also other taxa and ecosystem processes, and how these are likely to vary across ecosystem types, with implications for their effective management.

11.4 Conflict and Coexistence

As human populations expand and rates of land-use conversion intensify, wildlife populations will undoubtedly come into increasing contact with humans and livestock in the future. Nowhere is this problem likely to be more acute than in SSEA. In the majority of cases, such increased contact will almost certainly translate to increased levels of conflict, either actual or perceived, in the form of increased crop depredation by wildlife, heightened competition between livestock and native herbivores, increased hunting and persecution of wildlife by humans, and greater loss of property and potentially, human life. Such conflict is doubly problematic; it can seriously undermine conservation prospects by eroding the willingness of local communities to conserve wildlife, while simultaneously having negative consequences for large herbivores through loss of habitat, increased disease risk, reduce forage availability as a result of competition with livestock, and increased persecution by humans (Madhusudhan 2003, 2004; Mishra et al. 2004). The resolution of such conflict to ensure the long-term persistence of large herbivores in the region will ultimately hinge on successful cooperation between ecologists, sociologists, economists, politicians, managers and the public (Gordon et al. 2004). Ecologists, nevertheless, have much to contribute to conflict resolution by (i) identifying the extent to which native and domestic herbivores compete with or facilitate one another, the conditions under which each occurs, and the implications of such interactions for native herbivore populations (Bagchi et al. 2004; Madhusudhan 2004; Mishra et al. 2004; Bhatnagar et al. 2006; Odadi et al. 2011), (ii) identifying the ecological determinants of conflict such as crop raiding, including its spatial and temporal patterns of occurrence (Owen 2013), (iii) understanding the impacts in terms of disease prevalence and transmission of increasing contact between wild and domestic herbivores (Kilpatrick et al. 2009), (iv) understanding the population and genetic consequences of retaliatory killing of large herbivores by humans (Ginsberg and Milner-Gulland 1994; Mysterud 2011), and (v) helping local communities and land managers by determining the most effective ways of reducing conflict and managing large herbivore populations in the landscape (Mishra et al. 2003; Gordon et al. 2004). Such efforts are particularly critical given that significant wildlife populations currently persist outside protected areas, and conservation efforts must, by necessity, also extend to these human-dominated landscapes (Chazdon et al. 2009; Hoffmann et al. 2015).

11.5 Concluding Statements

The research themes and data requirements that we have highlighted in the previous sections represent but the bare minimum in terms of understanding the factors that limit herbivore populations in SSEA, and the roles that large herbivores play in influencing community and ecosystem processes in the region. The large herbivore guild of SSEA also provides a wealth of opportunity for addressing a diverse array of other basic, as well as applied, scientific questions relating to their evolutionary history and genetics (Chap. 2; Guha et al. 2007; James et al. 2008; Khaledi et al. 2009; Vidya et al. 2009), behavior (See Chap. 6; Isvaran 2005, 2007), body size and foraging ecology (Chap. 4; Ahrestani et al. 2012; English et al. 2014), habitat use and distribution patterns (see Chap. 3; Odden et al. 2005; Sridhara et al. 2013; Sharma et al. 2014), and their effects on plants (Chap. 7; Bagchi and Ritchie 2011), to name a few. In addition, a key area of research that has been largely neglected in SSEA, but one that is likely to become increasingly important for management in the future, is the role of disease in regulating wildlife populations. Most previous studies of disease have been carried out on captive populations of large herbivores (Fig. 1.2), and its role in regulating herbivore population numbers and community structure, as well as the factors that regulate the spatial dynamics of disease transmission in the wild including disease prevalence, persistence, and spread remain largely unknown.

The bulk of ecological research in SSEA has, thus far, been disproportionately centered on tropical forests and global biodiversity hotspots. However, hotspots of mammalian diversity are not necessarily congruent with 'general purpose' biodiversity hotspots, with less than 2 % of the prime area of large herbivore diversity estimated to overlap with 'global' biodiversity hotspots (Olff et al. 2002). Future efforts must, therefore, extend beyond the boundaries of global biodiversity hotspots to also encompass areas that are critical for large mammalian herbivores in the region (Olff et al. 2002; Ahrestani et al. 2011). There is also a need for sustained research efforts in 'priority' herbivore areas in order to generate much needed long-term time series data on herbivore population dynamics and ecology. The majority of research on large herbivores in SSEA thus far represent but a 'one-off' effort, addressing a specific question on a particular species in a particular location, with little to no follow-up research. As a result, for most species and sites, we lack even the most basic of foundations, built on a core body of past research, upon which to build and address questions of scientific and management relevance. Long-term studies of the blackbuck in Velavadar National Park (Chap. 6), mountain ungulates in the trans-Himalaya (Chap. 3) and monitoring of prey populations as part of tiger conservation efforts across SSEA (Chap. 9; Karanth and Nichols 1998; Karanth et al. 2004) represent welcome exceptions, but many more such efforts are currently needed. Finally, future research on large herbivores in SSEA must be set within the overarching context of climate change, as it is becoming increasingly evident that climatic variability and extreme events can exert significant, and often nonlinear, controls on herbivore populations (Coulson et al. 2001; Mysterud et al. 2001; Post and Forchhammer 2008; Post et al. 2009).

In closing, we hope that the understanding and the ongoing research on large herbivores in SSEA that this volume showcases serves to stimulate future research on this prominent but neglected guild. If we are to successfully manage and ensure the continued persistence of the diverse large herbivores of South and Southeast Asia, there is then an urgent need to immediately initiate research efforts that are ultimately sustained in the long term. At present, much remains unknown about the ecology of these species and the ecological roles they play in ecosystems, and the opportunities for future research are immense.

References

Ahrestani FS, Heitkönig IM, van Langevelde F et al (2011) Moisture and nutrients determine the distribution and richness of India's large herbivore species assemblage. Basic Appl Ecol 12:634–642

Ahrestani FS, Heitkönig I, Prins HHT (2012) Diet and habitat-niche relationships within an assemblage of large herbivores in a seasonal tropical forest. J Trop Ecol 28:385–394

Augustine DJ, McNaughton SJ (1998) Ungulate effects on the functional species composition of plant communities: herbivore selectivity and plant tolerance. J Wildl Manag 62:1165–1183

Augustine DJ, McNaughton SJ (2004) Regulation of shrub dynamics by native browsing ungulates on East African rangeland. J Appl Ecol 41:45–58

Bagchi S, Ritchie ME (2010a) Introduced grazers can restrict potential soil carbon sequestration through impacts on plant community composition. Ecol Lett 13:959–968
Bagchi S, Ritchie ME (2010b) Herbivore effects on above-and belowground plant production and soil nitrogen availability in the Trans-Himalayan shrub-steppes. Oecologia 164:1075–1082
Bagchi S, Ritchie ME (2011) Herbivory and plant tolerance: experimental tests of alternative hypotheses involving non-substitutable resources. Oikos 120:119–127
Bagchi S, Mishra C, Bhatnagar YV (2004) Conflicts between traditional pastoralism and conservation of Himalayan ibex (*Capra sibirica*) in the Trans-Himalayan mountains. Anim Conserv 7:121–128
Bagchi S, Bhatnagar YV, Ritchie ME (2012) Comparing the effects of livestock and native herbivores on plant production and vegetation composition in the Trans-Himalayas. Pastoralism 2:1–16
Bakker ES, Ritchie ME, Olff H et al (2006) Herbivore impact on grassland plant diversity depends on habitat productivity and herbivore size. Ecol Lett 9:780–788
Banks JE, Jackson C, Hannon LM et al (2010) The cascading effects of elephant presence/absence on arthropods and an Afrotropical thrush in Arabuko-Sokoke Forest, Kenya. Afr J Ecol 48:1030–1038
Beever EA, Woodward A (2011) Ecoregional-scale monitoring within conservation areas, in a rapidly changing climate. Biol Conserv 144:1255–1257
Beschta RL, Ripple WJ (2009) Large predators and trophic cascades in terrestrial ecosystems of the Western United States. Biol Conserv 142:2401–2414
Beschta RL, Ripple WJ (2012) The role of large predators in maintaining riparian plant communities and river morphology. Geomorphology 157:88–98
Bhatnagar YV, Wangchuk R, Prins HHT et al (2006) Perceived conflicts between pastoralism and conservation of the kiang Equus kiang in the Ladakh Trans-Himalaya, India. Environ Manage 38:934–941
Brander AAD (1923) Wild animals in Central India. Edward Arnold & Co, London
Ceballos G, Ehrlich PR (2002) Mammal population losses and the extinction crisis. Science 296:904–907
Ceballos G, Ehrlich PR, Soberón J et al (2005) Global mammal conservation: what must we manage? Science 309:603–607
Ceballos G, Ehrlich PR (2006) Global mammal distributions, biodiversity hotspots, and conservation. Proc Natl Acad Sci 103:19374–19379
Chazdon RL, Harvey CA, Komar O et al (2009) Beyond reserves: a research agenda for conserving biodiversity in human-modified tropical landscapes. Biotropica 4:142–153
Clutton-Brock TH, Coulson TN, Milner-Gulland EJ et al (2002) Sex differences in emigration and mortality affect optimal management of deer populations. Nature 415:633–637
Coulson T, Catchpole EA, Albon SD et al (2001) Age, sex, density, winter weather, and population crashes in soay sheep. Science 292:1528–1531
Diaz S, Lavorel S, McIntyre SUE et al (2007) Plant trait responses to grazing–a global synthesis. Glob Change Biol 13:313–341
Dublin HT, Sinclair ARE, McGlade J (1990) Elephants and fire as causes of multiple stable states in the Serengeti-Mara woodlands. J Anim Ecol 59:1147–1164
English M, GillespieG Ancrenaz M et al (2014) Plant selection and avoidance by the Bornean elephant (*Elephas maximus borneensis*) in tropical forest: does plant recovery rate after herbivory influence food choices? J Trop Ecol 30:371–379
Foster CN, Barton PS, Lindenmayer DB (2014) Effects of large native herbivores on other animals. J Appl Ecol 51:929–938
Frank DA (2008) Evidence for top predator control of a grazing ecosystem. Oikos 117:1718–1724
Fritz H, Loreau M, Chamaillé-Jammes S et al (2011) A food web perspective on large herbivore community limitation. Ecography 34:196–202
Gaillard JM, Festa-Bianchet M, Yoccoz NG (1998) Population dynamics of large herbivores: variable recruitment with constant adult survival. Trends Ecol Evol 13:58–63

Gaillard JM, Festa-Bianchet M, Yoccoz NG et al (2000) Temporal variation in fitness components and population dynamics of large herbivores. Annu Rev Ecol Syst 31:367–393
Gasaway WC, Gasaway KT, Berry HH (1996) Persistent low densities of plains ungulates in Etosha National Park, Namibia: testing the food-regulating hypothesis. Can J Zool 74:1556–1572
Ginsberg JR, Milner-Gulland EJ (1994) Sex-biased harvesting and population dynamics in ungulates: implications for conservation and sustainable use. Conserv Biol 8:157–166
Goheen JR, Young TP, Keesing F et al (2007) Consequences of herbivory by native ungulates for the reproduction of a savanna tree. J Ecol 95:129–138
Goheen JR, Palmer TM, Keesing F et al (2010) Large herbivores facilitate savanna tree establishment via diverse and indirect pathways. J Anim Ecol 79:372–382
Gordon IJ, Hester AJ, Festa-Bianchet M (2004) The management of wild large herbivores to meet economic, conservation and environmental objectives. J Appl Ecol 41:1021–1031
Grange S, Duncan P, Gaillard JM et al (2004) What limits the Serengeti zebra population? Oecologia 140:523–532
Grange S, Duncan P (2006) Bottom-up and top-down processes in African ungulate communities: resources and predation acting on the relative abundance of zebra and grazing bovids. Ecography 29:899–907
Greenwald KR, Petit LJ, Waite TA (2008) Indirect effects of a keystone herbivore elevate local animal diversity. J Wildl Manag 72:1318–1321
Guha S, Goyal SP, Kashyap VK (2007) Molecular phylogeny of musk deer: a genomic view with mitochondrial 16S rRNA and cytochrome b gene. Mol Phylogenet Evol 42:585–597
Guldemond R, van Aarde R (2008) A meta-analysis of the impact of African Elephants on savanna Vegetation. J Wildl Manag 72:892–899
Hebblewhite M, White CA, Nietvelt CG et al (2005) Human activity mediates a trophic cascade caused by wolves. Ecology 86:2135–2144
Hoffmann M, Belant JL, Chanson JS et al (2011) The changing fates of the world's mammals. Philos Trans R Soc B Biol Sci 366:2598–2610
Hoffmann M, Duckworth JW, Holmes K et al (2015) The difference conservation makes to extinction risk of the world's ungulates. Conserv Biol. doi:10.1111/cobi.12519
Holdo RM (2007) Elephants, fire, and frost can determine community structure and composition in Kalahari woodlands. Ecol Appl 17:558–568
Hopcraft JGC, Olff H, Sinclair ARE (2010) Herbivores, resources and risks: alternating regulation along primary environmental gradients in savannas. Trends Ecol Evol 25:119–128
Isvaran K (2005) Female grouping best predicts lekking in blackbuck (Antilope cervicapra). Behav Ecol Sociobiol 57:283–294
Isvaran K (2007) Intraspecific variation in group size in the blackbuck antelope: the roles of habitat structure and forage at different spatial scales. Oecologia 154:435–444
James J, Ramakrishnan U, Datta A (2008) Molecular evidence for the occurrence of the leaf deer *Muntiacus putaoensis* in Arunachal Pradesh, North-East India. Conserv Genet 9:927–931
Jerdon TC (1874) The mammals of India; a natural history of all the animals known to inhabit continental India. J Wheldon, London
Karanth KU, Nichols JD (1998) Estimation of tiger densities in India using photographic captures and recaptures. Ecology 79:2852–2862
Karanth KU, Nichols JD, Seidenstricker J et al (2003) Science deficiency in conservation practice: the monitoring of tiger populations in India. Anim Conserv 6:141–146
Karanth KU, Nichols JD, Kumar NS et al (2004) Tigers and their prey: predicting carnivore densities from prey abundance. Proc Natl Acad Sci USA 101:4854–4858
Karanth KK, Nichols JD, Karanth KU et al (2010) The shrinking ark: patterns of large mammal extinctions in India. Proc R Soc B Biol Sci. doi:10.1098/rspb.2010.0171
Khaledi KJ, Panandam JM, Siraj SS et al (2009) Estimation of inbreeding coefficient in Rusa Deer (Cervus timorensis) using Microsatellite Loci in Malaysia. J Appl Anim. Res 36:279–281
Kilpatrick AM, Gillin CM, Daszak P (2009) Wildlife–livestock conflict: the risk of pathogen transmission from bison to cattle outside Yellowstone National Park. J Appl Ecol 46:476–485

Madhusudan MD (2003) Living amidst large wildlife: livestock and crop depredation by large mammals in the interior villages of Bhadra Tiger Reserve, South India. Environ Manage 31:466–475

Madhusudan MD (2004) Recovery of wild large herbivores following livestock decline in a tropical Indian wildlife reserve. J Appl Ecol 41:858–869

Madhusudan MD, Karanth KU (2002) Local hunting and the conservation of large mammals in India. Ambio 31:49–54

McNaughton SJ (1985) Ecology of a grazing ecosystem: the Serengeti. Ecol Monogr 55:259–294

Mduma SA, Sinclair ARE, Hilborn R (1999) Food regulates the Serengeti wildebeest: a 40-year record. J Anim Ecol 68:1101–1122

Milchunas DG, Sala OE, Lauenroth W (1988) A generalized model of the effects of grazing by large herbivores on grassland community structure. Am Nat 132:87–106

Miller MF (1996) Dispersal of Acacia seeds by ungulates and ostriches in an African savanna. J Trop Ecol 12:345–356

Milner-Gulland EJ, Bukreeva OM, Coulson T et al (2003) Reproductive collapse in saiga antelope harems. Nature 422:135

Mishra C, Allen P, McCarthy TOM et al (2003) The role of incentive programs in conserving the snow leopard. Conserv Biol 17:1512–1520

Mishra C, Van Wieren SE, Ketner P et al (2004) Competition between domestic livestock and wild bharal *Pseudois nayaur* in the Indian Trans-Himalaya. J Appl Ecol 41:344–354

Moe SR, Wegge P (2008) Effects of deposition of deer dung on nutrient redistribution and on soil and plant nutrients on intensively grazed grasslands in lowland Nepal. Ecol Res 23:227–234

Morrison JC, Sechrest W, Dinerstein E et al (2007) Persistence of large mammal faunas as indicators of global human impacts. J Mammal 88:1363–1380

Mysterud A, Stenseth NC, Yoccoz NG et al (2001) Nonlinear effects of large-scale climatic variability on wild and domestic herbivores. Nature 410:1096–1099

Mysterud A (2011) Selective harvesting of large mammals: how often does it result in directional selection? J Appl Ecol 48:827–834

Odadi WO, Karachi MK, Abdulrazak SA et al (2011) African wild ungulates compete with or facilitate cattle depending on season. Science 333:1753–1755

Odden M, Wegge P, Storaas T (2005) Hog deer *Axis porcinus* need threatened tallgrass floodplains: a study of habitat selection in lowland Nepal. Anim Conserv 8:99–104

Olff H, Ritchie ME (1998) Effects of herbivores on grassland plant diversity. Trends Ecol Evol 13:261–265

Olff H, Ritchie ME, Prins HHT (2002) Global environmental controls of diversity in large herbivores. Nature 415:901–904

Owen NR (2013) Conservation, conflict and costs: living with large mammals in the Nilgiri Biosphere Reserve, India. Dissertation, University of Leeds

Pandey CB, Singh JS (1992) Influence of rainfall and grazing on herbage dynamics in a seasonally dry tropical savanna. Vegetatio 102:107–124

Peacock EH (1933) A game-book for Burma and adjoining territories. Witherby, London

Pillay R, Johnsingh AJT, Raghunath R et al (2011) Patterns of spatiotemporal change in large mammal distribution and abundance in the Southern Western Ghats, India. Biol Cons 144:1567–1576

Pillay R, Miller DA, Hines JE et al (2014) Accounting for false positives improves estimates of occupancy from key informant interviews. Divers Distrib 20:223–235

Post E, Forchhammer MC (2008) Climate change reduces reproductive success of an Arctic herbivore through trophic mismatch. Philos Trans R Soc Lond B Biol. Sci 363:2367–2373

Post E, Forchhammer MC, Bret-Harte MS et al (2009) Ecological dynamics across the Arctic associated with recent climate change. Science 325:1355–1358

Pringle RM, Young TP, Rubenstein DI et al (2007) Herbivore-initiated interaction cascades and their modulation by productivity in an African savanna. Proc Natl Acad Sci 104:193–197

Prins HHT, van der Jeugd HP (1993) Herbivore population crashes and woodland structure in East Africa. J Ecol 81:305–314

Proulx M, Mazumder A (1998) Reversal of grazing impact on plant species richness in nutrient-poor vs. nutrient-rich ecosystems. Ecology 79:2581–2592
Reynolds JH, Thompson WL, Russell B (2011) Planning for success: identifying effective and efficient survey designs for monitoring. Biol Conserv 144:1278–1284
Ripple WJ, Beschta RL (2006) Linking a cougar decline, trophic cascade, and catastrophic regime shift in Zion National Park. Biol Conserv 133:397–408
Ripple WJ, Beschta RL (2012) Trophic cascades in Yellowstone: the first 15 years after wolf reintroduction. Biol Conserv 145:205–213
Ripple WJ, Newsome TM, Wolf C et al (2015) Collapse of the world's largest herbivores. Sci Adv. doi:10.1126/sciadv.1400103
Ritchie ME, Tilman D, Knops JM (1998) Herbivore effects on plant and nitrogen dynamics in Oak savanna. Ecology 79:165–177
Sankaran M, McNaughton SJ (1999) Determinants of biodiversity regulate compositional stability of communities. Nat 401:691–693
Sankaran M (2005) Fire, grazing and the dynamics of tall-grass savannas in the Kalakad-Mundanthurai Tiger Reserve. Conserv Soc 3:4–25
Sankaran M, Ratnam J, Hanan NP (2004) Tree–grass coexistence in savannas revisited–insights from an examination of assumptions and mechanisms invoked in existing models. Ecol Lett 7:480–490
Sankaran M, Ratnam J, Hanan NP (2008) Woody cover in African savannas: the role of resources, fire and herbivory. Glob Ecol Biogeogr 17:236–245
Sankaran M, Augustine DJ, Ratnam J (2013) Native ungulates of diverse body sizes collectively regulate long-term woody plant demography and structure of a semi-arid savanna. J Ecol 101:1389–1399
Schipper J, Chanson JS, Chiozza F et al (2008) The status of the world's land and marine mammals: diversity, threat, and knowledge. Science 322:225–230
Sharma K, Chundawat RS, Van Gruisen J et al (2014) Understanding the patchy distribution of four-horned antelope *Tetracerus quadricornis* in a tropical dry deciduous forest in Central India. J Trop Ecol 30:45–54
Sinclair ARE (1985) Does interspecific competition or predation shape the African ungulate community? J Anim Ecol 54:899–918
Sridhara S, Edgaonkar A, Kumar A (2013) Understorey structure and refuges from predators influence habitat use by a small ungulate, the Indian chevrotain (*Moschiola indica*) in Western Ghats, India. Ecol Res 28:427–433
Staver AC, Bond WJ, Stock WD et al (2009) Browsing and fire interact to suppress tree density in an African savanna. Ecol Appl 19:1909–1919
Velho N, Karanth KK, Laurance WF (2012) Hunting: A serious and understudied threat in India, a globally significant conservation region. Biol Conserv 148:210–215
Vidya TNC, Sukumar R, Melnick DJ (2009) Range-wide mtDNA phylogeography yields insights into the origins of Asian elephants. Proc R Soc B Bio Sci 276:893–902

Species Index

© Springer Science+Business Media Dordrecht 2016
F.S. Ahrestani and M. Sankaran (eds.), *The Ecology of Large Herbivores in South and Southeast Asia*, Ecological Studies 225,
DOI 10.1007/978-94-017-7570-0

Subject Index

© Springer Science+Business Media Dordrecht 2016
F.S. Ahrestani and M. Sankaran (eds.), *The Ecology of Large Herbivores in South and Southeast Asia*, Ecological Studies 225,
DOI 10.1007/978-94-017-7570-0

M

N

O

P

R

The manufacturer's authorised representative in the EU is Springer Nature Customer Service Centre GmbH, Europaplatz 3, 69115 Heidelberg, Germany. If you have any concerns regarding our products, please contact ProductSafety@springernature.com

Printed and bound by CPI Group (UK) Ltd, Croydon, CR0 4YY
30/09/2026
02225492-0016